Biodiversity in India
Volume 6

The Editors

Prof. T. Pullaiah obtained his M.Sc. (1973) and Ph.D. degrees in Botany from Andhra University. He was a Post Doctoral Fellow at Moscow State University, Russia during 1976-78. He traveled widely in Europe and USA and visited Universities and Botanical Gardens in about 17 countries. Professor Pullaiah joined Sri Krishnadevaraya University as Lecturer in 1979 and became Professor in 1993. He held several positions in the University, which include Dean, Faculty of Life Sciences, Head of the Department of Botany, Chairman, BOS in Botany, Head of the Department of Sericulture, Co-ordinator and Chairman, BOS in Biotechnology, Vice Principal and Principal, S.K. University College. He retired from active service on 31st May 2011. He was selected by UGC as UGC-BSR Faculty Fellow and is working in Sri Krishnadevaraya University. Prof. Pullaiah has published 49 books, 265 research papers and 35 popular articles. His books include *Flora of Andhra Pradesh* (5 volumes), *Flora of Eastern Ghats* (Vols. 1-4), *Biodiversity in India* (Vols. 1-5), *Herbal Antioxidants* (3 volumes) *Taxonomy of Angiosperms* (presently in 3rd edition), *Plant Development, Plant Reproduction, Plant Tissue Culture, Flora of Guntur district, Flora of Kurnool, Flora of Anantapur district, Flora of Nizamabad, Flora of Ranga Reddi district* etc. He is Principal Investigator of 20 major Research Projects totaling more than a Crore of Rupees funded by DBT, DST, CSIR, UGC, BSI, WWF, GCC etc. Under his guidance 51 students obtained their Ph.D. degrees and 35 students their M.Phil. degrees. He is recipient of Best Teacher Award from Government of Andhra Pradesh, Prof. P. Maheswari Gold Medal and Prof. G. Panigrahi Memorial Lecture of Indian Botanical Society and Prof. Y.D. Tyagi Gold medal of Indian Association for Angiosperm Taxonomy. He was Vice President of Indian Botanical Society. He was Member of Species Survival Commission of International Union for Conservation of Nature and Natural Resources (IUCN).

Dr. K. Jaganmohan Reddy is Professor and Former Head, and presently Chairman, BOS in the Department of Botany, Kakatiya university, Warangal. He also Heading the Department of Botany, University Arts & Science College, and IQAC Coordinator same college. He has been engaged both in teaching and research for more than three decades and has published over 150 research articles in reputed National, International Journals and books. He has supervised 14 theses and 5 are in progress. He is a recipient of the eminent scientist award of the year, 2008 by NESA and honour's award by Academy of Plant Sciences India, 2006. He is also awarded Prof. V. Puri Medal 2012 from All India Botanical Society. He has visited research laboratories at Western Australia State Agricultural Biotechnology Centre, Murdoch University, Perth; Starch Biosynthesis and Molecular Biology and DNA Sequencing Laboratories, Thailand Science Park, Thailand. He is a member of the editorial boards of some reputed journals of India.

Biodiversity in India
Volume 6

—Editors—
T. Pullaiah
Department of Botany
Sri Krishnadevaraya University
Anantapur – 515 003
Andhra Pradesh

K. Jaganmohan Reddy
Department of Botany
Kakatiya University
Warangal – 506 009

2013
Regency Publications
A Division of
Astral International Pvt. Ltd.
New Delhi – 110 002

ISBN 9788170359722

Published by : **Daya Publishing House®**
A Division of
Astral International Pvt. Ltd.
– ISO 9001:2008 Certified Company –
4760-61/23, Ansari Road, Darya Ganj
New Delhi-110 002
Ph. 011-43549197, 23278134
E-mail: info@astralint.com
Website: www.astralint.com

Laser Typesetting : **Classic Computer Services**, Delhi - 110 035

Printed at : **Replika Press Pvt. Ltd.**

PRINTED IN INDIA

Preface

Since Rio Earth Summit Biodiversity has gained importance. Recent Conference of Parties on Biodiversity at Hyderabad has stressed the importance of Biodiversity, its sustainable utilization and conservation. India with its vas geographical area of about 329 m ha, ranks seventh largest country in the World. The diverse physical features and tropical monsoon climate supports rich biological diversity in different ecological habitats. India represents one of the 12 Megabiodiversity centre in the world. India is having 2.4per cent global area and exhibits 10 per cent of floral diversity and 6.67 per cent of faunal diversity. There is every necessity to document this Biodiversity. The present book is an attempt in this direction. Among the 34 Hot spots of Biodiversity in the world four are in India. Biodiversity of three of these hotspots is dealt with in the present book with accounts on one horned Rhinoceros in Assam, Frogs and *Impatiens* of Western Ghats and Flora of North Andamans. Bioresources is dealt with in two chapters. I congratulate all the contributors for their in depth studies.

We hope this will be useful to the conservationists and Biodiversity enthusiasts. Readers are requested to give suggestions for future volumes.

T. Pullaiah

K. Jaganmohan Reddy

Contents

Chapter 1

One Horned Rhinoceros Conservation in Assam, India

***Pranjal Bezbarua*[1,2], *Arundhati Bezbarua*[2], *Prasanna Boruah*[3] and *Chandra Kanta Baruah*[2]**

[1]*Grasshopper, Madhyam Khanda, North Guwahati, Guwahati – 781 030, Assam, India*

[2]*Botany Department, Gauhati University, Guwahati – 781 014, Assam, India*

[3]*Assam Remote Sensing Application Centre, Bigyan Bhawan, G.S. Road, Guwahati – 781 005, Assam, India*

ABSTRACT

The Indian one horned rhinoceros, having a global population of 2400 has been facing increasing threat from organized poaching and habitat loss. Despite 100 years of success in conservation of rhino in Kaziranga National Park (NP), the endangered species was locally eliminated in protected areas namely Laokhowa, Burachapari, Kochmara, Kurua and Manas tiger reserve of Assam, India. Besides threats of stochastic risk of rhinos in Kaziranga NP and Pabitora wildlife sanctuary, increasing man-rhino conflict; continuing poaching in all the three rhino areas having a population about 2000 was a cause of concern. Therefore, the forest department of Assam with support of all stakeholders designed a 15 years rhino conservation programme; Indian Rhino Vision-2020 that focused overall increasing security and rhino translocation to increase the population from 2000 to 3000 which will be distributed in six protected areas. Considering the gradual improvement of law and order problem and unique community participation in conservation through converting poachers to conservation volunteers, Manas tiger reserve, a former rhino habitat has been selected for rhino translocation.

The study aims to study the habitat suitability of the one horned rhinoceros in Manas tiger reserve, to rehabilitate ex-poachers/forest dependents cum conservation workers of the area and conservation education and capacity building in fringe villages of the reserve.

Remote sensing technique and GIS was used to generate maps considering all ground information like former rhino habitat in grassland and forests communities of Manas, aquatic bodies, topography, different anthropogenic factors namely past and present poaching threats, encroachment, weed invasion, grazing pressure from domestic cattle, burning intensity, awareness level in fringe villages, departmental infrastructure etc. The information helped in preparation of final map on rhino habitat suitability indicating most suitable (23.48 per cent), suitable (46.33 per cent), less suitable (24.16 per cent) and unsuitable zone (6.02 per cent) in Manas tiger reserve. Though, most suitable and suitable habitat still exists in western and Manas river ecosystem, the poaching threat scenario converted it to less suitable zone. The result confirmed the richness of rhino habitat in terms of diverse grassland associations and food plants. Of course the presence of different anthropogenic pressure caused degradation of habitat in southern zone of the reserve.

The rehabilitation programme on families of selected conservation volunteers (that included surrendered ex-poachers also) and forest dependent was started with introduction of capacity building on alternative livelihood on agro and veterinary products that varies from piggeries, zinger and citrus cultivation, vermi-compost, small business, bee keeping, handloom etc. Small in kind support was provided to the beneficiaries who formed Self Help Group (SHG) having 10 members. Professional experts, government departments and local NGOs were helping us in these efforts. The first thing, we observed was gradual behaviour change in terms of growing interest and confidence building which helped in achieving 50-60 percent success amongst different SHGs in next phases. They raised money amongst themselves to continue the work as well as lending it amongst themselves and others. More groups were coming to replicate this type of microfinance system. We also provided school bags and education material amongst the children of poor conservation volunteers and forest dependents. It is nice to see that the students (vulnerable to become future poachers) continue their school and a few of them secured top position in the examination.

Awareness and capacity building on rhino conservation was initiated with preparing a conservation leaflet in local language that was used in all conservation meetings, distributed amongst all local conservation groups, teachers and students, forest staff and other stakeholders. The material helped several local NGOs and ground staff to understand the mission of Indian Rhino Vision 2020 programme and need of the rhino conservation. We also focused on alternative livelihood generation programme to encourage the villagers of the fringe areas.

The PTES support in the programme helped to raise other fund to continue the work. We voluntarily helped local NGO in their ecotourism campaign and good amount of money was raised to supply ration for 80 conservation volunteers patrolling the Manas tiger reserve during critical period. We also monitored the translocated rhino using grassland habitat and also strayed to villages and helped forest official during the crisis. Threats to rhino habitat like flood in 2007 due to excess release of water by dam in Bhutan, weed invasion were also focussed. It is nice to see that the degraded rhino habitat has been restoring gradually due to hard work of forest department and local communities. The habitat and its use by translocated rhinos suggested more translocation operation with modified design.

The study recommended continuing community conservation programme to develop rural economy of the fringe villages of the Manas tiger reserve and restoring degraded habitat and population encouraging rhino research in gap areas, presence of well equipped trained front line staff and welfare schemes for them, increasing support in restoring rhino habitat in buffer zone also, better coordination amongst all stakeholders and strict vigilance on threats like movement of wildlife traders and forest mafias, involving biologist in habitat management, opening of blocked channel of Manas river including monitoring flood and dams in Bhutan. Despite two unfortunate poaching incidents and one natural death, the rhino reintroducton in Manas tiger reserve has been restored to a viable population of twenty two individuals during the 2005-2012 periods. This unique model will surely be replicated to other potential areas of Assam for long term survival of the species.

Introduction

Background of the Indian Rhinoceros Conservation Initiative

The Great Indian one horned rhinoceros: *Rhinoceros unicornis* Linnaeus 1758 has been facing great threat due to continuous organized poaching particularly for its horn and habitat shrinkage through different kinds of anthropogenic activities. During 2005, the global population of Indian rhino was estimated to be 2400 where about 2000 numbers were distributed in protected areas of Assam; the north eastern state of India. At that time Kaziranga National Park (NP) shared about 93 percent of rhinos in Assam and 75 percent of global population. Naturally, this can be regarded as a great success to the park as starting from a dozen of rhinos in 1905, the population lies at 1800 in hundred years through increasing security scenario. Besides Kaziranga National Park, Pabitora Wild Life Sanctuary (WLS) and Orang National Park have good number of rhinos due to constant security cover. Unfortunately despite 100 years of success in conservation history for rhino in Assam, the species was eliminated from other rhino habitat of the state namely Laokhowa WLS, Burachapari WLS, Kurua zone and Manas Tiger Reserve. In Manas, almost all 100 rhinos were eliminated during the political unrest period 1989-2001. The famous naturalist E.P. Gee estimated 15 rhinos in 1966 (Spillett, 1966). This population increased to 40 in ten years time in 1976 (Laurie, 1978). However, Sanjoy Debroy, the veteran forest officer estimated 75 rhinos in 1977 and the official number fluctuated from that to 100, till the early 1990s. The last rhino poaching was recorded in 2001 in Kokilabari area, south eastern boundary to Manas national park. The forest official could rescue only some blood and small amount of meat as local people took the meat and bone after poachers had taken away the horn. In Nepal also, there has been a dramatic decline in numbers (544 to 360) of rhino in Royal Chitwan NP (second largest population of the species on the planet) as a result of the Maoist insurgency in that country. Restriction of about 85 per cent of the rhinos to a single protected area like Kaziranga exposes the species to stochastic risks. There are currently only 9 breeding populations of *R.*

unicornis in 2 countries. Moreover, only two of the *R. unicornis* populations (Kaziranga and Chitwan National Park in Nepal) have more than 100 individuals. The population in Pabitora WLS has already exceeded the carrying capacity and the population needs to be reduced both to protect the habitat and to mitigate the increasing rhino-human conflicts occurring as animals move out of the park into agricultural areas. Therefore there is a need of long term practical management plan to find a solution to the threat to the rhino in Assam that share major population in the region.

As a consequence, the Assam Forest Department and the various NGOs and institutes interested and involved in conservation of *Rhinoceros unicornis* have agreed that range expansion of the rhino in Assam through translocations from Kaziranga and Pobitora to other appropriate protected areas is a prudent direction for future efforts. Such translocations have produced initial positive results in Nepal for *Rhinoceros unicornis* and longer term success in Africa for southern white rhino (Foose, pers comm. 2005).

This programme in Assam is globally known as Indian Rhino Vision-2020 where by rhino translocation and increasing security the population will be increased to 3000 from 2000 as well as to expand the distribution in at least 6 protected areas in 15 years period. To achieve the goal, there is a need to find suitable rhino habitats as well as increasing security through all means. The core of Manas tiger reserve (also adorned as Manas National Park) has been selected for the programme at first step being a former rhino habitat and has some hope of possibility of better security due to strong initiative taken by local conservation groups and forest department. With initiation of peace in 2003, local Bodo community developed a unique model of conservation in Manas landscape with support of the forest authority. In eastern part of Manas national park, one newly formed conservation group from local community helped more than 200 poachers to surrender and involved them in patrolling the park day and night to protect wildlife and their habitats. Significant success has been achieved to gather critical information on poachers, illegal wildlife trade and swift action. They extended support to forest department to rebuild park infrastructure. Active community participation helped lot in minimizing the management cost. Gradually this initiative was replicated throughout the southern fringe of the Manas tiger reserve and existing or new groups started their work in buffer areas. Encouraged by the development (since 2004) different regional, national and international NGOs, institutes have come forward to support the overall activities.

Though Manas was a former rhino habitat, it was not known scientifically the habitat suitability of the area or the changes occurred in the landscape due to continued anthropogenic pressure. The park will receive 20-30 rhinos from Kaziranga and Pabitora both situated in floodplain of Brahmaputra river whereas Manas is comparatively dry and located in high altitude bhabor terai zone in Indo-Bhutan border area (Bezbarua, 2007). On the otherhand there is very limited published literature on detail habitat suitability study on Indian rhinos. Most of the earlier publications were based on observations made by sportsmen or naturalists. Gee (1953,1959,1963) conducted a survey on rhinos in Kaziranga NP and in Royal Chitwan NP addressing the ecology and behaviour of rhinos. Laurie (1978) and

Dinerstein (1979) conducted landmark studies on the ecology of herbivores in the terai grasslands of Nepal. Laurie (1978) found much seasonal variation in the food availability and the resultant movement of rhinos between different habitat types. Rhinos were observed feeding on abundant fibrous food supplemented by a wide variety of other plants. Grasses constituted 70-89 per cent of their diet according to the season in Royal Chitwan NP (Laurie 1978). Jnawali (1995) also confirmed the diverse diet of the rhinoceros with more than 60 per cent of their diet being constituted of grasses. More recent studies addressing the ecology of the species includes the work by Dinerstein and Price (1991) on the demography and habitat use by rhino in Royal Chitwan NP. Sinha and Sawarkar (1993) studied the habitat use and seasonal variation in the range of movement in a case of reintroduced rhinos in Dudhwa NP, India. Jnawali and Wegge (1993) looked at the space and habitat use by the reintroduced rhino population in Royal Bardia NP in Nepal. Karki *et al.* (2000) looked at the effects of grazing and management intervention in Royal Bardia NP and concluded that ungulate concentration in habitat patches leads to the formation of grazing lawns that are characterized by growth of very short grasses differing in composition from the neighboring areas. These patches recovered once grazing pressures declined. Ghosh (1997) looked at the effect of fire in the chaurs of Corbett NP. Biswas (1999) looked at the habitat use by hog deer including sympatric species such as rhinos in the Duar grasslands of Jaldapara WLS across areas of natural grasslands and managed grasslands. We did a rapid assessment through remote sensing technique on rhino habitat which needs to be extended to microhabitat level considering all natural and anthropogenic factors. The grassland and forest dynamics model of Manas biosphere reserve (Bezbarua, 2007) indicated some vital information about the richness of habitat. The overall review of literature reveals that the rhino habitat suitability studies in Assam particularly in Manas have received scant attention. Therefore, the habitat suitability study at microhabitat level is very much important for the survival of the translocated rhinos and become a breeding population in the Manas tiger reserve.

Secondly for better protection of the species, it is also important to increase the level of involvement of local community and groups through conservation education. Despite being well protected, poaching continues in all the exiting rhino protected areas including Kaziranga. Presently, the international wildlife traders have built a better network with local poachers and anti-social elements and using most sophisticated methods to carry out operation. The absence or poor community participation may lead to local elimination of wildlife like rhino. Thirdly, rehabilitation of families of surrendered poachers now turned forest protectors of Manas is an urgent need through alternative livelihood to minimize the threats of their return to old profession *i.e.*, poaching profession. In past, it was noticed how surrendered poachers again had returned to their old profession due to lack of rehabilitation. The children of the area particularly from these poor families and neighbourhood have also a tendency to become poachers due to poor awareness and lack of school education.

Considering the above scenarios for effective rhino reintroduction and protection, this unique conservation initiative deals to find out the rhino habitat suitability status and strengthening greater community participation in rhino conservation in Manas tiger reserve. Detail objectives are as follows:

1. To study the habitat suitability of the one horned rhinoceros in Manas tiger reserve.
2. To rehabilitate ex-poachers/forest dependents cum conservation workers and poor villagers.
3. Capacity building in conservation education and extension in fringe villages of rhino habitat.

Methodology

Rhino Habitat Suitability Design

With support of the base map (satellite imageries IRS LISS III pan merge, 2006 and topomaps of survey of India scale 1: 50000) ground survey was conducted using GPS (Magellan model) to verify the land use cover of the area and locating past rhino distribution. We randomly placed 26 belt transects of length 500 meters each having 10 quadrats of size 1m × 1m in the grassland. In forest, 8 belt transects were placed with 10 numbers of quadrats in each of them and size 10m × 10m for tree species. Nested quadrats of size 2m × 2m were also placed for undergrowth vegetation. Data on cover of grass and non grass species in grassland areas and number and basal area for trees and undergrowth species in the forest were collected. Data on threats like grassland invasive species, human entry routes, illegal burning, and resource collection including poaching threat were also recorded. A key was prepared to categorise suitability criteria of rhino habitat. It covered type of vegetation characteristics at micro level, availability of water bodies, nearby forest cover and its undergrowth vegetation, food plants availability all type of threat perspective like habitat disturbance and poaching aspect that also linked present infrastructure and antipoaching measures, awareness level and socio-economic-political situation in fringe villages.

The data was analysed through biological software TWINSPAN (version 2.3, 2005) and R-Console (Oksanen, 2006) to find out the grassland associations at microhabitat level. For forest community, importance value index (Mishra, 1968) of the species was considered to assess habitat characteristics. We modified the methods of Kushwaha *et al.* (2000) to generate a habitat suitability map with support of GIS software Arc/Info version 9.1. Presence of plant associations in preferred rhino habitat of Orang (Hazarika, 2007), Kaziranga (Baruah, 1998) and former rhino areas of Manas were considered in spatial interpolation of geospatial field data to improve the design of the suitability model.

Rehabilitation of Ex-poachers, Forest Dependents and Fringe Villagers

Free training sessions were arranged on available agro-veterinary resources to promote alternative livelihood of families of ex-poacher turned conservation distribution of volunteers and forest dependents. This was followed in by kind support to beneficiaries to start their work by formation of self help groups. Time to time

Table 1.1: Rule for Designing Rhino Habitat Suitability in Manas Tiger Reserve

Suitability Category	*Soil*	*Distance from Aquatic Bodies in Meters*		*Grassland Species Type*	*Distance of Forest Near Grassland in Meters*	*Biotic Pressure*			*Poaching Threat*
		Grass Area	*Forest Area*			*Weed Invasion*	*Grassland Burning**	*Resource Collection*	*Present Situation*
Most suitable	Wet sandy-loam and mud	0-1000	0-500	Wet and lowland	0-500	0	41-60%	0	0
Suitable	River sand	1001-2000	501-1200	Semi wet	501-1000	0.1-20%	30-40%, 61-70%	0.1-5%	Low
Less Suitable	River rock	2001-10000	1201-10000	Upland dry	1001-7000	21-80%	1-29%, 71-80%	6-30%	Moderate
Unsuitable	Dry rock	Above 10000	Above 10000	Degraded or Human habitation	Above 7000	>80%	>80%	>30%	High

* The burning percentage is applicable on those dry sites that prone to annual severe burning, lack of mosaic burning, heavy fuel load due to lack of burning for three years. It is not applicable for wetland or wet grass community having perennial water source. The wet grass community may not be burnt for several years yet fuel load is minimized due to water logging condition to encourage microbial decomposition and deposition of silt in the base of the grass and naturally managed.

monitoring, counselling was done by local NGOs and project personnel to assess the development.

To promote education, school kits were provided to the children of selected ex poachers turned conservation volunteers and forest dependents. We requested the school administrations through our local partners to take care of these children to get proper education.

Capacity Building on Conservation Education and Extension in Fringe Areas

A Rhino conservation education material was prepared in local language. Different information resulted from literature and ongoing research on rhino was incorporated in the preparation of the education material in simple language to reach all section of mass. The material included information about the importance of rhinos and their ecological services as well as about Indian Rhino Vision-2020. Local NGOs and groups working in core of the tiger reserve and adjacent buffer including school teachers and student groups, local personalities, village heads were involved in organizing public meetings targeting sensitive villages. We distributed the conservation material to the participants of the awareness meetings. The benefits of rhino conservation as well as biodiversity conservation, local example of ecological boomerang and ecological services including eco-developmental works and scope of ecotourism were discussed to aware the local people and appeal to the mass to involve in the conservation and protection of rhinos. The design of discussion was thoroughly demonstrated before the local NGOs and other stakeholders as well as free distribution of rhino education material as part of the capacity building activities.

Study Site

Manas tiger reserve, comprising an area of 2837.12 sq. km. is situated in the foothills of Himalayas, flanked by Indo-Bhutan international boundary in the north and more than 200 fringe villages of different tribal and plain communities of Assam in the south. The western boundary extends to river Sonkosh, adjacent to Buxa tiger reserve of West Bengal; while the eastern boundary is marked by Dhansiri river of Udalguri district of Assam. Geographically the area is located in between 89°51′45″E-92°07′00″E longitude and 26°30′00″ N-26°56′43″N latitude and altitude ranging between 40-720 meters from mean sea level. The core zone of Manas tiger reserve having an area of 519.77 sq. km. and lies in between 90°48′00″ E-91°15′00″E longitude and 26°36′00″ N-26°49′00″ latitude was selected for habitat suitability study. Of course, apart from adjacent village of the core zone, the community conservation works was extended to near by buffer villages. The climate of the tiger reserve was found to be subtropical in nature, with temperature lies in between 6°-37° Celsius. The coolest period was from late December to early January while hottest season was recorded in the month of May. The average annual rainfall lies in between 3000 mm-4000 mm, attaining peak during the month of July and August that often causes flash flood for brief period. The driest period of the season was in between late February to early March. The topography of the Manas tiger reserve is a flat plain with gently slopes towards southern areas in core zone while some undulating hilly areas at the

Indo-Bhutan border led to different kinds of ecosystem formation. There is a well-marked bhabor tract of rocky soil towards the northern boundary that contains sandstone, limestone and shale having very little moisture retaining capacity. The bhabor tract generally remains dry and is characterized by maximum relief, dissection and drainage density amongst all the major streams in this area and is deeply entrenched within it. The flood plains abut against the dissected mountain front of the sub Himalayas. The streams and rivers that become turbulent in the rainy season frequently flood the riparian areas. Apart from bifurcating Manas river in Indo-Bhutan border, the core area also comprised of streams like Sukhanjan in west border, Burisuti, Gabharukhonda, Jongrong-Songrong Gyati, Pahumara, Garusora, Rabang, Suti Rabang Doimari, Teklai, Tangunmara, Jia simbu, Mora simbu, Kalpani, Sikari, Madulijora, Sarphuli and Pota at the eastern boundary. These streams and rivers are running through the reserve, carry enormous amount of silt and rock from the foothills resulting in the development of alluvial terraces comprising of thick deposits of mineral particles and detritus overlain with sand and soil of varying depth with shifting river channels and swamps. Highly porous bhabor formation in the northern part as well as terai tract and fine alluvial deposits with underlying pans in the southern parts is typical of the reserve. Except for a few perennial rivers/streams most of the other smaller and shallow water bodies or streams become dry during winter months. This bhabor-terai zone shows characters of dryness of most of the streams even in monsoon period and formation of natural springs in the southern side of the reserve. Of course during rains, there are flows of high speed water through these natural drainages for short period. Besides there are number of standing water bodies, locally known as 'beels' which act as reservoirs for extensive rainy water during monsoon. Most of the water bodies were created by the changing of the river and streams and blockade of the channels. Besides, several small ponds were created by the management authority in different parts of the reserve.

Results and Discussion

Habitat Suitability of Rhinoceros in Manas Tiger Reserve

Habitat Characteristics

The remote sensing satellite imagery based map generated from the study has indicated the presence of all types of plant communities namely tall and short grassland (37.19 per cent), evergreen and semi-evergreen forest (0.18 per cent), mixed moist dense and open deciduous forest (47.34 per cent), water bodies like perennial river, streams, pond and beels (4.41 per cent), dry river (4.08 per cent), river sand (4.202 per cent). The investigation observed that 3.725 per cent area is under encroachment covering two zones in Panbari and Bhuyanpara ranges.

The study identified nine grassland assemblages or associations in rhino habitat namely *Phragmites karka – Themeda villosa - Saccharum ravennae, Arundo donax, Saccharum ravennae, Phragmites karka – Saccharum spontaneum – Saccharum ravennae-Saccharum narenga, Phragmites karka, Imperata cylindrica, Imperata cylindrica-Saccharum narenga, Saccharum narenga, Saccharum spontaneum* and one non grass assemblage *Alpinia allughas.* The investigation showed that the grassland assemblages were distributed in diverse altitude ranging from 50 meters in southern boundary to 280 meters in northern upland. Well developed upland grassland are covered by *Imperata cylindrica* and *Saccharum narenga* based associations where as preferred rhino habitats in the low lying grassland are comprised of associations dominated by *Phragmites karka, Themeda villosa, Saccharum ravennae, Arundo donax* and *Alpinia allughas* type species. Riparian grassland is mainly covered by *Saccharum spontaneum*. The majority of rhino populations were distributed mainly in the eastern part *i.e.*, Bhuyanpara range dominated by low-lying tall grassland assemblage having a height of 3-4.5 meters comprised of *Phragmites karka, Phragmites-Themeda villosa-Saccharum ravennae, Phragmites-Themeda villosa-Alpinia allughas, Arundo doax* and *Saccharum spontaneum* including pure *Alpinia* patch, swampy forest and wetlands located in remote forest areas.

Although one horned rhinoceros prefer grassland community, it also roams forest patches for cover to get rid of hot sun, using aquatic bodies of the forest as preferred wallowing sites during summer and for undergrowth food plants. Sometimes it consumes fruit, even leaves of young trees. In Manas, the evergreen forest is comprised of *Michelia champaca, Dysoxylum binectariferum, Aphanomixis polystachya, Premna bengalensis, Premna latifolia* along with moderate to low population of *Cinnamomum tamala, Madhuka longifolia, Morus laevigata, Eriobotrya bengalensis, Mangifera sylvatica, Artocarpus chaplasha, Spondias axillaris.* The semievergreen forest is dominated mainly by *Aphanamixis polystachya, Anthocephalus chinensis, Lagerstroemia parviflora, Dysoxylum binectariferum, Stereospermum chelonoides, Trewia nudiflora, Syzygium cumini, Syzygium formusum, Michelia champaca, Bauhinia purpurea, Mallotus philippensis, Careya arborea, Toona ciliata, Bombax ceiba.* The mixed deciduous forest is covered by *Lagerstroemia parviflora, Dillenia pentagyna, Bombax ceiba, Careya arborea* along with moderately grown *Trewia nudiflora, Bridelia retusa, Holaruhena antidysenterica, Lagerstroemia parviflora, Terminalia bellerica, Terminalia chebula, Toona ciliata, Sterculia villosa, Oroxylum indicum, Albizzia procera, Albizzia lebbeck, Bischofia javanica, Gmelina arborea*. The riparian bed is also covered by *Bombax ceiba* and *Dalbergia sissoo* assemblage.

The undergrowth forest community of Manas is mainly comprised of sciophytic species like *Phlogocanthus thyrsiflorus, Litsea salicifolia, Adhatoda vasica, Piper mullesua, Mimosa himalayana, Murraya koenigii, Clerodendron viscosum, Glycosmis arborea, Costus speciosus, Blechnum orientale, Piper diffusum, Zingiber zerumbet, Setaria pallida* in the moist forest. The forest floor of the dry deciduous and disturbed forest was covered by mainly heliophytes like *Chromolaena odorata, Leea asiatica, Ageratum conyzoides, Tephrosia candida, Crotalaria sessilifera* etc. The forest floor with high moisture content when opened by moderate logging pressure was quickly covered by fern *Dryopteris paleacea*. On the other hand, the swampy forest floor is mainly covered by cane like *Calamus floribundus* as well as wetland non grass *Alpinia allughas*. The presence of grass *Paspalum conjugatum* indicated bhabor forest having moderate moisture in the forest floor. The riparian bed was naturally covered by *Saccharum spontaneum, Leea asiatica, Dryopteris palaecea, Crotalaria* sp., along with tree sapling like *Bombax ceiba, Lagerstroemia parviflora, Dillenia pentagyna, Callicarpa arborea* and *Dalbergia sissoo.*

Several perennial water bodies generally adjacent to streams were characterized by complex hydro-seral stages even in small areas. These habitats were comprised of free floating hydrophytes, suspended submerged hydrophytes, anchored submerged hydrophytes, anchored hydrophytes with floating leaves, anchored hydrophytes with floating shoots, emergent amphibian hydrophytes and wetland species. The common aquatic plants in the former rhino habitat are *Vallisnaria spiralis, Ottelia alismoides, Eichhornia crassipes, Trapa natans, Pistia stratiotes, Ceratophyllum demersum, Limnophila sessiliflora, Nymphoides cristatum, Nelumbo nucifera.* With increase of terrace height, several emergent amphibian hydrophytes like *Polygonum posumbu, Typha elephantina, Monocharia hastata, Hygroryza aristata, Leersia hexandra, Ipomoea aquatica* spread in the area. Besides several wetland sedges and short grass like *Cyperus brevifolius, C. digitatus, C. killingia, C. iria, Scirpus juncoides, Scripus grossus, Panicum paludosum, Paspalum conjugatum, Carex spiculata,* young *Phragmites karka, Arundo donax,* and non grass species *Lasia spinosa, Polygonum hydropiper, P. barbatum, Alpinia allughas* are distributed in the system. The wet tall grass *Phragmites karka* and *Arundo donax* along with non grass hydrophytes are highly competitive in the aquatic ecosystem present in the grassland areas. They even invaded on the dead remains of the different hydro-seral stages. *Arundo donax* even found to grow with submerged and free floating hydrophytes. The distribution of aquatic species is based on the depth of the water as well as microhabitat condition. Some small water bodies even streams become dry in the winter season. The hydro-seral stages of the aquatic body could be related to the formation of wet grass community and swampy forest as a result of very long term temporal change. While presence of swampy forest was indicated by presence of wetland species like *Alpinia* and cane species *Calamus* as undergrowth vegetation, dominant tall wet grassland formation was confirmed by presence of key grass species covering adjacent to aquatic community as well within it. There are presences of several low lying areas and water bodies in Bhuyanpara range that become shallow due to heavy siltation in fluvial period.

Food Plants

Based on the field observation in Orang national park (Buddhin, 2007) in the north bank of Brahmaputra river, having a rhino population of 68 numbers in 2006 and earlier information, several plant species were identified as potential food plants for one horned rhinoceros in Manas. It may be mentioned that Orang NP is very much similar with Pabitora WLS and Kaziranga NP (both are donor sources for rhino translocation) in terms of habitat condition *i.e.,* located in floodplain of Brahmaputra river. Even there are reports of migration of rhinos from Pabitora to Orang crossing the Brahmaputra river and its sandbars. The preferred food plants are namely *Arundo donax, Phragmites karka, Saccharum ravennae, Cynodon dactylon, Imperata cylindrica, Hymenachne assamica, Hemarthria comprersa, Leersia hexandra, Saccharum narenga, Saccharum spontaneum, Alpinia allughas, Vetiveria zizanioides, Andropogon* sp, *Sclerestachya fusca, Cyperus rotandus, Pallinia ciliata, Ipomoea reptans, Enhydra fluctuans, Solanum carotenens.* Besides they selected young shoots of various trees like *Dalbergia sissoo, Ziziphus jujuba,* fruit of *Trewia nudiflora, Gmelina arborea* etc. Under stress condition creeper weed like *Mikania micrantha* was consumed by translocated rhino that strayed from Manas several times. During flood season even *Eichhornia crassipes* another problematic weed is taken by rhinos that suffered them loose motion (Hazarika, pers comm.). The rhinos also strayed to raid rice crops in fringe villages as observed in all protected areas of Assam.

Management Constrains and Prescription Followed

Exploitation of Bio-resource

There is still some logging pressure in Panbari range, while river channels pf Manas have been used for smuggling of the timber and fuel wood. The smugglers even collected timber from Royal Manas national park, Bhutan and used the western buffer of Panabri range as exit route. The grass collection was confined in boundary areas of Manas. Poachers were still active in some pockets although the overall security of the park found to be better in comparison to 1989-2003 periods. The recent recovery of wildlife items including fresh parts in southern village of Bhuyanpara range and nearby town proved that wildlife traders are still active in some pockets.

There is a good network of different stakeholders and agencies working in the area that helped in minimizing the wildlife trade in the area. The incident of recent recovery of wildlife item is a good example of such good activities. The incident of grass and fuel wood collection has been lowered with active participation of conservation volunteers of local NGOs. More poachers from Kahitema near Manas – Beki river system surrendered to the authority. This was termed as a replication and extension of the community initiative to help poachers and wood cutters to surrender and work as conservation volunteers if they wish. The security forces and local administration also extended support to the forest department in antipoaching operations.

Status of Infrastructure

The southern boundary needs to be completely restored immediately by construction of permanent bridges and restoration of degraded roads. There is also need to develop the remaining camps and beats with all facilities before the major rhino translocation. It may be mentioned that despite well designed security system sudden increase of poaching incidents in Kaziranga national park was a cause of concern in last 2-3 years.

Majority of the camps have been restored while the construction work of remaining camps in some points delayed due to adverse situation and technical difficulty. The bridges have been restored to facilitate two wheelers. There is a need of strong permanent bridge so that vehicle can be easily passed regularly particularly during crisis period.

Flood and Threat of Dam in Bhutan

While low to moderate nature of flood favour in improvement of rhino habitats by maintaining successional grassland, creating new wallowing sites and recharging aquatic bodies, heavy flood created a great threat to the biodiversity of Manas. The heavy siltation by devastating flood blocked one major channel of Manas river system in 2004. The other channel bearing heavy water washed away several fringe villages of the reserve due to breaching of an embankment. This was due to excess release of water from Kurichu dam located in a tributary of Manas river in Bhutan. The same incident was repeated on 31st July, 2007 resulting breaching of the same Narayanguri embankment. It was reported that excess water released at a rate of 10500 cubic units per second from the Kurishu dam. The breaching site was very near to the central range office Bansbari. The main road between Borpeta road field director office and Bansbari range office was cut off due to this severe flood. Badly effected fringe villages of Manas tiger reserve were namely Narayanguri, Soonbari, Alangabari, Bispani, Kalpani, Churaguri, Khagrabari. The excess water through a channel called Kalpani moved to southeast direction and met another stream Palla that approaches to south west direction. This also threatened to the fringe villages in eastern side of the park and road approaching to Bhuyanpara range as well as alternative road to the Bansbari range. This flood also posed a threat to some areas inside the park. The riparian camps were washed away. The gradual increasing of flood intensity in recent years also threatened the beautiful Mothanguri bunglow in Indo-Bhutan border. The crop fields in the fringes became unsuitable due to heavy siltation. Naturally the human dependence on resources of Manas will be increased if a practical strategy is not taken to rehabilitate the affected villagers.

The local administration extended some support of flood relief and organized medical camps, later repaired and restored the Narayanguri embankment and road communication.

Man Animal Conflict

The man animal conflict in the form of crop raids, household and property damage and injury and death by Asian elephant in the southern fringe is a continued threat. There is no buffer cover on the southern side of the Manas tiger reserve.

Some study on the conflict problem was done by some organizations. Department staff, volunteers and villagers have a tough time during the conflict period.

Man Power

There was shortage of 150 forest staff besides lack of well trained young officials in Manas during starting of Indian Rhino Vision-2020. Though presently volunteers of local NGOs have been extending support to the authority in patrolling the reserve, the shortage of well trained permanent forest guards must be fulfilled before the major translocation programme.

Gradually the forest department with support of other stakeholders provided fixed remuneration to conservation volunteers of local NGOs. Initially armed staff from other departments have been engaged to fill up the gaps.

Grassland Burning Perspective

Annual grassland burning pattern has been playing a major role in the temporal dynamics of the rhino habitat. The investigation indicated that the intensity of the burning percentage of grassland depended on the soil moisture of the area. The soil moisture in the upland areas reached significantly low level in dry season causing higher to severe burning (65-100 per cent) in different patches of *Saccharum narenga* and *Imperata cylindrica* dominated assemblages whereas grassland assemblages of the low lying and swampy habitat showed low percentage of burning (10-40 per cent) or remained unburned. The grass associations of riparian areas showed moderate burning (45-55 per cent) due to diverse water holding capacity of the sand bars. The presence of control burning is limited. Better patrolling activities restricted illegal burning in some places. The presence of low moisture in Panbari range and low awareness of villagers might resulted severe burning just like other years observed by us since 2000. There has been improvement to keep the burning percentage low in central zone *i.e.*, Bansbari range but areas adjacent to southern boundary could not be saved of anthropogenic burning while the low lying Bhuyanpara range naturally observed very low percentage burning due to high moisture content of the biomass. We also looked into the matter of recent forest fire caused in Manas landscape. While it is anthropogenic in nature, the drought observed in last winter caused heavy fire in eastern buffer areas of Manas.

The forest guards and volunteers were instructed by officials to check the anthropogenic burning in the grassland. We also suggested control mosaic burning pattern as well as early, mid and late burning depending on the perspective of grassland habitat. During awareness and capacity building programme, we always cover this issue.

Grassland Invasion

High intensity of invasion by tree species *Bombax ceiba* was observed in *Imperata cylindrica, Saccharum narenga, Imperata cylindrica-Saccharum narenga, Saccharum spontaneum* grassland assemblages in response to various disturbance situations like burn, burn-cut and burn-cut-external grazing while in disturbance free condition there was slow growth of the already existing saplings of the invasive species (Bezbarua, 2007). Maximum increase of other non grass species like undershrubs *Leea asiatica* and *Chromolaena odorata* were noticed in the gap areas created by different disturbed conditions in the grassland of the Manas tiger reserve. The *Leea asiatica* is generally known to spread through underground propagules and locally cover the gap areas formed each year by continued disturbances, while *Bombax* has advantage of seeds attached to cotton ball that spreads through high speed wind. The preferred rhino areas i. e., Bhuyanpara range mainly dominated by *Phragmites karka, Themeda villosa, Alpinia allughas* type species was also gradually invaded by *Chromolaena odorata, Leea asiatica* and *Bombax ceiba* in adjacent boundary areas due to excess external biotic pressure. Apart from presence of some buffalo Khuti/Bathan in boundary areas of Manas, most of villagers have domestic cattle that graze on the grassland of the tiger reserve. The *Alpinia allughas* growing in wet condition also has been decreased due to overexploitation and subsequent invasion by weed species *Chromolaena odorata*. The most threatened aquatic weed *Eichhornia crassipes* was observed in southern boundary zone.

Awareness on the impact of weed in Manas was started by us and message was gradually well understood by local, national and international stakeholders. During preparation of management plan, practical prescription to minimize weed problem was delivered to the authority of Manas biosphere reserve and subsequently followed in some pockets.

Medical Facility

Malaria is the main disease in the foot hills of Indo-Bhutan border. Each year, both forest staff and villagers suffered badly in monsoon period due to malaria. In 2006, a few cases of death were also reported from the fringe villages. There was no good hospital in the region and villagers have the extra burden of expenditure for better treatment in outside hospital.

Recently death cases declined significantly due to better awareness, timely identification of malaria and necessary medication provided by medical camps, health centres and pharmacies.

Tourist Pressure

Considering the improvement of law and order situation, excessive domestic tourist flow in the park was observed in 2005-2006. At least 450 plus vehicles entered on 1st January, 2006. Average 50-100 numbers of vehicle entered for successive 2 weeks. Tourists were not aware of the rule and regulation of the national park although forest guards tried their best to maintain the law inside the park. A better management of tourist was observed since 2007.

Socio Economic and Political Perspective

The political situation was significantly improved that resulted restoration of Manas and its biodiversity attributes. Different local NGOs in fringes of the Manas landscape converted the poachers and forest dependents to forest protectors. Yet some groups/individuals are trying to mislead one section of poor villagers having low awareness plus different political ideology. It may be mentioned that Bodoland Territorial Council was created after one militant group laid arms and formed the tribal council. While a major faction of another group is maintaining ceasefire with the government. The problem arises with some incidents related to law and order to get powers particularly during elections. The local NGOs working in different zones have member with different political ideology. It has been observed that opportunists are trying to increase difference between these conservation groups. The level of success of different groups is also different. We agree that healthy competition is necessary. But there should be better coordination between all the groups particularly to deal with antipoaching operations and wildlife crime issues and always take support of forest authority and local administration.

Our partner NGO faced tough situations in eastern and western fringe villages in different periods. On western side of Kokilabari local villagers (having poor awareness) near Satrapa camp destroyed one forest camp on 13th April, 2007 with the encouragement of an owner of Baffalo Bathan. The situation was later controlled by the forest authority and local administration. Similarly in eastern buffer area they faced threat from local villagers due to restriction of timber felling and in a few incidents there was creation of law and order situation. Local administration and forest department controlled the situation in time.

The type of problem has been gradually minimized with continuous awareness programme by different stakeholders and formation of village based local conservation groups. We talked repeatedly to increase community conservation activities in these areas for long term conservation.

Stray of Rhinos

There was history of stray incidents of rhinos in southern eastern fringes of Manas tiger reserve (Core zone). A detailed survey identified the different sensitive points. A strong monitoring group that included forest staff and officials including administrative and police, conservation volunteers of all the local NGOs and field biologists was formed to provide security to animal in such situations. Awareness programme was conducted in those areas.

Geo-spatial Habitat Suitability Modeling of Rhinoceros

Apart from land cover map showing complexity of the rhino habitat, poaching threat map and habitat disturbance (integrating anthropogenic burning, resource exploitation, weed, and external grazing pressure) maps were generated. The final integration was done in two scenarios. The land cover map was subjected to only habitat disturbance map in scenario I while both habitat disturbance and poaching threat maps in Scenario II to design four classes of rhino habitat suitability. The suitability model categories the habitat in most suitable, suitable, less suitable and unsuitable zone. The analysis without poaching threat indicated the most suitable zone (42.35 per cent) in grassland and adjacent swamp forest of eastern Bhuyanpara range and extending to central Bansbari range covering perennial water bodies including Manas river ecosystem and part of Panbari range. The buffer of the most suitable habitat comprises suitable habitat (38.72 per cent) located in fragmented island grassland, unnoticed small ponds under the shade of deciduous forest in Indo-Bhutan border zone with several small streams, moderately moist grassland in western Panbari range and southern wet grass communities in low-lying zone. The GIS software identified a portion of forest in northeastern zone and grassland in western zone in same suitable category. Though grassland is more preferred by rhinos, the software gave more weight on the well protection measures by tribal community and authority in north eastern forest habitat with isolated aquatic bodies and island grass patches to place with the good grassland in western Panbari range but with moderate protection measure. The dryness of bhabor forest patches in north and different grassland areas in south particularly in south western side subjected to biotic pressure like resource collection, unscheduled burning, external grazing pressure by domestic cattle and increasing invasive species like *Chromolaena*, *Leea* and *Bombax* converted the suitable area to less suitable (13.35 per cent) category. The encroached area in Panbari range and Bhuyapara range including degraded habitat are categorized as unsuitable (5.57 per cent) zone. When we incorporated the poaching threat also linked with low awareness level and current socio-economic-political situation, the habitat suitability scenario was significantly changed. The most suitable area was decreased to 23.48 per cent and there was slight increase of suitable area (46.33 per cent) and unsuitable (6.02 per cent) areas and moderate increase of less suitable area (24.16 per cent). The most suitable and suitable areas in different pockets in Manas river ecosystem and western zone are identified as less suitable even unsuitable due to poaching threat.

We designed the suitability considering both annual dry winter and flood season. Short time food scarcity for herbivores was noticed in Panbari range during dry season (February–March 10) and repeated burning of grassland. Secondly considering growth pattern of grassland (20 percent of sampling sites) influenced by excess water and drought condition, the current habitat suitability may show some change. We

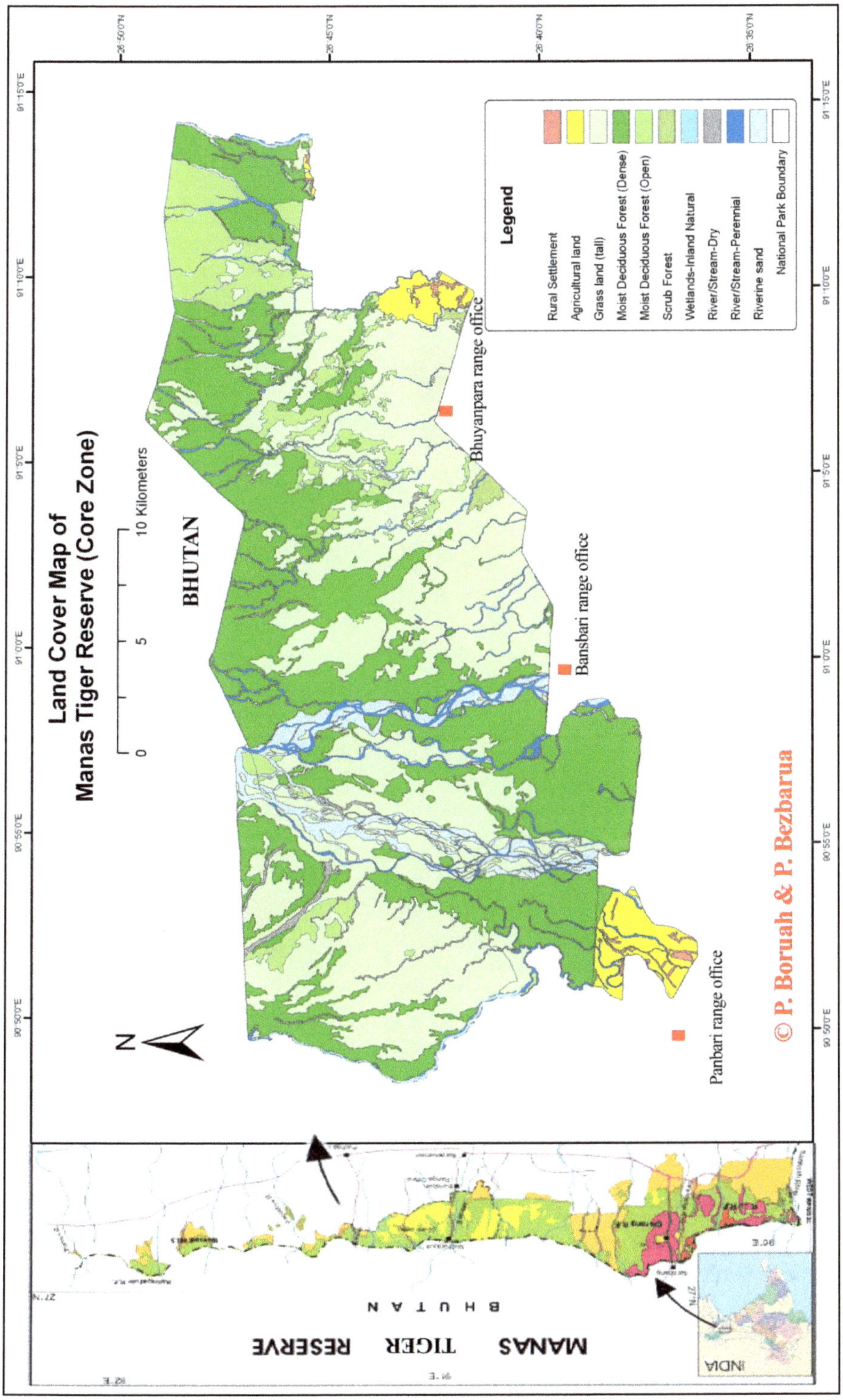

Figure 1.1: Land Cover Map of Manas Tiger Reserve (Core Zone)

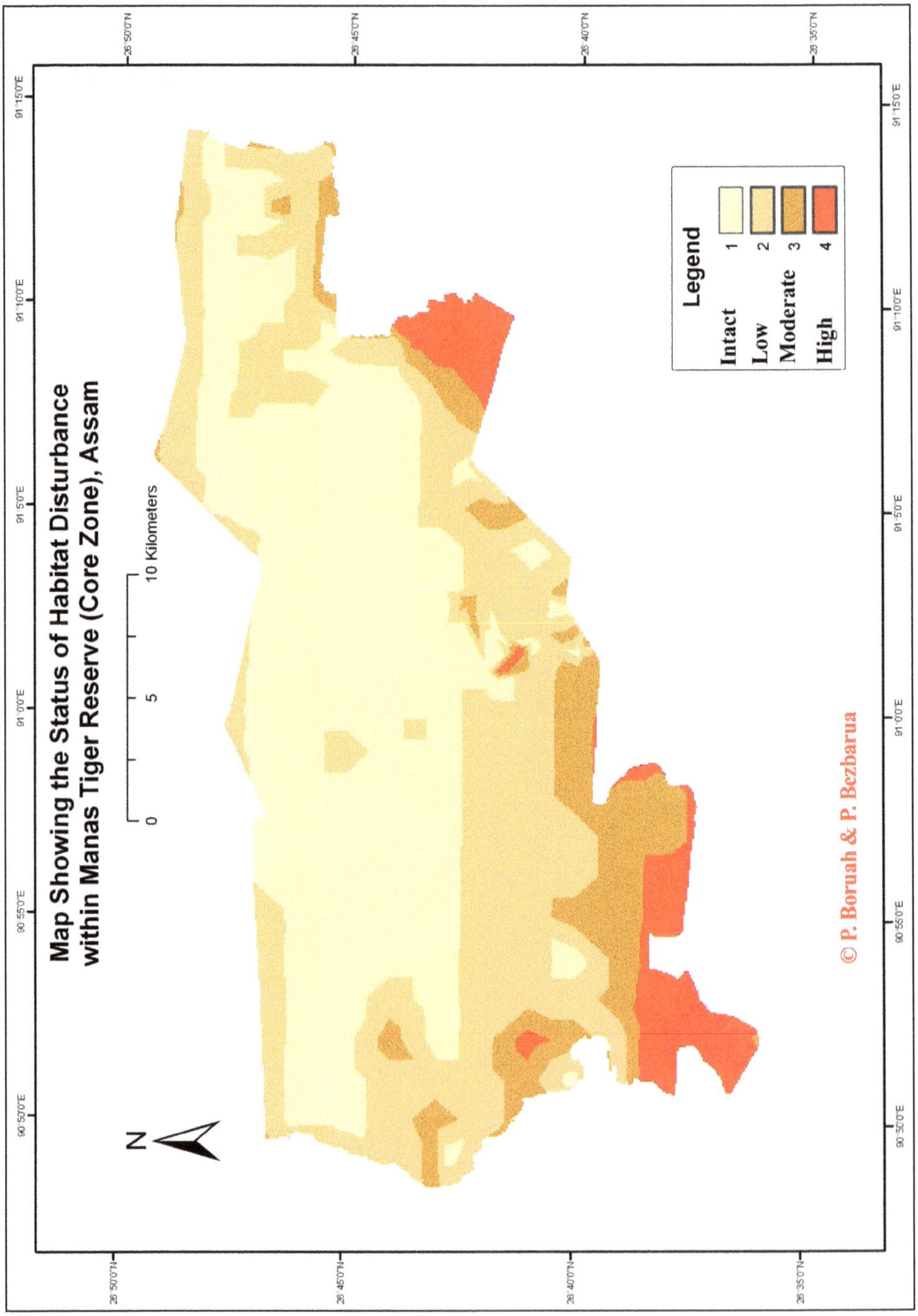

Figure 1.2: Habitat Disturbance Status in Manas Tiger Reserve

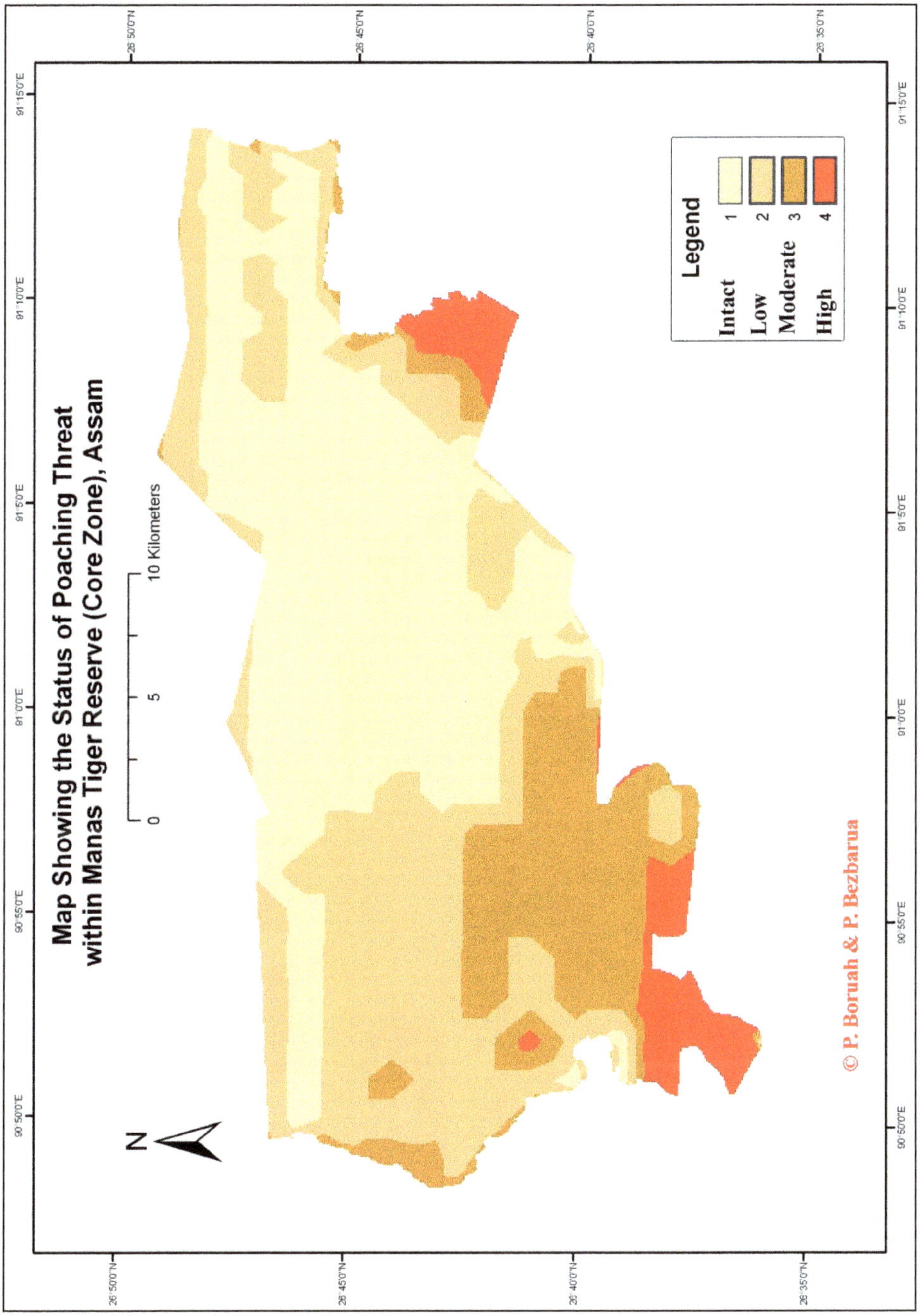

Figure 1.3: Poaching Threat Status in Manas Tiger Reserve

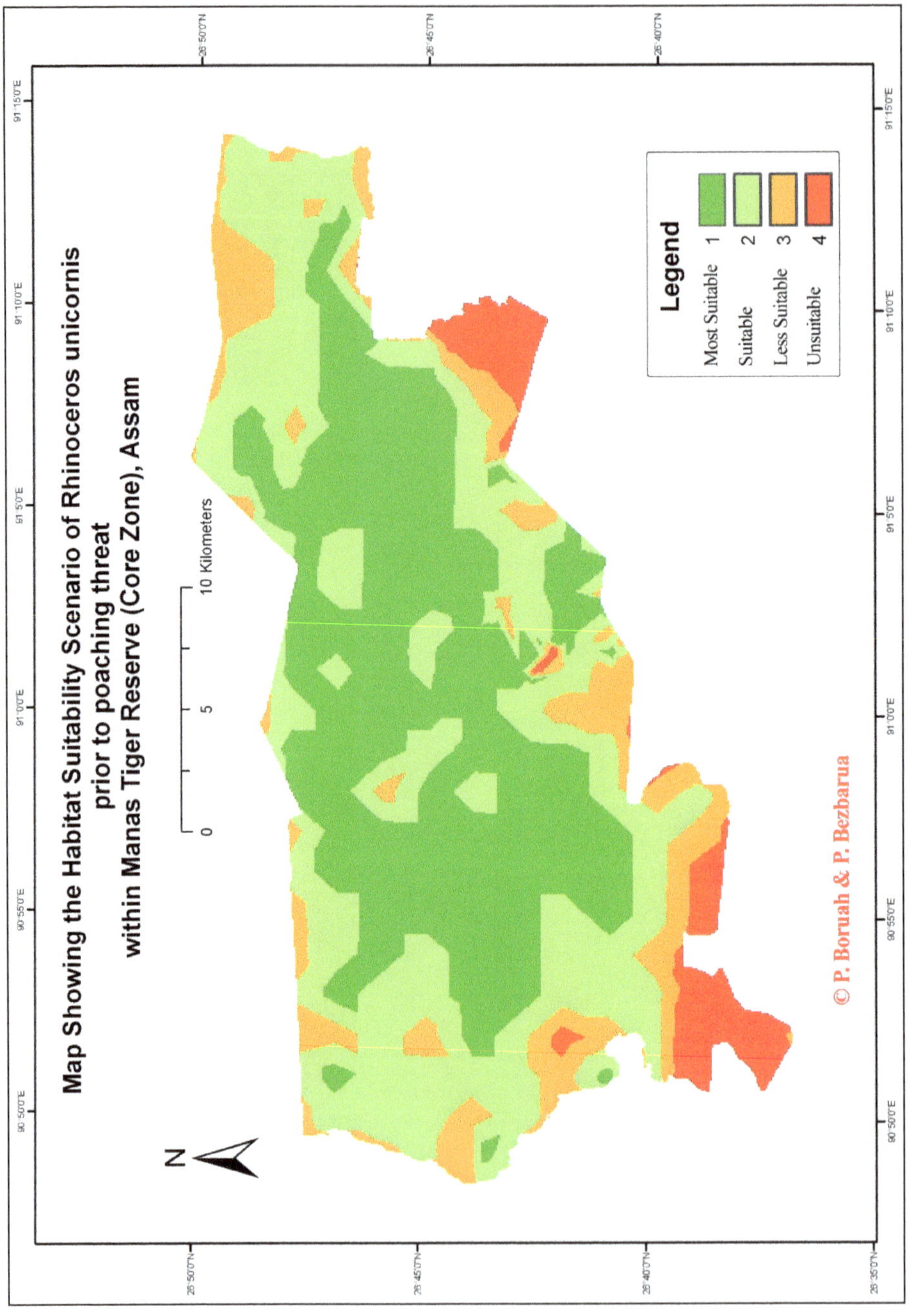

Figure 1.4: Habitat Suitability for *Rhinoceros unicornis* in Manas Tiger Reserve (Scenario I only habitat disturbance were considered)

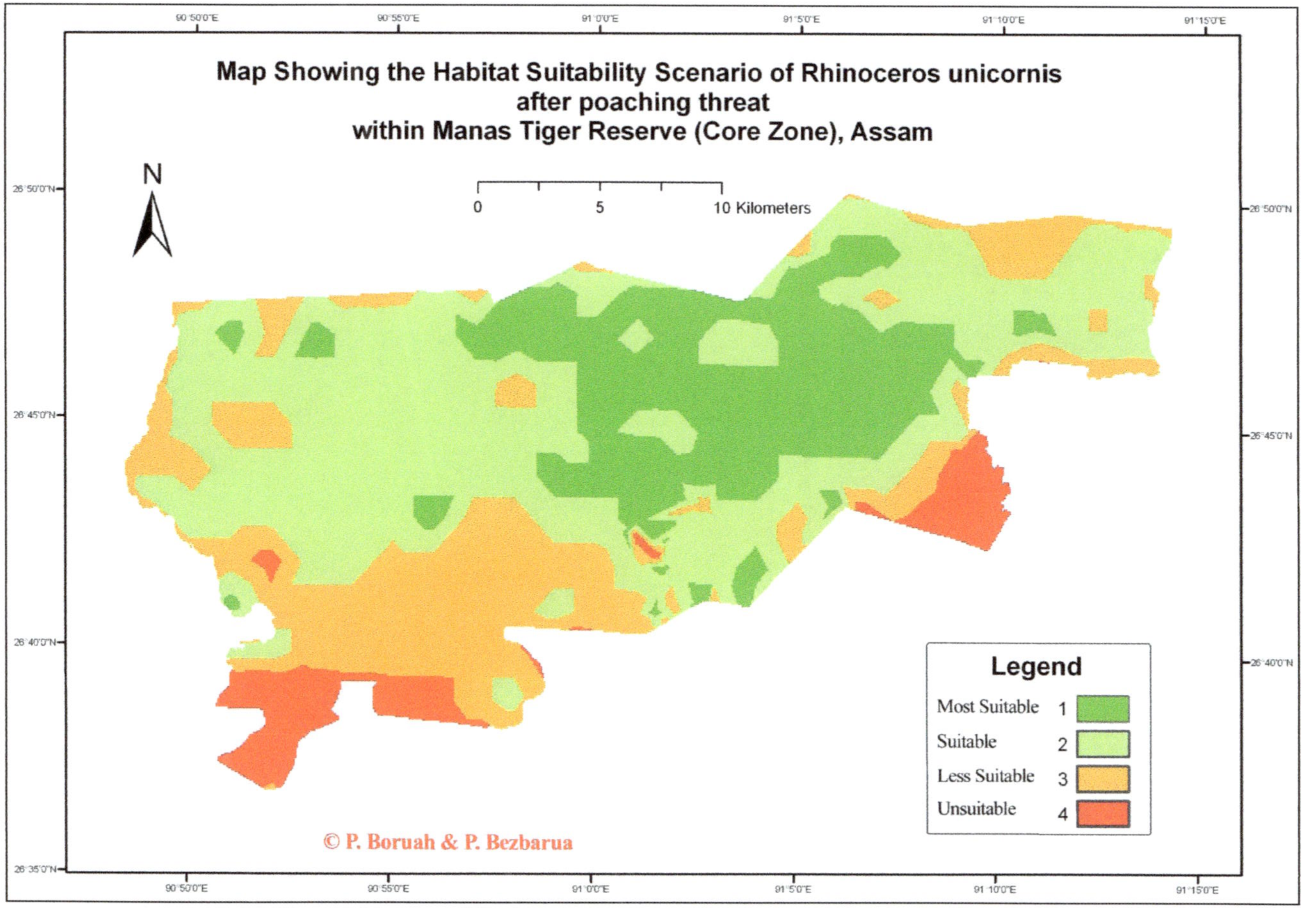

Figure 1.5: Habitat Suitability for *Rhinoceros unicornis* in Manas Tiger Reserve (Scenario II Both, Poaching threat and habitat disturbance were considered)

Figure 1.6: Community Awareness on Manas Biodiversity

Figure 1.7: Translocated Rhinos in Manas Tiger Reserve

predicted that the most suitable, suitable and less suitable areas would show fluctuation to other category and vice versa about ±5-10 per cent depending upon the severe drought or heavy flood caused by low or excess rainfall including melting rate of glaciers in Bhutan Himalayas due to global warming and climate change.

Experimental Rhino Reintroduction in Manas

About 2 sq. km. of plot under Bansbari range with solar fencing and near to the boundary was specially surveyed before the first experimental rhino reintroduction in Manas. We found two major grassland associations namely *Saccharum narenga* (*2.5* meter tall) and *Imperata cylindrica* in uplands. It also included perennial stream - Kasimdoha known to be a former rhino breeding site in Manas. It is near to a forest camp (Gahari farm) having two plots initially (250m long and 150m wide, the connecting side of each plot has a wideness of 70 m, GPS point–26°40′43″N, 91°00′37″E and 26°40′31″N and 91°00′48″E) connected by a small corridor. It was prepared to keep three rescued young rhinos treated and cared in the Rescue and Rehabilitation Center near Kaziranga NP. The grassland in the plot as well as surrounding areas is mainly dominated by grass *Saccharum narenga* while *Themada villosa, Alpinia allughas* are also scattered in aquatic environment. A small seasonal drain (26°40′32"N 91°00′51"E) linked to the stream Gyati (approximate 5 m wide, 260m long inside the plot and 1 meter deep of water in December) is passing through the first plot. One rescued rhino was translocated on 2006 followed by another two numbers on 2007. The rhinos were monitored round the clock for finding its adaptation to new sites and better security in a park that experienced total elimination of the pachyderm during violent agitation and insurgency problem. From the security point of view (in 2006) the area was suitable for introduced rhino calfs. But it was noticed that there were very few wallowing sites as well as fresh food plants during dry winter season. Therefore selective grassland burning was suggested for growing of new palatable grass after observing the poor health condition of the animal during winter season. The area has been further extended covering some low lying areas. Different sections of the people living in fringe areas were interviewed to know peoples' view regaring the rhino reintroduction and future of Manas which were incorporated with inputs of intelligence source to assess the threats.

Considering the adaptation of released rhino calfs in Manas, the forest department took careful initiative for wild to wild rhino translocation. The initiatial plan to translocate 2 male and 2 female adult rhinos from Pabitora wildlife sanctuary to Manas tiger reserve was restricted to only 2 male rhinos due to security and other technical reasons apart from fourth rescued young rhino from rehabilitation centre of Kaziranga NP. In 2008, two radio collared wild rhinos from Pabitora wildlife sanctuary was released in open grassland (26°41′46″ N and 91°00′37″ E) areas of Manas. The post release monitoring through radio collar signal and direct sighting indicated that the pachyderm spent most of the time in most suitable and suitable habitat designed. Of course one male rhino prefer to stay in the fenced grassland for long period while the second one strayed several times. One rhino calf died last year due to unknown reason (suspected to be infighting and infection of wound) and there were 5 rhinos in Manas in 2008.

Stray of Translocated Rhino: Chronology of the Crisis and Critical Observation

On 1st September, 2008 morning, we came to know from one member of a local NGO that one translocated male rhino again had crossed the southern boundary through Betbari area of the Manas tiger reserve on the night of 31st August. The forest guard and conservation volunteers of all local conservation groups tried to push back the animal to the reserve. But the animal proceeded towards southern direction through paddy fields and located in Silbari area near Anchali which was aerially around 5 km south east from Betbari. The incident was immediately reported to the top forest officials of Assam Forest department as well as Manas tiger reserve. The rhino rested the night in the ponds covered by community bamboo forest. Domestic elephants were on move towards the area from the central range office Bansbari. There was large gathering of villagers that disturbed the animal. It further moved to southern direction and located in muddy crop field near Baghmara, aerially 10km south to nearest park boundary on 2nd September. The forest official and conservation volunteers has been trying to push back it towards the reserve including ensure protection of the animal while local administration and security forces of the area also extended help for overall control of the situation raised from gathering villagers. Possibility of tranquilizing the animal was discussed although the unfavorable ground condition *i.e.*, continuous rainy periods causing inundated the low lying areas, muddy grounds and bad roads for transportation of the animal were focused. Meanwhile the animal roamed around Soibari near Nityananda on 3rd September. We supported top forest official in Guwahati in sharing information from ground and GPS data was plotted in satellite imageries based map to get a better perspective of the movement pattern of the rhino. The information generated from the map has been immediately delivered to the concerned authority at ground that helped in better understanding of the situation and taking appropriate measures. The map indicated that it was aerially 14 km south from Kokilabari beat, 16km south east from Bhuyanpara range (the eastern range of Manas national park) and 22 km south east from Bansbari range (central range). The animal crossed the Kaldia river and found roaming towards east direction. In the meantime, it was decided to tranquilize the animal despite bad weather. On 4th September, the animal moved further east and roamed around Pamapathar - Dalbil area near Musalpur- the headquarter of Baska district. One tranquilizer expert visited the site with other veterinary experts. While the forest official and conservation NGOs continuously monitoring the situation another NGO- Manas Souci Konkhor Ecotourism Society (working the eastern buffer of Manas tiger reserve) joined the team. On 5th September, One biologist from department of Botany, Gauhati University, Guwahati rushed to the area via Hajo Nalbari road as the 31st National Highway near Rangia has been under water. Meanwhile the animal already crossed Morapagladia (one branch of Pagladia river system) and moved in the paddy field of Kadamtola village.

We observed that the rhino was very in stress condition. It was lying in a rice field having approximately 0.675 sq km that was surrounded mainly by thin community bamboo forest including *Areca catechu* plantation. We found that the rhino consumed *Mikania micrantha* a creeping weed, young bamboo (*Bambusa tulda*) leaves and rice leaves. Large gathering of people numbering near about 700 were watching the animal. Meanwhile we found tired conservation volunteers and forest officials. But the team continued to monitor and protecting the animal day and night. Elephants were moved to the place to cover the animal. Local administration discussed with the

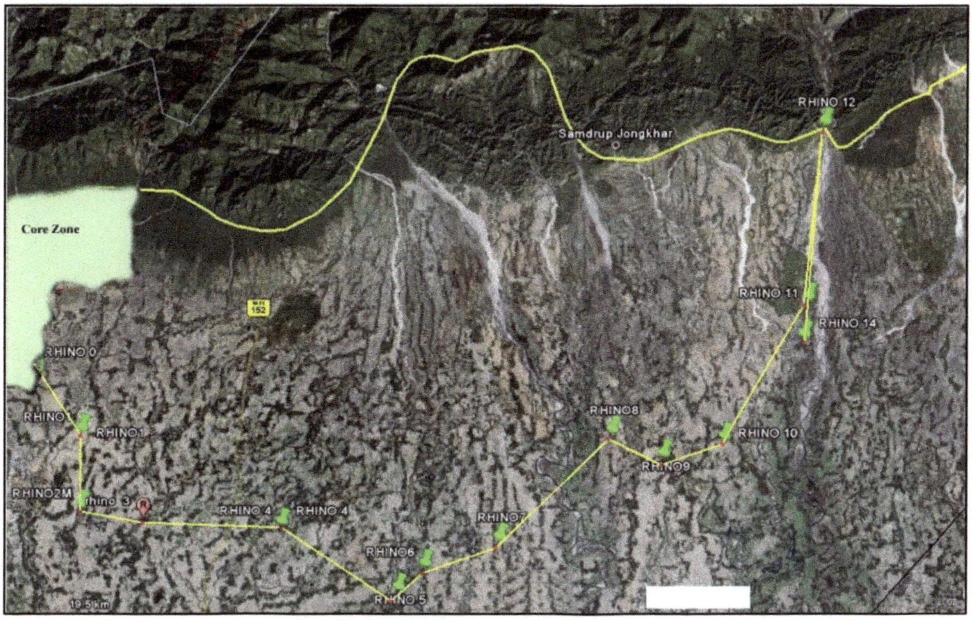

Figure 1.8: Map Showing Rhino Stray Path in Eastern Fringes of the Manas Tiger Reserve

Figure 1.9: Surrendered Arms of Poacher in Community Museum

Plate 1: Wallowing site

Plate 5: Suitable rhino habitat

Plate 2: Riparian habitat

Plate 6: Grassland

Plate 3: Anthropogenic burning

Plate 7: Grazing pressure and *Leea asiatica* invasion

Plate 4: *Chromaelaena odorata* invasion

Plate 8: Encroachment area in Panbari range

Plate 9 : Fuelwood market near Manas TR

Plate 13: *Bombax ceiba* invasion

Plate 10: Fishing and *Eichhornia* invasion in rhino habitat

Plate 14 : Flood cut off road communication

Plate 11 : Surrendered Arms by poachers

Plate 15 : Hardcore ex-poachers turned to conservation volunteers

Plate 12 : Restoration of forest camp

Plate 16: Restoration of boundary bridge

Plate 17 : School bag distribution to children of conservation volunteer

Plate 18 : Rhino mask attracted young students

Plate 19 : Conservation education programme

Plate 20 : Participants of capacity building programme on alternative livelyhood generation

Plate 21 : Students with school bags

Plate 22 : Rhino conservation awareness campaign

Plate 23 : Distribution of Citrus saplings to self help groups

Plate 24: Piggery developed by SHG

Plate 25 : PTES tourists with conservation volunteers

Plate 26 : PTES official interacting with SHG

Plate 27 : Wild adult rhino translocated to Manas

Plate 28 : Stray of translocated rhino

Plate 29 : Forest guards & conservation volunteers patrolling in rhino habitat

Plate 30 : PTES tourists with members of SHGs

Plate 31 : Rescued translocated young rhino in fenced grassland in Manas

Plate 32: Principal investigator collecting data in rhino habitat

community leaders to maintain the law and order and safety of the animal. While we found people were very much interested to get a view of the animal, some enthusiastic young boys teased the pachyderm, came close even touched it. The result was injury of a couple of young boys since the rhino entered in human habitation. On 6th September strategies made to tranquilize the animal. The animal was found in a place called Khagrabari in the evening. On 7th September it moved to 5 km north east from the previous day's position. On 8th September it was located 2 km north to Dhamdhama. Forest official and other experts discussed that it would be risky to tranquilize the highly exhausted rhino. Decision was taken to continue the monitoring of the animal. On 8th September it moved further east by crossing Pagladia river. It moveed further south east direction during the period of 9th -11th September. On 12th September it was located in north side near another river Barnadi that was also close to the Barnadi wildlife sanctuary. At last on 13th September the radio-collared rhino was tranquilized and back to the Manas tiger reserve.

It may be mentioned the same rhino strayed several times in past. On 6th July we saw how the conservation volunteers and forest guards spent the whole night to prevent the animal to cross the eastern, southern and northern boundary of the park. That time one source informed us that sometimes signal from the radio collar had not been received properly due to technical problem which increased the risk to locate the exact location of the animal at night time.

There is a need to find out the stray behaviours of the rhino. First we think, all should agree that wild animal does not follow a boundary or area decided by human. In this particular case, this stray behaviour of the translocated rhino might be linked to insecuredness in the new environment. Though rhino is a solitary animal it shared common wallowing site. Further it needs mating partner. This wild rhino was from Pabitora wildlife sanctuary, having highest density of rhinos and high intensity of rhino stray incidents. The other male rhino released at the same time did not show the stray behabiour. As it found its partner in the fencing plot since mid June where four rescued female rhinos (one or two attained sexual maturity) were already translocated by the department. The dominant wild male rhino in the fencing plot also droved away the other rhino (strayed rhino) when it tried to enter the fencing plot. During early part of our study, we came to know about several incidents of stray of rhinos in the south eastern boundary of Manas during paddy season. As rice is the primary crop of the region rhinos graze rice during summer period but generally confined to short distance from the boundary.

Therefore it can be concluded that the wild rhino should be translocated in cluster having both male and female. The initial plan was two male and two female at first phase. Though it was restricted to two rhinos at first phase yet one male and one female would naturally have less chances of stray in Manas. Further we must not forget if a major force, both conservation volunteers and forest staff engaged in such incident then security of the park will be at risk when there would be more rhinos. The common mass should have awarness about the wildlife stray while allowing them to view from a safe distance, avoid teasing the animal and extend support to the department.

Rehabilitation of Ex-poachers-cum-Conservation Volunteers and Forest Dependents

Promotion of School Education for the Children of Conservation Volunteers

School bags with necessary education accessories were initially distributed to the children of 20 ex-poachers residing in the fringe areas of Bhuyanpara range of Manas tiger reserve. It was distributed during the Sahitya Sabha (an annual large gathering or meeting to discuss various aspect of language and literature) of the Bodo community. The children are from eight villages and from six schools having age ranging from 3-10 years. The aim and objectives of Indian Rhino Vision-2020 was focused during the distribution of the school bags. This innovative method to promote education for the children of ex-poachers turned conservation volunteers has a good impact in the local society to give more importance in education to the young generation. It helped to get additional support from other organization to replicate the model and we distributed school bags to the children of another 60 families identified as forest dependents and ex-poachers cum conservation workers from 23 fringe villages. All the 80 students are continuing their education. It was a satisfaction to us when a proud father and a hardcore ex rhino poacher cum hardcore conservation volunteer informed the visiting officials of a funding organization that his girl child had stood first in the annual examination. Now local NGO informed us that a few big organizations have come forward to provide scholarship to the school students of the forest dependents. A small beginning to save the children from being a future poacher or anti-social element, once unknown and neglected issue in the wildlife conservation movement, gradually get momentum and surely have long term impact in rhino conservation.

Alternative Livelihood Resource Generation Programme

Based on the potential of the area we organized series of training programme on alternative livelihood generation to benefit the families of conservation volunteers. In the first phase training on plantation crop management was provided to 19 numbers of female and 3 numbers of male participants in Kokilabari area. Each trainee belonged to a member of either a family of ex poacher or forest dependent. This design was followed as per the discussion with the local NGO. In kind support like rhizomes of Zinger (*Zingiber officinale*), seeds of Pigeon pea (*Cajanas cajan*) was provided to three self help groups (female) having members of 10 numbers. The most encouraging observation was that the participants had come forward for joint effort and working for the resource generation. Each member deposited a small amount to raise a common fund which was used for preparation of soil, bio-fencing and other related cost. One group even got allotment (one year lease in Kokilabari farm) of land for one year for rice cultivation. These activities although very simple but a new beginning by forest dependents/families of ex-poachers cum conservation volunteers significantly encouraged the other forest dependents. Observing the changing behaviour, we planned to involve more forest dependents in the programme in phase by phase. In the second phase we supported in raising piggeries to three self help groups. All the groups found this option more easy and continue the works. The profit from the piggeries has been used by one group to cultivate rice as well as expenditure for the education of their children. In the third phase we also got additional support from another organization to continue the community conservation work. Last year a training programme on agricultural and veterinary resources and their disease control were organized. About 1000 numbers of Citrus saplings were distributed to the 15 self help groups (having 10 members in each group) of the area. The resource person also explained the potentiality of multiple horticultural items in the same land for citrus, vermi-compost to get organic manure and raising duck amongst the beneficiaries. Apart from supplying natural vitamin C to the children and other family members, the citrus fruit has good demand in the market to increase the income level of the SHGs. Now we found more people coming for citrus plantation in the area. The average success rate of all beneficiary SHGs found to be increased from 35 percent (in first phase) to 50-60 percent, while some groups did extremely well in all three items beside increased the income significantly and involve in rice cultivation by taking annual lease of land. Thus our small community conservation work has also experimented with other traditional items like advanced handloom, bee keeping etc. and provided linkage to rural development department to get more support to continue the work. We jointly organized a seven days training programme on bee keeping and allied subjects like sericulture in eastern buffer of Manas with collaboration with another local NGO, Government rural development department and Botany department, Gauhati University. The area is a former rhino habitat and one translocated rhino strayed there for several days. Amongst the total 30 participants from 16 fringe villages of buffer zones, 20 persons selected bee keeping as preferred item. After the training, bee boxes and one live bee hive captured were distributed amongst the 10 SHGs formed by beneficiaries. The model of alternative livelihood generation programme gradually empowered the women around Manas and played a pivotal role in better involvement of society in rhino conservation.

The funding organization People's Trust for Endangered species (PTES), UK visited Manas to monitor our work. We arranged a trip that helped our local partner NGO in a crucial period of fund scarcity to supply ration for 80 conservation volunteers patrolling the eastern range of Manas tiger reserve. This trip raised an amount about Indian Rupees 100,000 equal to 1370 pound plus global campaign by travel writers to promote ecotourism of the NGO and to continue their good work.

Conservation Education and Capacity Building in Fringe Villages of Rhino Habitat

We traveled and extensively visited the fringe areas of the Manas tiger reserve in January 2006 to evaluate the present perspective of conservation awareness level on Indian rhinoceros and threat. We met local stakeholders that included village heads, villagers, local conservation groups, ex poachers turned conservation volunteers, teachers, students and forest officials including guards and officials of local administration. It helped in finding out gap areas that need immediate attention regarding rhino conservation awareness. It had been noticed that although there was some improvement of security scenario; disturbance in terms of external grazing

Figure 1.10: Rhino Conservation Education Material in Local Language

pressure from local cattle farms near the southern boundary of the park, collection of thatch grass, logging and fuel wood exploitation mainly in the riparian and western range was continued to some extent. There were reports of some poachers in some areas though concrete information was not shared by local people at that time. The survey also indicated that most of the local people including teachers and local tribal leaders even some ground forest staff and conservation NGOs and volunteers were also unaware about actual aim Indian Rhino Vision 2020. A few persons knew that rhino might be brought to Manas from Kaziranga.

Therefore we started our conservation campaign through discussion with the local stakeholders in their area. We learned the local problem faced by the stakeholders and tried to find solution to minimize with an aim to link with ecological service provided by Manas landscape and its rhinos. We also explained how their problem was originated due to ecological boomerang in the area. We prepared a rhino conservation material in local language that focused status of rhino and detail of Indian Rhino Vision 2020 and distributed it. Next with support of local NGOs, we campaigned for rhino conservation in small as well as big meetings. Focus on the needs of Manas and its rhinos as well as other wildlife were discussed during the awareness meetings where large numbers of local villagers, teachers, students, and community leaders actively participated. These conservation meetings found to be unique in the sense that we gave more importance on community conservation activities to motivate poor forest dependents to take alternative livelihood methods instead of forest resources. In village level meetings we covered 27 villages of eastern and central range including buffer zones considering the immediate need to improve the awareness level as these were prime rhino habitat. Of course the participants from 49 villages attended our different programmes. The opening and closing Manas centenary celebration and Bodo Savitya Sabha (Bodo Language and literature Meeting) gave us the scope to reach local community of most of the fringe of the tiger reserve even the buffer zone. Besides in later part, we discussed conservation issues with two local NGOs working in western range and distributed leaflets as part of capacity building in their area. The limited fund was another cause we could not invest more time in western range. So far, we provided large number of leaflets to the six front line conservation NGOs working in core zone. We also monitored how these activities improve the rhino conservation awareness level amongst students and villagers. We observed that local conservation leaders and student replicated the rhino awareness campaign in gap areas besides discussing alternative livelihood generation works, ecosystem service by Manas.

The initiative of community conservation activities in the form of promoting alternative livelihood amongst community and conservation education helped in assesssing the habitat and poaching threats. We got information about roaming of two rhinos in the Indo-Bhutan border areas in the February, 2006. Information also revealed that one active poacher tried to get a suitable gun during that period. We already informed the matter to the field director and deputy field director of Manas TR. Meanwhile, on eastern buffer in a site near Disima river stream one boy claimed to have seen a rhino in 2006. The pug mark was photographed by the authority. These incidents indicated that small population of rhinos still might exist in Bhutan.

Meanwhile another source indicated that one person got information about 10-20 rhinos in a safe riparian zone of Royal Manas national park, Bhutan. There is a need of joint effort of authorities of India and Bhutan for surveying the existing rhino population in Bhutan for possible gene flow between the Kaziranga and Pabitora population introduced in Manas in near future as well as maintaining genetic diversity of the species for long term survival. One source also revealed that he counted poaching of 106 rhinos since starting of the political turmoil in the region. We also gathered information from old ex poachers and even some top forest officials believed that earlier actual rhino population might exceeded 100 numbers in the landscape covering some buffer forests also.

Management Recommendation

1. The security scenario should be improved through community participation and strict vigilance by the forest authority to encourage converting less suitable rhino habitats in riparian, boundary areas to suitable and most suitable category.
2. Community conservation work should be continued in fringes of Manas tiger reserve. The encroached area should be gradually restored involving local community by encouraging alternative livelihood options which will be providing more income to villagers.
3. Rehabilitate all conservation volunteers (ex-poachers) including education to their children.
4. Numbers of well trained young permanent forest staff should be increased. Government should provide remuneration to all conservation volunteers at higher rate. There should be well equipped facilities (ration, travel, antipoaching kits, night vision facility, arms etc.) for forest staff and volunteers involved in patrolling activities.
5. Organize well designed training to all staff for using communication facility.
6. Family benefits like good quarters, good schools, and medical facilities should be developed and improved in fringe areas.
7. While encouraging ecotourism, rural village economy should be also supported to minimize the dependence on resources of Manas tiger reserve.
8. The buffer areas on east and west side of Manas should be developed as part of long term rhino conservation.
9. The bridges in boundary road should be built as permanent structure so that four wheelers can move immediately during crisis period. Inside road and bridges should be also maintained in all seasons. Of course for low lying areas, departmental elephant will do the purpose. Therefore number of patrolling elephant needs to be increased.
10. Regular vaccination of domestic cattle should be organized in all fringe villages.
11. Initiative should be taken to minimize the man-animal conflict particularly the man-elephant conflict in the fringe village. If possible, adjacent 500

meter wide fringe crop field should be converted to spiny plants like citrus species which may minimize the wild elephant raid. Alternative crops which are avoided by elephants should be practiced. In buffer areas enough food plants of elephant need to be grown.

12. There should be an alternative to cattle and buffalo farm. Advanced methodology of dairy farm can be introduced where people will get more production and cattle need not to be pushed to grassland of Manas.
13. Rain water harvesting in dry bhabor areas and drinking water facility for all forest camps of Manas in Indo-Bhutan borders should be available.
14. A balance sex ratio of rhinos needs to be maintained in future translocation. Absence of mating partner and competition might be the cause of recent stray.
15. There is a need of better balance in overall political situation including long term peace in the area.
16. The silted areas in rhino habitats need to be restored to maintain wetland grass community.
17. Control burning practice is a must and better involvement of grassland ecologists will help in maintaining the grassland habitat.
18. Eradicate the invasive weed before it covers large areas. Apart from exiting *Chromalaena odorata, Leea asiatica,* and tree species *Bombax ceiba* covering large areas; *Eichhornia crassipes*-the most problematic weed in water bodies is a new threat to Manas. Biomass sharing of weed, encouraging biological control through large wild herbivores like elephant and control burning prescribed by grassland ecologist should follow. The succession of *Bombax ceiba* in some prime rhino habitat should be prevented by eradicating the saplings at earliest.
19. Early warning systems of flood in Manas need to be developed through more cooperation with Bhutan. Stop developing of more dams in rivers that pass through Manas. Instead solar energy is a potential source of electricity.
20. Immediate opening of blocked channel of Manas river is necessary. Other wise the Panbari range- the western rhino habitats where once 30-40 rhinos roamed will be changed to unsuitable habitat. It will increase ecological density of rhinos in Manas and there would be a more threat to the whole population including frequent stray incidents.
21. The Makhibaha area a most suitable habitat of Bhuyanpara range can be a potential release site of translocated rhino by just improving the road communication.
22. Detail research in gap areas like ecology of translocated rhino, buffer habitat status etc. need to be carried out.
23. Strong cooperation between all the conservation NGOs is needed. Forest department is doing good job by involving all in conservation works. Authority must design a strategy that benefits all the local NGOs involved

in ecotourism. This umbrella needs to be strong enough for hundred percent successes to rhino conservation in Manas landscape.

Conclusion

- ☆ The rhino habitat suitability model indicated that Manas possess 122 sq. km. area as most suitable habitat, followed by 240.80 sq. km. as suitable, 125.57 sq. km as less suitable and 31.29 sq. km. as unsuitable. The most suitable area can be increased to 220 sq km. by minimizing the poaching threat and law and order problem near zero level. Currently the suitable area without poaching threat is covering 201.25 sq km. followed by less suitable 69.38 sq km. and unsuitable as 28.951 sq km.
- ☆ The microhabitat condition indicated presence of rich biophysical heterogeneity in grassland association having 11 types and complex hydro-seral stages of aquatic plant community. The forest covers 47.53 percent as deciduous forest including very less (less than 1 per cent) area as close canopy evergreen and semievergreen forest.
- ☆ The suitability model proved that habitat diversity and availability is satisfactory in most of the area.
- ☆ The initial release of rescued and wild rhinos and post release monitoring helped forest authority to deal with rhino stray, habitat management, local community participation and security as well as modification of design for future rhino translocation.
- ☆ The management constraints like incompletion of some infrastructure with all modern facility, poaching threat in some points, weed invasion, illegal burning, resource exploitation influenced the habitat suitability. Of course the current prescription - joint work of forest department and local community found to be encouraging in restoring major areas of the Manas. The dam in Bhutan is a growing threat to the tiger reserve.
- ☆ The community conservation work related to rehabilitation of ex-poachers and forest dependent families showed positive result with 50-60 percent success.
- ☆ The capacity building on rhino conservation education and replication in different areas found to be satisfactory and need further extension in buffer areas.
- ☆ The overall habitat improvement and adaptation of one horned rhinoceros helped forest authority to translocate more rhinos to Manas, though two poaching incidents in 2011 and 2012 and a death related to suspected infighting of rhinos restricted the population to 22 individuals in Manas tiger reserve till June, 2012. The forest authority, with partnership of all stakeholders are working for better management of the introduced rhino population in Manas so that it can be replicated to other potential areas of Assam.

Acknowledgements

We would like to thank Peoples' Trust for Endangered Species of UK for the financial support of the project. We also express our sincere gratitude to former Director Late Dr. Thomas J. Foose of International Rhino Foundation (IRF) for his personal initiative in transferring the PTES grant which was later personally handed over by Late Dr. Nico Van Strein of IRF to Grasshopper in Guwahati, India. We would also like to thank Dr. Susie Ellis of IRF for her help to get matching grant from other organization to continue the rhino conservation work. Dr. Cathy Dean of EAZA rhino campaign initiated to introduce our work with IRF before starting of the Indian Rhino conservation programme. Prof. C.K. Baruah and Dr. G.C. Sarma of Botany department, Gauhati University helped us at different periods including identification of problematic plant species. Special thanks goes to our colleague Dr. Buddhin Hazarika for the part of study carried out on rhino habitat use and food plants in Orang national park. This information helps us to compare with the data on Manas to design habitat suitability model. We would like to thank Mr. N. Bhagoboti of IASST, Guwahati and Mr. Prasanna Baruah of ARSEC, Guwahati for their help in data analysis and map preparation respectively. We also would like to thank the local forest authority for the necessary permission and infrastructural facilities. We specially like to mention the help by the present field director A. Sargiary, deputy field director, C.R. Bhobora, past field director A. Rabha, range officers M. Brahma, L. Ramchiary and front line forest staff. We also extend our gratitude to different local conservation groups namely Manas Maozigendri Ecotourism Society, Manas Souci Konkhor Ecotourism Society under NGO (Prateedhani), Agrang Ecotourism Society, Bhuyanpara Ecotourism Society, Manas Ever Welfare Society, Panbari Manas National Park Protection and Ecotourism Society, Raiguli Ecotourism Society, Green Manas during the project period. Besides different officials of organizations like Aranyak, WTI and WWF extended cooperation during the project period. We would like to specially mention the support provided by conservation volunteers, village heads, local Bodo leaders, teachers and students and villagers of the area. We would like to extend our gratitude to members of Grasshopper and my family for their kind support during the project period.

References

Baruah P. P. 1998. Study of vegetation dynamics and periodic migration of animal populations in Kaziranga National Park in relation to flood and fire. Unpublished Ph.D. thesis. Gauhati University, Guwahati, India.

Bezbarua, P. 2007. Study of plant diversity status in Manas biosphere reserve with special reference to community structure. Unpublished Ph.D. thesis Gauhati University, Guwahati, India.

Biswas, T. 1999. Habitat utilization by hog deer (*Axis porcinus*) in relation to other sympatric species in Jaldapara Wildlife sanctuary. M.Sc. dissertation. Saurashtra University, Rajkot, India.

Dinerstein, E. 1979. An ecological study of the Royal Karnali- Bardia Wildlife Reserve, Nepal II. Habitat animal interactions. Biol. Conserv. **16**: 265-300.

Dinerstein, E and Price L. 1991. Demography and habitat use by greater one horned rhinoceros in Nepal. J. Wildlife Management. **55**: 401-411.

Gee, E. P. 1953. Further observations on the Great One horned Rhinoceros. J. Bombay Nat. Hist. Soc. **51**: 765-772.

Gee, E. P. 1959. Report on the survey of the rhinoceros areas of Nepal. Oryx. 5: 59-85.

Gee, E. P. 1963. Report on a brief survey of the wildlife resources including the rhinoceros. Oryx. **7**: 67-76.

Ghosh, S. 1997. Ecological impacts of prescribed burning in Corbett Tiger Reserve, Uttar Pradesh. M.Sc. dissertation. Saurashtra university, Rajkot, India.

Hazarika, B. 2007. Studies in Eco-behavioural aspects of *Rhinoceros unicornis* Linn. 1758 in the Orang national park, Assam, India. Unpublished Ph.D. thesis Gauhati University, Gauhati, India.

Jnawali, S.R. 1995. Population ecology of the Great One horned Rhinoceros (*Rhinoceros unicornis*) with particular emphasis on habitat preference, food ecology and ranging behaviour of the reintroduced population in Royal Bardia National Park in lowland Nepal. Ph. D. Thesis. Deptt. of Biology and Nature Conservation, Agricultural Univ. of Norway.

Jnawali, S.R. and P.Wegge. 1993. Space and habitat use by a small reintroduced population of greater one horned rhinoceros (*Rhinoceros unicornis*) in Royal Bardia National Park-A preliminary Report. In: Ryder, O. A. (Ed.) Rhinoceros biology and conservation. Proc. of an international conference. Zool. Soc, San Diego, USA. pp. 208-217.

Karki, J.K. and Y.V. Jhala and P.P. Khanna. 2000. Grazing lawns in Terai Grasslands, Royal Bardia National Park, Nepal. Biotropica. **32 (3)** : 423-429.

Kushwaha, S. P. S.; S. Munkhtuya and P. S. Roy 2000. Geospatial modeling for Goral habitat evaluation. J. Indian Soc. Remote Sensing **28(4)**: 296-297.

Laurie, W, A. 1978. The ecology and the behaviour of the Great one horned rhinoceros (*Rhinoceros unicornis*). Ph.D. dissertation, University of Cambridge, Cambridge.UK.

Mishra, R. 1968. Ecology work book. Oxford and I.B.H. publishing Co. Calcutta, India.

Oksanen, J. 2006. Multivariate analysis of ecological communities in R. Vegan tutorial. Manual for software R. Console. R Development core team. Available online at http://cran.r-project.org.

Sinha, S.P. and V.B. Sawarkar. 1993. Management of the reintroduced greater one horned rhinoceros (*Rhinoceros unicornis*) in Dudhwa National Park, Uttar Pradesh, India. In: Ryder, O. A. (Ed.) Rhinoceros biology and conservation. Proc. of an international conference. Zool. Soc, San Diego, USA. pp. 218-227.

Spillett, J.J. 1966. A report on wildlife surveys in North India and Southern Nepal, January - June 1996. J. Bombay Nat. Hist. Soc. **63(3)**: 492-628.

Appendix I: Local NGOs involved in the Programme

Sl.No.	*Name of Conservation Group*	*Name of the Area Covered*	*Zone*
1.	Manas Souci Konkhor Ecotourism Society under NGO (Pratidhani)	Covering the eastern buffer namely Subonkhata, Uttarkuchi, Batabari R.F.	Eastern buffer
2.	Manas Maozigendri Ecotourism Society	Major part of Bhuyanpara range, some parts of eastern buffer	Eastern buffer and core
3.	Agrang Ecotourism Society	Bhuyanpara range southern eastern side	Core
4.	Bhuyapara Ecotourism Society	Bhuyanpara and some parts of eastern Bansbari	Core
5.	Manas Ever Welfare Society	Bansbari range covering the riparian side	Core
6.	Panbari Manas National Park Protection and Ecotourism Society	Panbari range	Core and western buffer
7.	Raiguli Ecotourism Society	West to Panbari *i.e.* Manas R.F. near Kukloong	Western buffer
8.	Green Manas	Stated working in Kahitema riparian zone of Manas-Beki river in Bansbari range	Core zone

Appendix II: Name of the Villages Involved in Fringes of Manas Tiger Reserve

Bhuyanpara range Core of Manas TR	Dongpar, Agrang, Kumguri, Khusatari Bhuyanpara, New Betbari, Thaijwoguri, Majormakha, Rangidara, New Borgaon, New Randhanipara, Simlaguri, Chalchali North Khamardwisa, Randhanipara, South Khamardwisa, Hazirapara, Rangapani, Dwimguri (Khamardwisa), Garumara, East Khamardwisa, 2 no. Kokilabari, Langdangpara, Daohara, Bhalaguri, Thebarmur, Hatijan, Daoharu.
Bansbari Range Core of Mans TR	Mayagpara, Palsiguri, Bishpani, Narayanguri, Katajhar, Gyati, Barengabari, Kahitema.
Panbari Range Core of Mans TR	Provided education material to local NGO Panbari Manas National Park Protection and Ecotourism Society.
Eastern buffer of Manas TR	Uttarkuchi, Subankhata, Amlaiguri, Dimapur, Bandagui, Pachim ambari, Bhabanipur, Katabari, Bhujelsepa, Aliguri, Deghelipam, Bhutankhuti.
Western buffer of Manas TR	Provided education material to local NGO Raiguli Ecotourism Society.

Chapter 2

Prioritisation of Conservation Areas in the Central Western Ghats, India through Biological Indicators–Amphibians

T.V. Ramachandra, M.D. Subash Chandran, N.V. Joshi, K.V. Gururaja and Vishnu D. Mukri

Energy and Wetlands Research Group, Centre for Ecological Sciences, Indian Institute of Science, Bangalore – 560 012, Karnataka, India
URL: http://ces.iisc.ernet.in/energy; http://ces.iisc.ernet.in/biodiversity
E-Mail: cestvr@ces.iisc.ernet.in

ABSTRACT

Amphibians are considered as biological indicators for their susceptibility to even very small changes in the surrounding environment and their habitats typically spread across the interface between terrestrial and aquatic habitats. They are the only vertebrate group with dual life stages (*i.e.,* tadpoles and adults) and perform vital ecological functions. Semi-permeable skin, anamniotic eggs and biphasic life style make them particularly vulnerable to changes and contamination of their habitats on land and in water. Habitat destruction and overexploitation are the major threat for amphibians, apart from *Chytrid* fungus and other synergistic effects of human induced changes. Presence of a diverse population of amphibians in a region is indication of a healthy environment. They are being used as surrogates in conservation and management practices. Monitoring amphibian diversity and their distribution would provide insights to the prevailing conditions of an ecosystem and its health, which in turn helps in prioritizing the region for conservation and management action in the Western Ghats.

River basins/catchments are topographically and hydrologically well defined unit of space and the present study has been carried out in five river basins namely Sharavathi, Aghanashini, Bedti and Kali of Uttara Kannada district using amphibians as biological indicators to arrive at conservation priority regions in the district. Forty six species were recorded from the five river basins. Sharavathi river basin forms species rich and endemic rich, while Venkatapura is species poor and endemic poor. Kathalekan of Sharavathi river with 34 species of which 24 of are endemic, is an apt candidate for the status of heritage site of biological diversity as per biodiversity act 2002 (Chapter IX, Biodiversity heritage sites).

Keywords: *Biological indicators, Western Ghats, River catchment, Amphibians, Heritage sites.*

Introduction

The Western Ghats of the Indian peninsula constitute one of the 34 global biodiversity hotspots along with Sri Lanka, on account of exceptional levels of plant endemism and by serious levels of habitat loss (Conservation International, 2005). The rugged range of hills stretching for about 1600 km along the west coast from south of Gujarat to the end of the peninsula (lat. 8° and 21° N and long. 73° and 78° E), is interrupted only by a 30 km break in Kerala, the Palghat Gap (Radhakrishna, 2001). Covering a geographical area of about 160,000 km^2, the Western Ghats have an average height of 900 m, with several cliffs rising over 1000 m. The Nilgiri Plateau to the north and Anamalais to the south of the Palghat Gap exceed 2000 m in many places. Towards the eastern side the Ghats merge with the Deccan Plateau which gradually slopes towards the Bay of Bengal. The northern half of the Western Ghats is covered with basaltic rocks of volcanic origin whereas the southern half is of Pre-Cambrian rocks of different kinds like the crystalline rocks, the peninsular gneisses and the charnokites. Nearly a hundred rivers originate from these mountains and most run their westward courses towards the Arabian Sea that is close-by. Only three major rivers, joined by many of their tributaries flow eastward, longer distances, towards the Bay of Bengal (Dikshit, 2001; Radhakrishna, 2001). The Western Ghat rivers are very critical resources for peninsular India's drinking water, irrigation and electricity (Subash Chandran *et al.*, 2010; Ramachandra *et al.*, 2007). The region has varied forest types from tropical evergreen to deciduous to high altitude sholas. It is also an important watershed for the peninsular India with as many as 37 west flowing rivers, three major east flowing rivers and innumerable tributaries. The richness and endemism in flora and fauna of this region is well established with over 4,000 species of flowering plants (38 per cent endemics), 334 butterflies (11 per cent endemics) [Kunte, in Press], 290 fishes (65 per cent endemics) [Dahanukar *et al.*, 2011], 157 amphibians (86 per cent endemics) [Biju *et al.*, 2010], 157 reptiles (62 per cent endemics) [Hegde, 2011; Ganesh *et al.*, 2008; Ganesh *et al.*, 2009; Chandramouli and Ganesha, 2010; Dasa *et al.*, 2006], 508 birds (4 per cent endemics) [Molur *et al.*, 2011] and 140 mammals (12 per cent endemics) [Karanth *et al.*, 2009]. This mountain stretch has influenced regional tropical climate, hydrology and vegetation and endemic plant species.

The entire region is reeling under tremendous pressure from human induced changes in terms of developmental projects like hydroelectric or thermal power plants, big dams, mining activities, unplanned agriculture practices, monoculture plantations, illegal timber logging, etc. This has led once contiguous forest habitats to fragmented patches, which in turn led to shrinkage of original habitat for the wildlife, change in the hydrological regime of the catchment, decreased inflow in streams, human-animal conflicts, etc. Under such circumstances, a proper management practice is called for requiring suitable biological indicators to show the impact of these changes, set priority regions and in developing models for conservation planning.

Uttara Kannada district with a spatial extent of 10,291sq.km is the second largest district in south India with good vegetation cover (68 per cent). Dandeli Anshi tiger reserve (814.89sq.km), which is about 8 per cent of the total area is the only protected area in the district. It is a clear indication that a single large protected area is not sufficient for the conservation and management of biodiversity of this district. The rationale for protected area networks or any region to consider for conservation and management must be based on a well-defined functional unit. A river basin (catchment) is topographically and hydrologically well defined unit, which can be very well considered for conservation management. Across various spatial scales a river basin helps to understand the ecological processes and landscape influence on biodiversity. The current focus is to prioritise conservation regions in five river basins in Uttara Kannada district, using biological indicator - amphibians (surrogate for many other species). Objectives of the current research are:

1. Mapping of diversity and distribution of amphibian species - river basin wise in the district.
2. Prioritise areas for conservation based on amphibian richness and association to habitat characteristics.

Materials and Methods

Study Area

Five river basins, namely Kali, Bedti, Aghanashini, Sharavathi and Venkatapura of Uttara Kannada district in the Central Western Ghats (between 12°–16°N) were considered for this study as depicted in Figure 2.1. These rivers are west flowing rivers and form the part of Uttara Kannada, the district with highest forest cover (78 per cent) in Karnataka.

River Kali

It is the northern most river in coastal Karnataka, originates at Diggi in Supa Taluk, Uttara Kannada district and traverses for about 184km before joining Arabian Sea at Karwar. For the initial 100 km, the river flows south eastwards and at Thattihalla due to geological fault it flows towards southwest. The river has four major and two minor dams constructed across river Kali. Major dams are at Supa, Bommanalli, Kodasalli and Kadra, while minor ones are at Tattihalla and Kaneri. Pandri, Ujli, Nujji, Thananala, Kaneri and Vaki are the other streams that joins Kali at various places. The entire river basin is about 5,104 sq. km, encompassing dry deciduous-

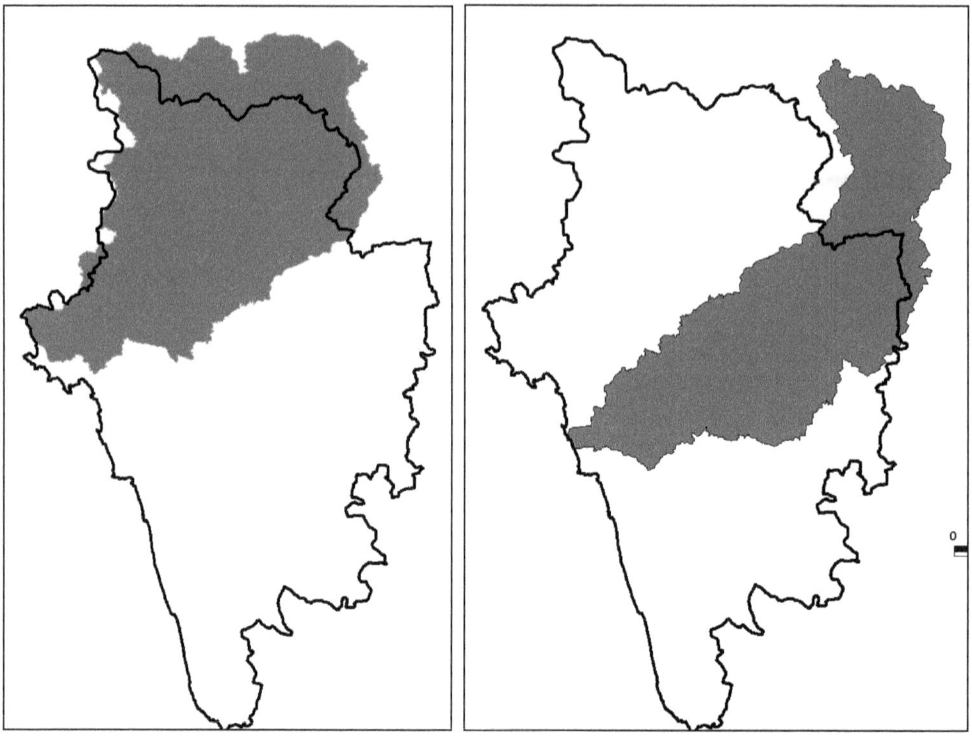

River Kali River Bedti

evergreen-mangroove vegetation of the Western Ghats. The Kali river basin receives on an annual rainfall between 850-3200 mm. Figure 2.2 illustrates drainage network and sampling sites in River Kali.

River Bedti

River Bedti (also called Gangavalli) originates at Dharwad District as Shalmala and confluences at Kalghatgi with another stream from Hubli, flows westward for about 161km to merge with Arabian sea. It has a catchment of about 3878sq.km, the second largest catchment in Uttara Kannada district. There are two tributaries to this river Shalmala and Sonda. The river forms a fall at Magod from about 220m. The river has dense evergreen, semi-evergreen to deciduous forests along its path. Soils are mainly lateritic. Annual rainfall ranges from 1,700 - 6,000 mm. Figure 2.3 depicts sampling sites and drainage network in River Bedti.

River Aghanashini

River Aghanashini having a catchment of about 1390.52 sq.km traverses westward for about 121km from the origin at Manjguni of Sirsi Taluk, and confluences with Arabian Sea at Tadri. Estuarine part of Aghanashini is 13km long making it as the longest among Uttara Kannada rivers. Unchalli falls (Lashington falls) forms a major water fall of this river. Figure 2.4 shows drainage map and sampling sites in River Aghanashini. Despite being a small river basin, Aghanashini has diverse vegetation predominates with evergreen-semi-evergreen to mangrove.

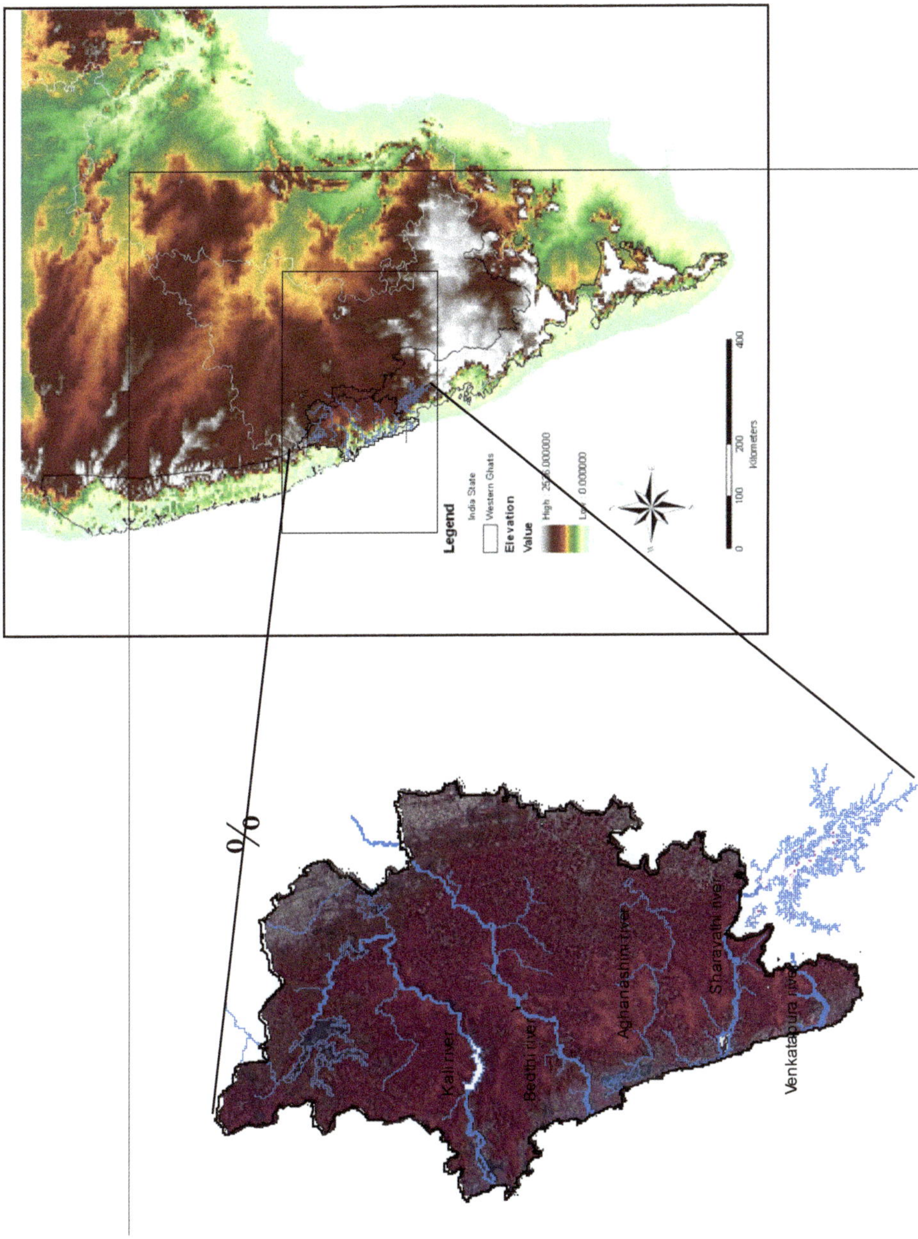

Figure 2.1: False Colour Composite Image of Uttara Kannada District

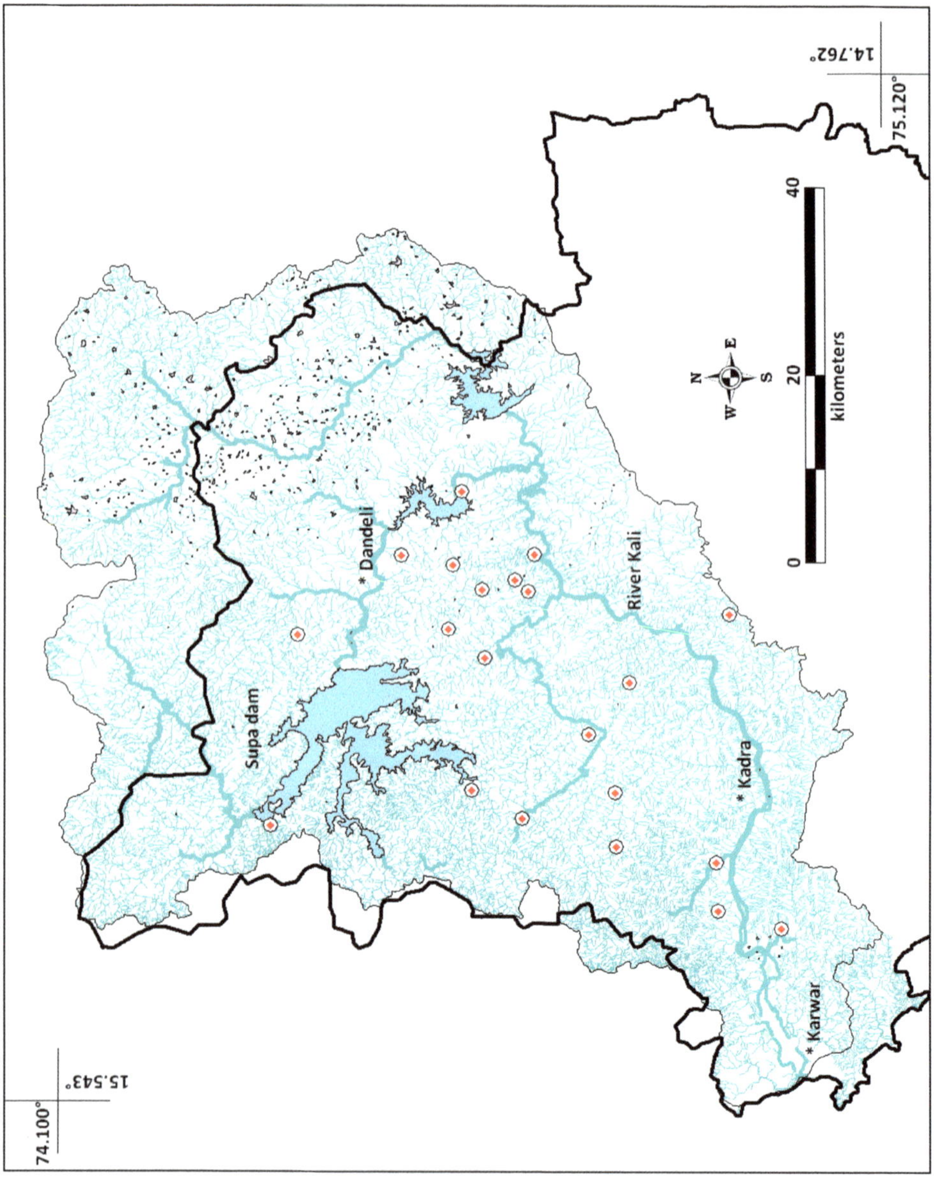

Figure 2.2: Drainage Network and Sampling Sites in Kali River Basin

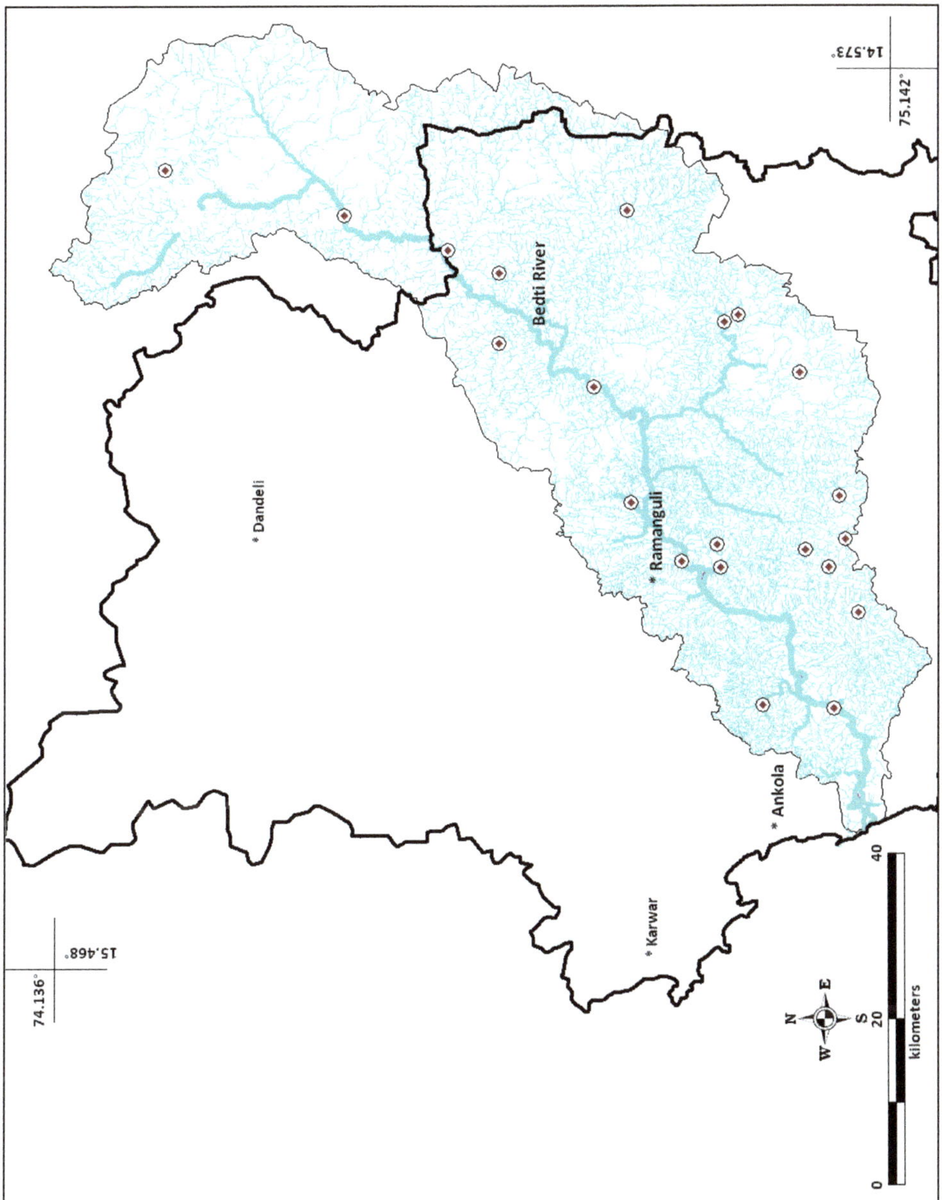

Figure 2.3: Drainage Network and Sampling Sites in Bedti River Basin

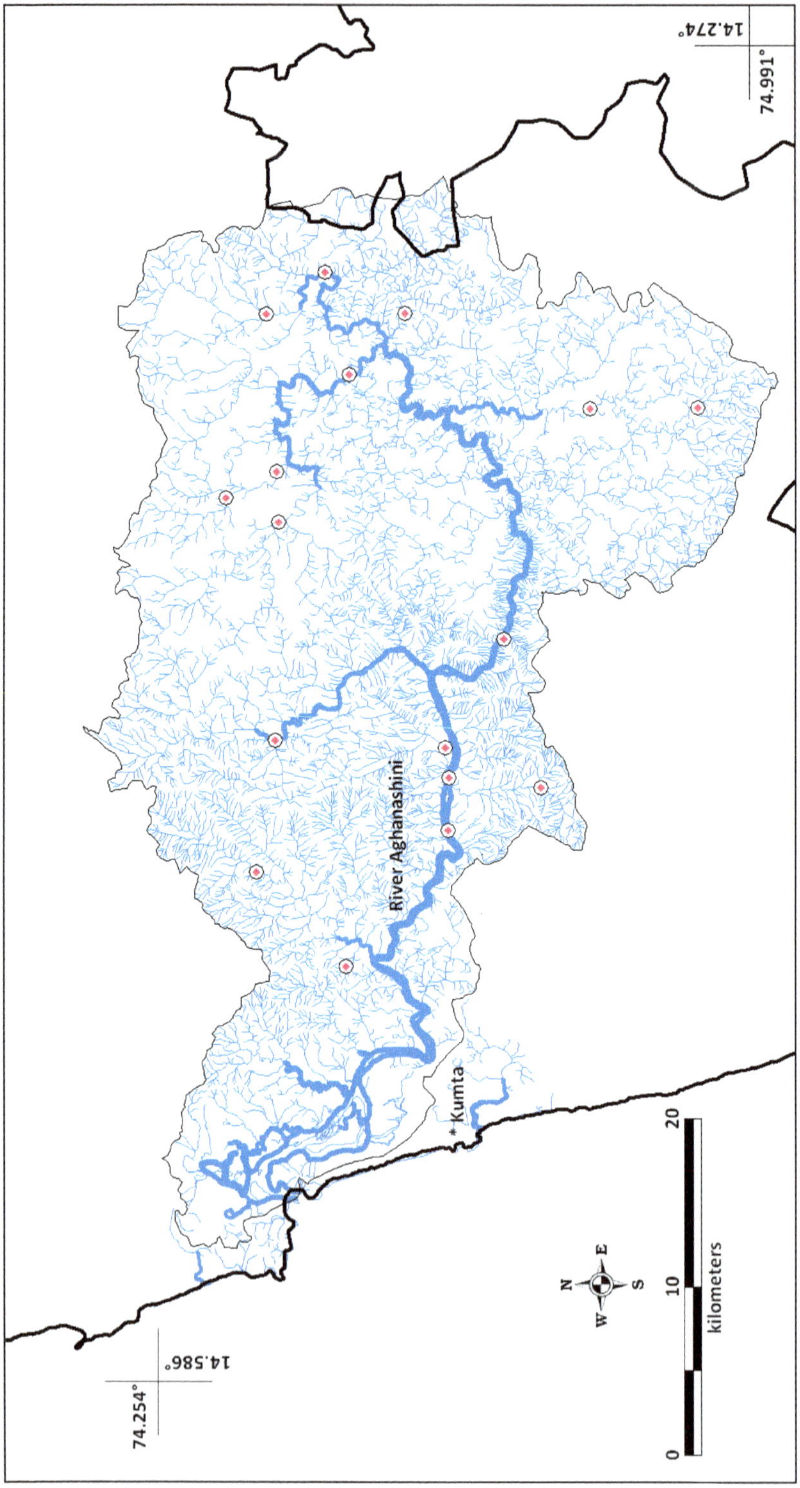

Figure 2.4: Drainage Network and Sampling Sites in Aghanashini River Basin

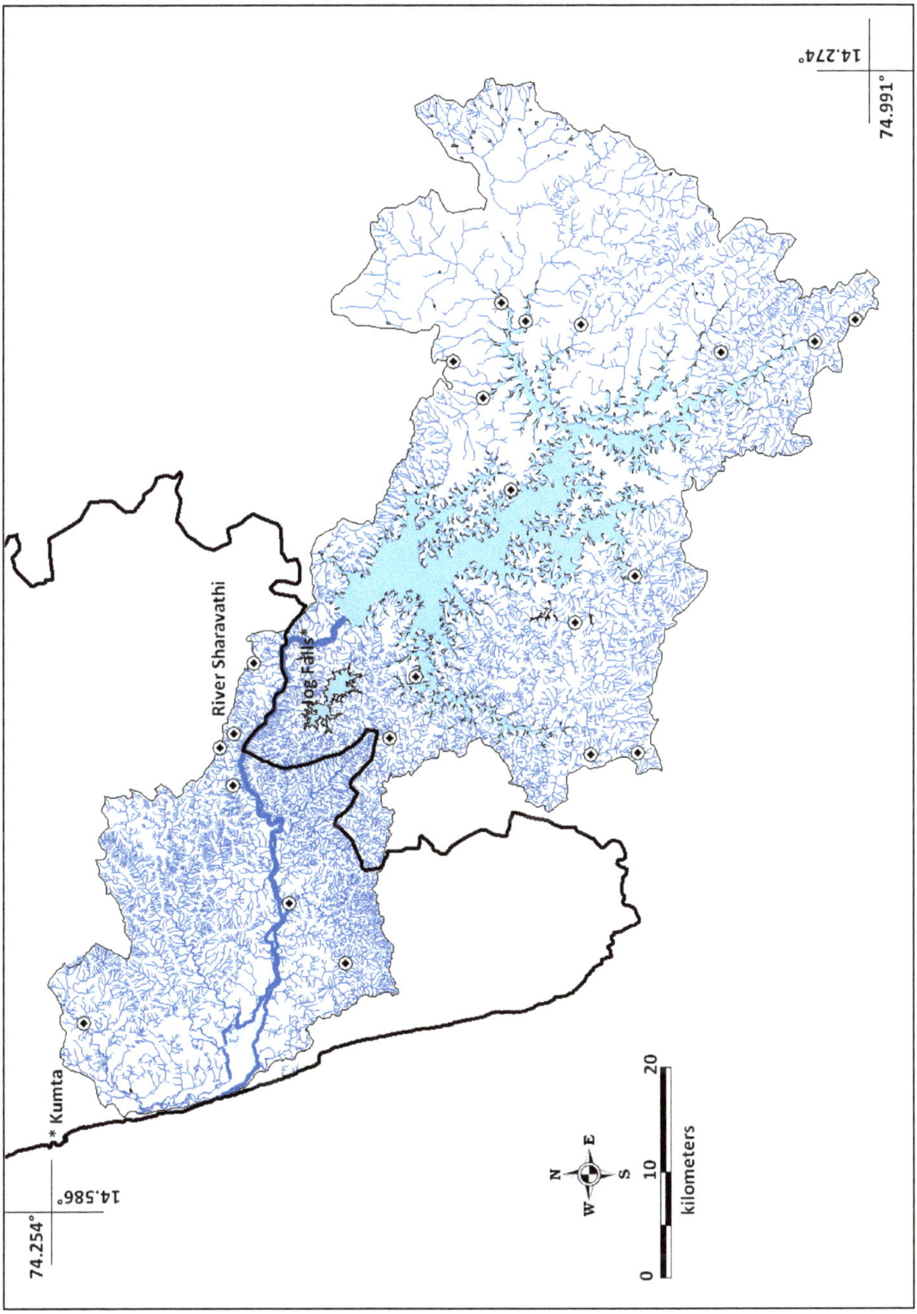

Figure 2.5: Drainage Network and Sampling Sites in Sharavathi River Basin

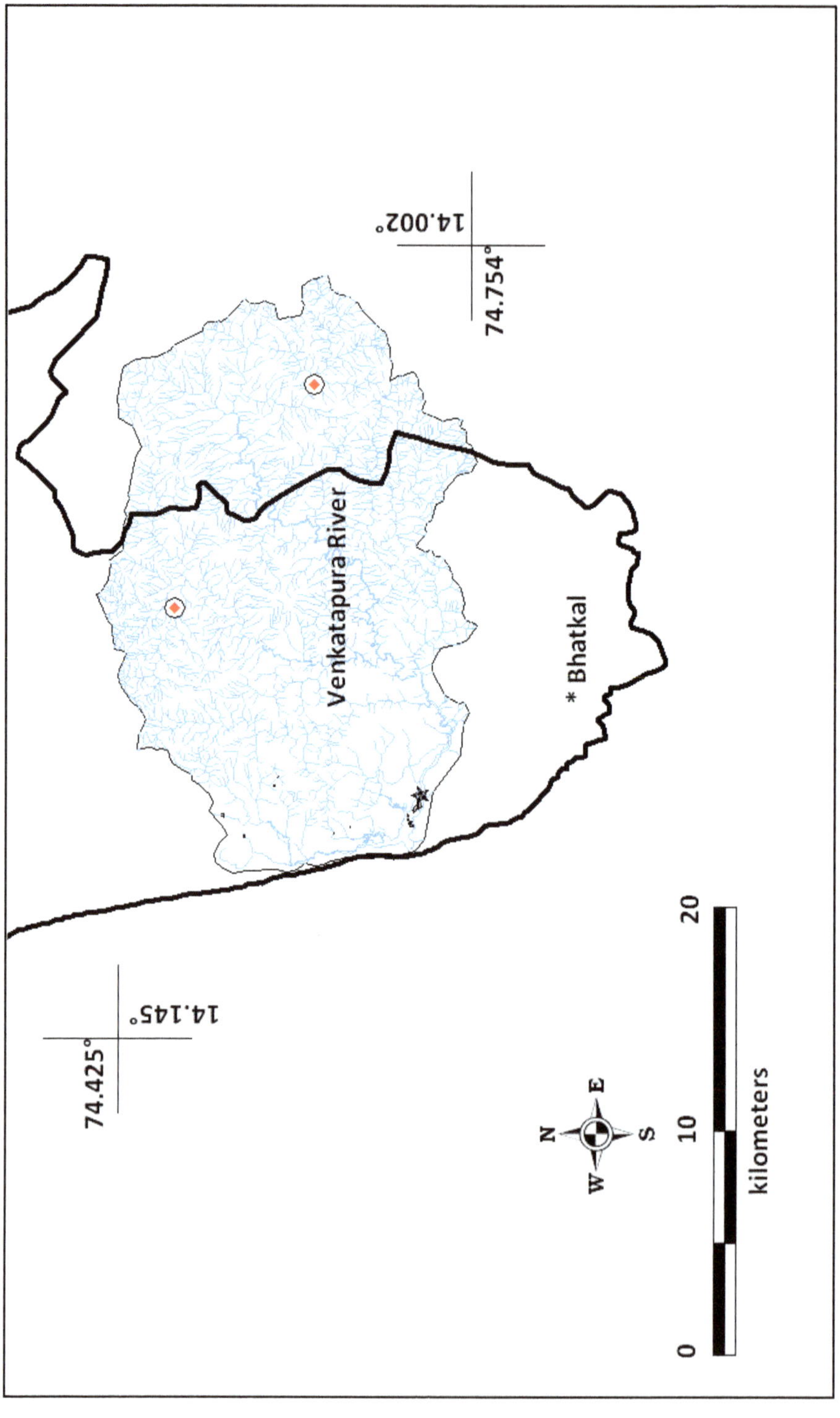

Figure 2.6: Drainage Network and Sampling Sites in Venkatapura River Basin

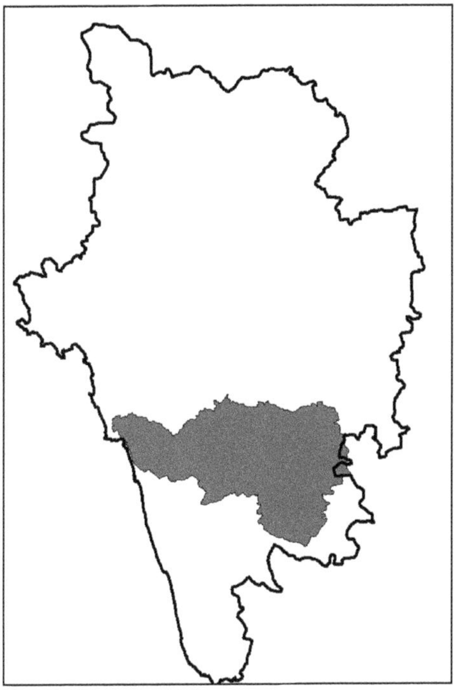

River Aghanashini

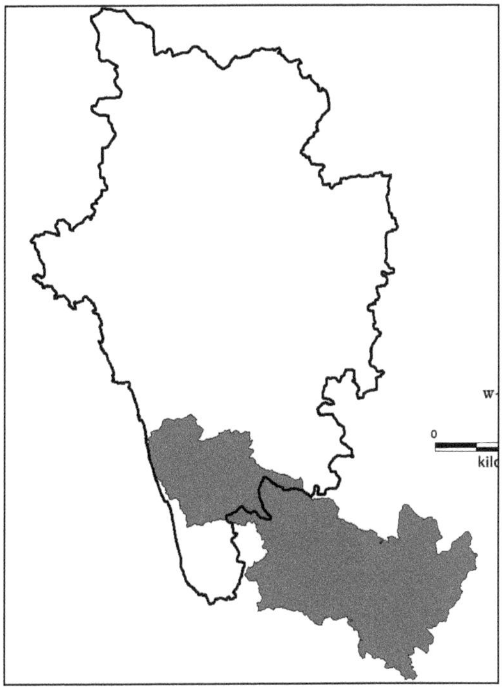

River Sharavathi

River Sharavathi

It originates near Ambuthirtha of Shimoga district, traverses for about 132km and confluences at Honnavar to the Arabian sea. The magnificent waterfall, Jog, is situated in the course of this river. The catchment area of this river is about 3005 sq.km. Only down stream part of the river is within Uttara Kannada district. This river has four dams across its catchment. The first dam was built way back in 1940 at Hirebhaskar, which got submerged after the construction of Linganmakki Dam in 1964. It was among the largest in Asia at that time submerging an area of 326sq.km. Figure 2.5 illustrates catchment area and sampling sites in Sharavathi River.

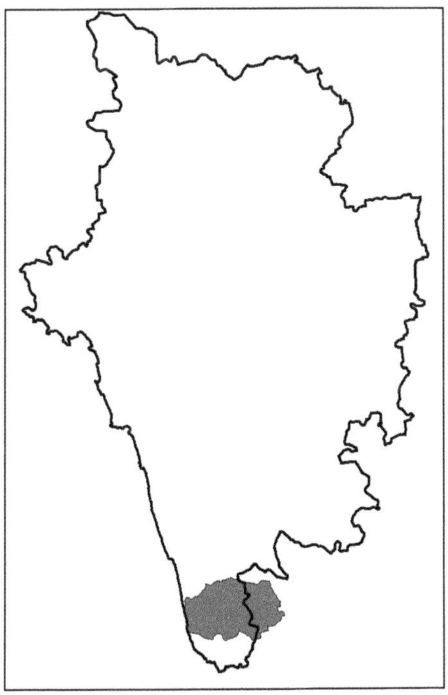

River Venkatapura

River Venkatapura

Smallest of the rivers in Uttara Kannada, Venkatapura has very small catchment area of about 326.2sq.km. The vegetation is predominantly low land evergreen forests. Figure 2.6 illustrates the sampling sites and catchment of River Venkatapura.

Sampling Methods

Site Selection

Based on the land use and the extent of catchment area in each river basin, 83 sampling sites were selected: two from Venkatapura, 22 from Sharavathi, 15 from Aghanashini, 22 each from Bedti and Kali river basins.

Habitat Variables

Altitude (meters above sea level), rainfall (annual, mm), stream perenniality (seasonal or perennial) and predominating land-use (from the LULC analysis) are used to determine anuran distribution. These habitat variables were graded from lowest to highest. For the analysis, these grades are used than the actual values. River basin wise sampling sites and habitat variables are given in Table 2.1.

Table 2.1: Sampling Sites and Respective Habitat Variables

Sl.No.	*River*	*Place*	*Altitude (m)*	*Rainfall (mm)*	*Landuse**	*Stream***
1.	Venkatapura	Ondalasu	94	4429	1	1
2.	Venkatapura	Kelanur	231	4662	4	2
3.	Sharavathi	Malemane	309	4171	4	2
4.	Sharavathi	Kathalekan	599	4079	4	2
5.	Sharavathi	Watehalla	698	4119	4	2
6.	Sharavathi	Mavingundi	611	3859	4	2
7.	Sharavathi	Dabbe	609	4302	4	2
8.	Sharavathi	Hosagadde	41	4240	1	1
9.	Sharavathi	Magod	8	4145	1	1
10.	Sharavathi	Chandavar	22	3693	3	2
11.	Sharavathi	Nandihole	561	2198	1	1
12.	Sharavathi	Haridravathi	563	2339	1	1
13.	Sharavathi	Mavinhole	593	2583	3	2
14.	Sharavathi	Sharavathi	610	3655	4	2
15.	Sharavathi	Hilkunji	601	4402	3	2
16.	Sharavathi	Nagodi	572	4040	4	2
17.	Sharavathi	Hurli	605	3999	3	2
18.	Sharavathi	Karni	657	4877	4	2
19.	Sharavathi	Yennehole	578	4744	4	2
20.	Sharavathi	Muppane	575	4069	3	2
21.	Sharavathi	Mundigesara	657	2284	3	2
22.	Sharavathi	Niluvase	763	4124	4	2

Contd...

Table 2.1–*Contd...*

Sl.No.	*River*	*Place*	*Altitude (m)*	*Rainfall (mm)*	*Landuse**	*Stream***
23.	Sharavathi	Sampekai	578	2516	1	1
24.	Sharavathi	Hubse	586	2959	1	1
25.	Aghanashini	Kathagal	45	3662	3	2
26.	Aghanashini	Ullur	69	3793	1	1
27.	Aghanashini	Sapurti	512	3196	3	2
28.	Aghanashini	Baillalli	534	3769	4	2
29.	Aghanashini	Yanahole	72	3553	2	1
30.	Aghanashini	Bennehole	418	3769	3	2
31.	Aghanashini	Revankatta	520	3300	1	1
32.	Aghanashini	Bolurhole	495	3054	3	2
33.	Aghanashini	Bilgihole	489	3655	1	1
34.	Aghanashini	Hulidevaragadde	55	3694	1	1
35.	Aghanashini	Donnehole	26	3641	3	2
36.	Aghanashini	Nellimadke	526	2757	1	1
37.	Aghanashini	Neralamane	504	3019	1	1
38.	Aghanashini	Deevalli	30	3676	3	2
39.	Aghanashini	Mudagi	31	3617	3	1
40.	Bedti	Abageri	405	3582	4	2
41.	Bedti	Andhalli	491	2642	2	1
42.	Bedti	Angadibailu	84	3481	3	2
43.	Bedti	Chitageri	524	1574	2	1
44.	Bedti	Daanandi	492	2245	2	1
45.	Bedti	Dabguli	89	2742	4	2
46.	Bedti	Devnalli	541	3322	1	1
47.	Bedti	Gundabala	8	3478	1	1
48.	Bedti	Gunjavathi	505	1747	2	1
49.	Bedti	Hasehalla	71	3044	3	2
50.	Bedti	Hemmadi	485	2278	2	1
51.	Bedti	Kalghatgi	517	1153	1	1
52.	Bedti	Kalleshwara	104	3131	4	2
53.	Bedti	Karadrolli	490	2025	2	1
54.	Bedti	Kelginkeri	344	3481	1	1
55.	Bedti	Makkigadde	52	3427	3	2
56.	Bedti	Manchikeri	424	2506	2	1
57.	Bedti	Melinkeri	428	3496	4	2
58.	Bedti	Nyctisite	531	3496	4	2

Contd...

Table 2.1–*Contd...*

Sl.No.	*River*	*Place*	*Altitude (m)*	*Rainfall (mm)*	*Landuse**	*Stream***
59.	Bedti	Tarihal	647	849	1	1
60.	Bedti	Vajagadde	563	3518	4	2
61.	Bedti	Yerebail	497	1555	2	1
62.	Kali	Goira	63	3074	1	1
63.	Kali	Badapoli	532	3407	2	1
64.	Kali	Anshi	529	3332	3	2
65.	Kali	Jhalavali	546	3191	2	1
66.	Kali	Kaneri	489	3149	2	2
67.	Kali	Cyntheri	445	2812	2	1
68.	Kali	Ulvi	645	3093	1	1
69.	Kali	Gunda road	483	2635	2	1
70.	Kali	Water1	579	2430	2	1
71.	Kali	Amgaon	557	2370	1	1
72.	Kali	Water 2	564	2452	2	1
73.	Kali	Mines	661	2531	2	1
74.	Kali	Nagzhari 1	180	2414	2	1
75.	Kali	Nagzhari 2	391	2224	2	1
76.	Kali	Kulgi	502	2085	2	1
77.	Kali	Virnolii	457	1932	2	1
78.	Kali	Sakatihalla	17	3055	3	1
79.	Kali	Beegaru	257	3014	4	2
80.	Kali	Kanshirda	473	2012	2	1
81.	Kali	Deriye	640	3015	3	2
82.	Kali	Castlerock	571	2894	3	2
83.	Kali	Gowliwada	534	2125	2	1

*Landuse–1: Agriculture, 2: Deciduous, 3: Semi-evergreen, 4: Evergreen.

**Stream–1: Seasonal, 2: Perennial.

Ecological Guilds

Anuran amphibians in this study were classified into ecological guilds (Table 2.2) on the basis of four variables describing their functional ecology, namely, endemism (referring to their spatial extent of occurrence), threat status (IUCN, 2009), habitat specificity (aquatic to arboreal, considering aquatic to be very primitive) and finally tadpole habitat (direct development being considered most advanced).

Table 2.2: Ecological Guild Gradation Used in the Present Study

Variable	*Range*	*Grade*
Endemism	Non-endemic	1
	Endemic to Western Ghats-Sri Lanka Hotspot	2
	Endemic to the Western Ghats	3
Threat status	Data deficient	1
	Least concerned	2
	Near threatened	3
	Vulnerable	4
	Endangered	5
Habitat specific	Aquatic	1
	Semi-aquatic	2
	Terrestrial	3
	Fussorial	4
	Arboreal	5
Tadpole	Aquatic	1
	Semi-aquatic	2
	Arboreal/Direct development	3

Sampling of Amphibians

Amphibian sampling was carried out very systematically covering all seasons. Visual encounters, calls, tadpoles, foam nests, spawn are used to record the amphibians in the field. Two man hours of searching is made using torch lights between 19:00-20:00 hr, by walking across the streams, forest floors, gleaning leaf litters, prodding bushes, wood logs, rock crevices etc. All the species encountered are identified up to species level using the keys of Bossuyt and Dubois (2001), and Daniels (2005). New species names are based on literature by Biju *et al.* (2010) and Dinesh *et al.* (2010). Opportunistic encounters are also recorded to enlist the species of the region.

Statistical Analysis

Using presence data for species, grading in ecological guilds and environmental variables, non-metric multidimensional analysis (NMDS) is carried out. Non-metric multidimensional scaling is based on Bray-Curtis distance matrix. In NMDS, data points are placed in 2 or 3 dimensional coordinates system preserving ranked differences. Absolute distances are not taken into consideration. Spatial interpolation by Krigging technique is used to produce map with continuous spatial estimate of species endemism based on scattered data points.

Results

Amphibian Diversity and Distribution Across Uttara Kannada District

Forty seven species of amphibians were recorded from Uttara Kannada district (Table 2.4). This is nearly 30 per cent of observed amphibians from the Western Ghats (157 species). These species belonged to two orders, nine families and 20 genera. Two families, namely, Nyctibatrachidae and Micrixalidae are among the oldest frog families found in the Western Ghats and are Gondwanan relicts. Of the 46 species recorded, 67 per cent of them are endemic to the Western Ghats (31 species). Family Dicroglossidae has highest species (15) followed by Rhacophoridae with 10 species. Least species were recorded in Ranixalidae and Ichthyophiidae with two each (Table 2.3).

Table 2.3: Family-wise Species Recorded in Uttara Kannada

Family	*Genera*	*Species*
Bufonidae	2	4
Dicroglossidae	5	15
Micrixalidae	1	3
Microhylidae	3	4
Nyctibatrachidae	1	3
Ranidae	2	4
Ranixalidae	1	2
Rhacophoridae	4	10
Ichthyophiidae	1	2

River Basin-wise Diversity of Amphibians

Amphibian species recorded in each of the river basin in given in Table 2.5. There were 45 species from Sharavathi, 32 from Aghanashini, 29 each from Bedti and Kali river basins and five species in Venkatapura river.

Sharavathi River Basin

Forty five species were recorded from Sharavathi river basin listed in Table 2.5. Majority of the species recorded for the entire Uttara Kannada district is known from Sharavathi river basin, except for *Raorchestes bombayensis*. Sharavathi harbours nearly 69 per cent of endemic species of the Western Ghats. The species rich sites (> 10 species) are Kathalekan (34 species), Watehalla and Muppane (14 species each), Hurli (12 species) and Niluvase (11 species). Kathalekan is a well known Myristica swamp having ancient origin (Chandran *et al.*, 2010), and provides habitat for uniquely breeding species (Gururaja, 2010). It also has highest number of endemic species (24). *Raorchestes ponmudi*, an endangered species is also recorded from this site.

Table 2.4: Species Recorded and their Ecological Status in the Four River Basins of Uttara Kannada

Species	*Common Name*	*Endemic*	*IUCN*
CLASS: AMPHIBIA Gray			
ORDER: ANURA Fischer von Waldheim			
Family: Bufonidae Gray			
Duttaphrynus melanostictus (Schneider 1799)	Common Indian toad	Non endemic	LC
Duttaphrynus scaber (Schneider, 1799)	Ferguson's toad	Non endemic	LC
Duttaphrynus stomaticus (Lutken, 1862)	Assam toad	Non endemic	LC
Pedostibes tuberculosus Günther 1875	Malabar Tree toad	Western Ghats	EN
Family: Dicroglossidae Anderson			
Euphlyctis aloysii Joshy, Alam, Kurabayashi, Sumida and Kuramoto, 2009	Aloys' skittering frog	Western Ghats	DD
Euphlyctis cyanophlyctis (Schneider, 1799)	Skittering frog	Non endemic	LC
Euphlyctis hexadactylus (Lesson, 1834)	Indian Pond frog	Non endemic	LC
Fejervarya brevipalmata (Peters, 1871)	Peter's frog	Western Ghats	DD
Fejervarya caperata Kuramoto, Joshy, Kurabayashi and Sumida, 2007	Wrinkled Fejervarya	Western Ghats	DD
Fejervarya granosa Kuramoto, Joshy, Kurabayashi and Sumida, 2007	Granular Fejervarya	Western Ghats	DD
Fejervarya kudremukhensis Kuramoto, Joshy, Kurabayashi and Sumida, 2007	Kudremukha Fejervarya	Western Ghats	DD
Fejervarya mudduraja Kuramoto, Joshy, Kurabayashi and Sumida, 2007	Mudduraja Fejervarya	Western Ghats	DD
Fejervarya rufescens (Jerdon, 1853)	Reddish burrowing frog	Western Ghats	LC
Hoplobatrachus crassus (Jerdon, 1853)	Jerdon's bull frog	Non-endemic	LC
Hoplobatrachus tigerinus (Daudin, 1803)	Indian bull frog	Non-endemic	LC
Minervarya sahyadris Dubois, Ohler and Biju, 2001	Minervarya frog	Western Ghats	EN
Sphaerotheca aff. leucorhychus (Rao, 1937)	Rao's burrowing frog	Western Ghats	DD
Sphaerotheca breviceps (Schneider, 1799)	Indian Burrowing frog	Non-endemic	LC
Sphaerotheca dobsonii (Boulenger, 1882)	Dobson's burrowing frog	Western Ghats	LC

Contd...

Table 2.4–*Contd...*

Species	*Common Name*	*Endemic*	*IUCN*
Family: Micrixalidae Dubois, Ohler and Biju			
Micrixalus aff. elegans (Rao, 1937)	Elegant torrent frog	Western Ghats	DD
Micrixalus saxicola (Jerdon, 1853)	Small torrent frog	Western Ghats	VU
Family: Microhylidae Günther			
Kaloula taprobanica Parker, 1934	Painted frog	Non-endemic	LC
Microhyla ornata (Dumeril and Bibron, 1841)	Ornate narrow mouthed frog	Non-endemic	LC
Microhyla rubra (Jerdon, 1854)	Red narrow mouthed frog	Non-endemic	LC
Ramanella aff. montana (Jerdon, 1854)	Jerdon's Ramanella	Western Ghats	NT
Family: Nyctibatrachidae Blommers Schlösser			
Nyctibatrachus cf. aliciae Inger, Shaffer, Koshy and Bakde, 1984	Alice's Night frog	Western Ghats	EN
Nyctibatrachus cf. major Boulenger, 1882	Malabar Night frog	Western Ghats	VU
Nyctibatrachus cf. petraeus Das and Kunte, 2005	Castle rock night frog	Western Ghats	LC
Family: Ranidae Rafinesque			
Clinotarsus curtipes (Jerdon, 1853)	Bicoloured frog	Non-endemic	NT
Hylarana aurantiaca (Boulenger, 1904)	Golden frog	Western Ghats	VU
Hylarana malabaricus (Tschudi,1838)	Fungoid frog	Non-endemic	LC
Hylarana temporalis (Günther,1864)	Bronzed frog	Non-endemic	NT
Family: Ranixalidae Dubois			
Indirana beddomii (Günther, 1875)	Beddome's Indian frog	Western Ghats	LC
Indirana semipalmatus (Boulenger,1882)	South Indian frog	Western Ghats	LC

Contd...

Table 2.4–*Contd...*

Species	*Common Name*	*Endemic*	*IUCN*
Family: Rhacophoridae Hoffman			
Polypedates maculatus (Gray, 1834)	Chunam frog	Non-endemic	LC
Polypedates occidentalis Das and Dutta, 2006	Charpa tree frog	Western Ghats	DD
Polypedates pseudocruciger Das and Ravichandran,1998	False hour glass tree frog	Western Ghats	LC
Pseudophilautus amboli (Biju and Bossuyt, 2009)	Amboli bush frog	Western Ghats	DD
Pseudophilautus wynaadensis (Jerdon, 1853)	Wynaad bush frog	Western Ghats	EN
Raorchestes bombayensis (Annandale, 1919)	Maharashtra bush frog	Western Ghats	VU
Raorchestes luteolus (Kuramoto and Joshy, 2003)	Coorg yellow bush frog	Western Ghats	DD
Raorchestes ponmudi (Biju and Bossuyt, 2005)	Large Ponmudi bush frog	Western Ghats	CE
Raorchestes tuberohumerus (Kuramoto and Joshy, 2003)	Kudremukh bush frog	Western Ghats	DD
Rhacophorus malabaricus Jerdon, 1870	Malabar gliding frog	Western Ghats	LC
ORDER: GYMNOPHIONA Müller			
Family: Ichthyophiidae Taylor			
Ichthyophis beddomi Peters, 1879	Beddome's caecilian	Western Ghats	LC
Ichthyophis bombayensis Taylor, 1960	Bombay caecilian	Western Ghats	DD

Notes: E: Endemic; NE: Non-endemic; GAA: Global Amphibian Assessment; EX: Extinct from type locality; EN: Endangered; Vu: Vulnerable; NT: Near threatened; LC: Least concerned, DD: Data deficient.

Aghanashini River Basin

Thirty two species are recorded from Aghanashini river basin (Table 2.5). None of the caecilians were recorded in this river basin. Kathagal with 17 species is species rich site in Aghanashini followed by Sapurthi (15 species) and Baillalli (13 species). Sapurthi harbours higher endemic species in the entire river basin.

Bedti River Basin

Twenty nine species were recorded from Bedti river basin. Species rich regions are Makkigadde (13 species), Devnalli (12 species), Kelginkeri and Daanandi (10 species each). Endemism is highest is Makkigadde and Devnalli (8 species each). It is interesting note that Yerebail has no endemic species despite having 6 species.

Kali River Basin

Kali river basin has 29 species (Listed in Table 2.5). Castle rock, Virnolli and Ulvi are species rich (13 species each) followed Gowliwada (10 species). Castle rock leads with higher endemism than Ulvi (6 endmic species) Virnolli (9 endemic species). The species recorded from Deriye were all endemic to Western Ghats.

Table 2.5: River Basin-wise Species Record in Uttara Kannada

	Shara-vathi	*Aghana-shini*	*Bedti*	*Kali*	*Venkata-pura*
Family: Bufonidae					
Duttaphrynus melanostictus	+	+	+	+	
Duttaphrynus scaber	+	+	+	+	
Duttaphrynus stomaticus	+		+		
Pedostibes tuberculosus	+	+	+	+	
Family: Dicroglossidae					
Euphlyctis aloysii	+			+	
Euphlyctis cyanophlyctis	+	+	+	+	+
Euphlyctis hexadactylus	+	+			
Fejervarya brevipalmatus	+				+
Fejervarya caperata	+	+	+	+	
Fejervarya granosa	+	+	+		
Fejervarya kudremukhensis	+		+	+	
Fejervarya mudduraja	+	+			
Fejervarya rufescens	+	+	+	+	
Hoplobatrachus crassus	+	+			
Hoplobatrachus tigerinus	+	+	+	+	
Minervarya sahyadris	+	+	+	+	+
Sphaerotheca aff. leucorhychus	+	+		+	
Sphaerotheca breviceps	+	+	+	+	
Sphaerotheca dobsonii	+				

Contd...

Table 2.5–*Contd...*

	Sharavathi	*Aghanashini*	*Bedti*	*Kali*	*Venkatapura*
Family: Micrixalidae					
Micrixalus aff. elegans	+				
Micrixalus saxicola	+	+	+		
Family: Microhylidae					
Kaloula pulchra	+			+	
Microhyla ornata	+	+	+	+	
Microhyla rubra	+	+	+	+	
Ramanella aff. montana	+			+	
Family: Nyctibatrachidae					
Nyctibatrachus cf. aliciae	+	+	+		+
Nyctibatrachus cf. major	+	+			
Nyctibatrachus cf. petraeus	+	+	+	+	
Family: Ranidae					
Clinotarsus curtipes	+	+	+	+	
Hylarana aurantiaca	+	+	+	+	
Hylarana malabaricus	+	+	+	+	
Hylarana temporalis	+	+	+	+	
Family: Ranixalidae					
Indirana beddomii	+	+	+	+	
Indirana semipalmatus	+	+	+	+	+
Family: Rhacophoridae					
Polypedates maculatus	+	+	+	+	
Polypedates occidentalis	+	+			
Polypedates pseudocruciger	+				
Pseudophilautus amboli	+		+	+	
Pseudophilautus wynaadensis	+	+		+	
Raorchestes bombayensis			+	+	
Raorchestes luteolus	+	+	+		
Raorchestes ponmudi	+				
Raorchestes tuberohumerus	+	+		+	
Rhacophorus malabaricus	+	+	+	+	
Family: Ichthyophiidae					
Ichthyophis beddomi	+				
Ichthyophis malabaricus	+		+		
Species richness	**45**	**32**	**29**	**29**	**5**

Cluster analysis based on species richness and endemics using Bray-curtis distance measure is given in Figure 2.7. There are four clear groups with Kathalekan standing out exceptionally. Group I is species poor, while Group IV is species rich. Non-metric multidimensional scaling analysis based on habitat variables and species parameters are given in Figure 2.8. All the 83 sampling sites can be categorically grouped into four. Each quadrant in the graph are representing a unique group. First quadrant carries species rich but endemic poor group, II Quadrant having both richness and endemism very poor, III Quadrant with species poor but endemic rich and IV Quadrant with species rich and endemic rich sites. Kathalekan, Watehalla and Muppanne in Sharavathi, Baillalli in Aghanashini, Kelginkeri in Bedti are in Quadrant IV, needing immediate conservation measures. These sites were also influenced by arboreal species richness, direct developing species, critically endangered and endangered species.

Krigging based on species endmism is given in Figure 2.9. Kathalekan is being deliberately removed from the analysis to know other sites in the Uttara Kannada district with higher endemism. The warm coloration and contours joining them indicates the endemism value. Sites like Castle rock in Kali, Makkigadde in Bedti, Kathagal, Sapurthi and Bailalli in Aghanashini and Maleman, Nagodi and Hurli in Sharavathi river basin along with Kathalekan are species rich and endemics rich.

Conclusion and Recommendation

River basin based studies provide insights on species distribution and diversity as catchments are topographically and hydrologically well defined. Among five rivers, Venkatapura is least rich in terms of amphibians, which could be attributed to its smaller catchment. Among the other four river basins, Sharavathi is species rich and also endemics. One of the sampling site, Kathalekan, a very well known Myristica swamp, harbours 34 species, attributed to the vegetation and seasonality of stream. This site certainly needs immediate attention from decision makers as surrounding areas are used for agriculture purpose and there are instances of human activities inside the region. Muppane also has higher richness, but is already inside Sharavathi valley wildlife sanctuary. As one proceeds further north in Uttara Kannada district, the vegetation also changes to semi-evergreen to deciduous, which could be the reason of less diversity in Kali and Bedti river (29 species each). Kathagal in Aghanashini river basin is relatively closer to coast among the sites is an example for 'refugia' concept, where in amphibian species were found despite a small area surrounded by agricultural activities. Bailalli and Sapurthi also harbor higher species in Aghanashini. In Bedthi, Makkigadde and Kelginkeri are the sites with high amphibian richness and endemism. Similarly, Castlerock, Virnolli, Ulvi and Gowliwad are in Kali with amphibian richness and endemism. However, all these sites with high richness and endemism in all the river basin face the threat, directly are indirectly from human activities such as diversion of streams, encroachment for agriculture, illegal felling and collection of forest yields. Kathalekan in Uttara Kannada district is an ideal heritage sites from Biodiversity perspective. The other sites mentioned needs the attention of forest managers for better conservation and management of biodiversity in Uttara Kannada district.

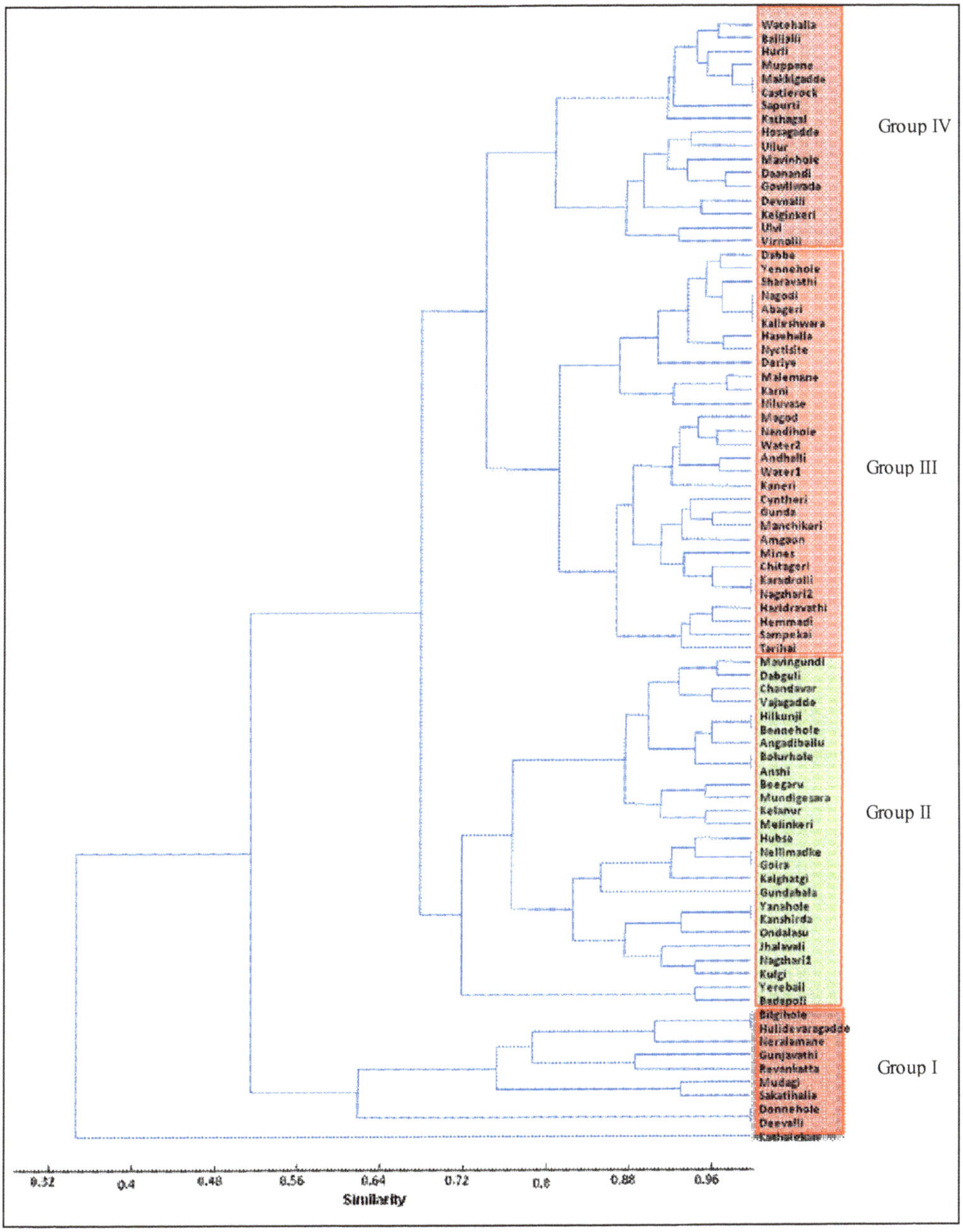

Figure 2.7: Cluster Analysis Based on Species Richness and Endemism

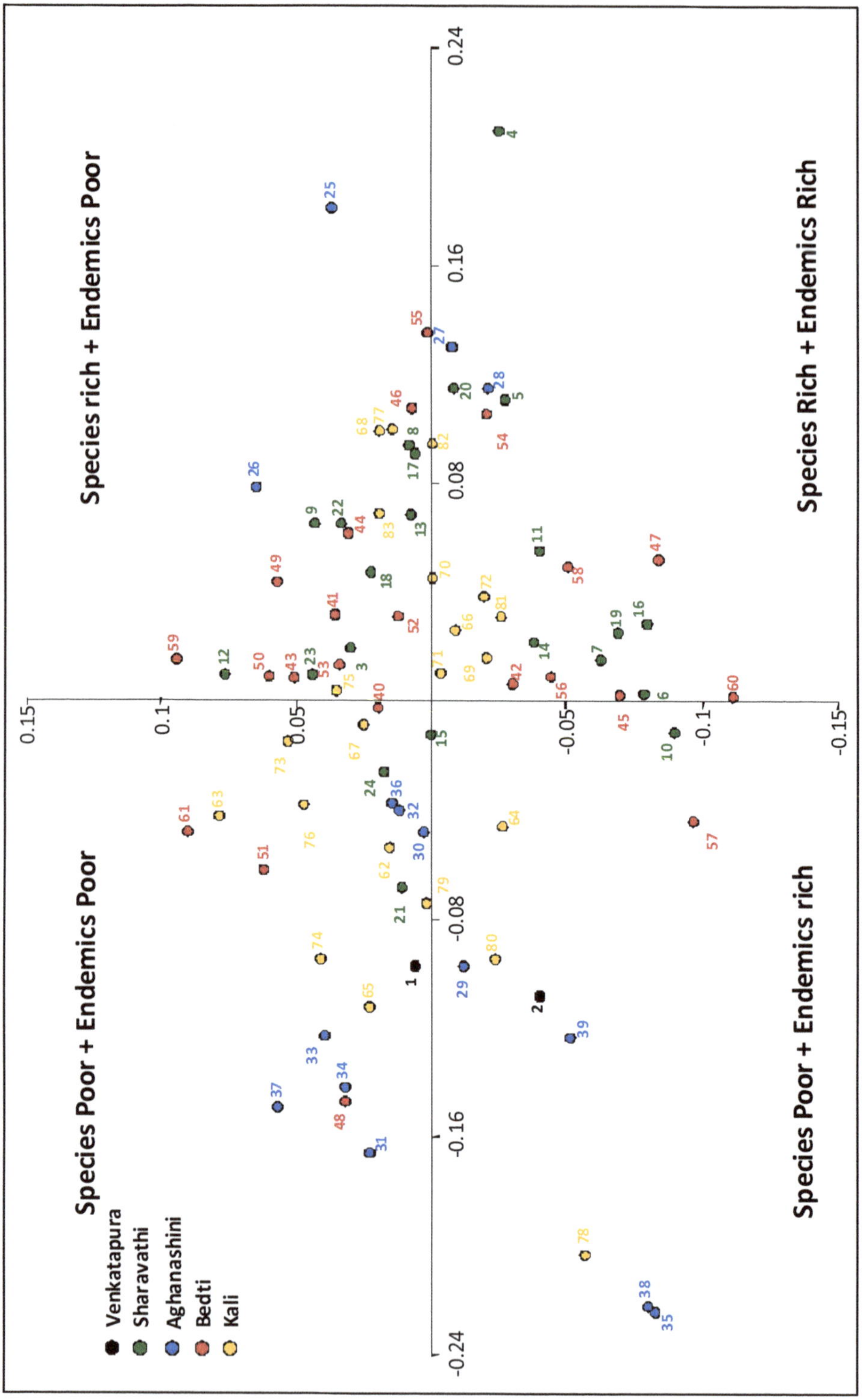

Figure 2.8: Non-Metric Multidimensional Analysis of Amphibians of Uttara Kannada

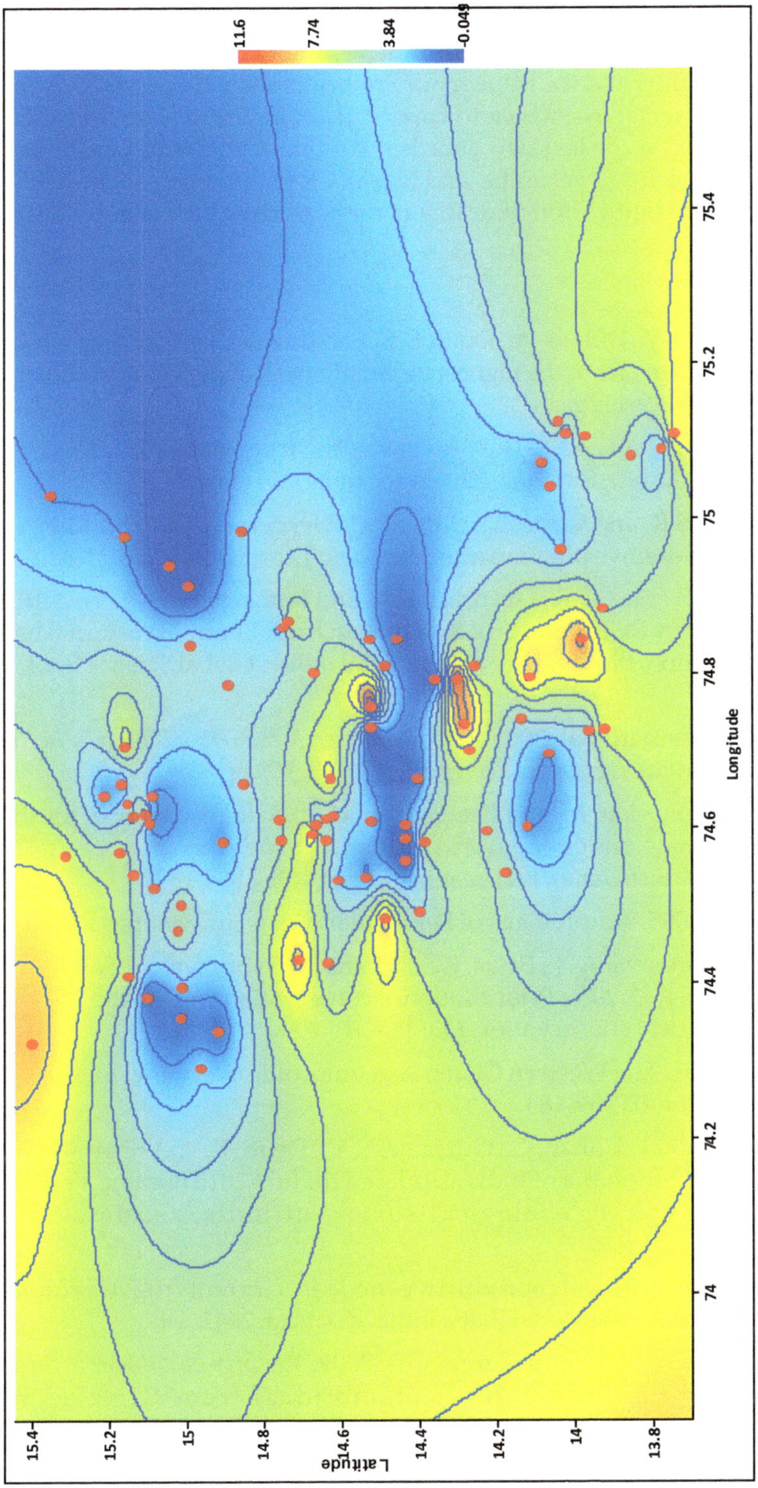

Figure 2.9: Spatial Interpolation Using Kriging Based on Species Endemism

Acknowledgements

We are grateful to NRDMS division, The Ministry of Science and Technology, Government of India and the Ministry of Environment and Forests, Government of India for the financial assistance to carry out ecological research in the Western Ghats region during the last two decades. We thank Shrikant, Lakshminarayana, Sameer Ali, Karthick B, Sreekantha, and Nayak CR for their help in the field. Avinash KG helped in the digitization of drainage maps of river basins of Uttara Kannada district.

References

Biju, S.D., Shouche, Y., Dubois, A., Dutta, S. K. and Bossuyt, F. 2010. A ground-dwelling rhacophorid frog from the highest mountain peak of the Western Ghats of India. Curr. Sci. **98**:119-1125.

Bossuyt, F. and Dubois, A. 2001. A review of the frog genus *Philautus* Gistel, 1848 (Amphibia, Anura, Ranidae, Rhacophorinae). Zeylanica, **6**:1-112.

Chandramouli, S.R. and Ganesha, S.R., 2010. Herpetofauna of Southern Western Ghats, India-Reinvestigation after decades, Taprobanica **2(2)**:21-32.

Chandran, M.D.S., Rao, G.R., Gururaja, K.V. and Ramachandra, T.V., 2010. Ecology of the Swampy Relic Forests of Kathalekan from Central Western Ghats, India. Bioremediation, Biodiversity and Bioavailability, Global Science Book Journals, 4: 54-68.

Conservation International, 2005. *Hotspots Revisited: Earth's Biologically Richest and Most Endangered Terrestrial Ecoregions*, CI, US, 392 pp.

Dahanukar, N., Diwekar, M. and Paingankar, M. 2011. Rediscovery of the threatened Western Ghats endemic sisorid catfish *Glyptothorax poonaensis* (Teleostei: Siluriformes: Sisoridae). J. Threatened Taxa **3(7):** 1885-1898.

Daniels, R.J.R., 2005. Amphibians of Peninsular India. Universities Press.

Dasa, A.C., Krishnaswamy, J., Bawa, K.S., Kirana, M.S., Srinivas, V, Samba Kumar, N. and Karanth, K.U. 2006. Prioritisation of conservation areas in the Western Ghats, India. Biological conservation **133**: 16 –31.

Dikshit, K.R. 2001. The Western Ghats: A geomorphic overview. *Memoir Geological Society of India* **47**: 159-183.

Dinesh, K.P., Radhakrishnan, C., Gururaja, K.V., Deuti, K. and Bhatta, G.K. 2010. A Checklist of Amphibia of India, Updated till July 2010 (Online Version). pp 11 http://zsi.gov.in/zoological-survey-of-india/zsi-data/checklist/Amphibia_final.pdf.

Gururaja, K.V. 2010. Novel reproductive mode in a torrent frog *Micrixalus saxicola* (Jerdon) from the Western Ghats, India. Zootaxa, **2642**: 1-8.

Ganesh. S.R., Asokan, S. and Kannan, P. 2009. Record of *Oligidon travancoricus* Beddome, 1877 (Serpentes; Columbridae) from Grizzled Squirred Santuary,Western Ghat,Tamil Nadu. The Herpetological Bulletin,**109**:25-28.

Hegde, B. 2011.Applying landscape Ecology Grieswald approach for conservation in Western Ghat.www.botanik.uni.grieswald.de/thesis/2011.

Karanth, K. K., Nichols, J. D., Hines, J. E., Karanth, K. U. and Christensen, N. L., 2009. Patterns and determinants of species occurrence, J. Appl. Ecol. **46:** 1189–1200.

Kunte, K. (in press.) Checklist of the Butterflies of the Western Ghats, Southwestern India. In K. A. Subramanian (ed.) Diversity and Conservation of Invertebrates in the Western Ghats.

Molur, S., Smith, K.G., Daniel, B.A. and Darwell, W.R.T. 2011.The Status and Distribution of Freshwater Biodiversity of Western Ghats, India. IUCN Red list of Threatened Species.

Radhakrishna, B.P., 2001. The Western Ghats of the Indian peninsula. Memoir Geological Society of India **47**: 133-144.

Ramachandra, T.V., Chandran, M.D.S., Gururaja, K.V. and Sreekantha. 2006. Cumulative Environmental Impact Assessment, Nova Science Publishers, New York, 371 pp.

Chapter 3

Vegetation and Floristic Diversity of Nawarangpur District, Odisha: A Repository of Threatened and Industrial Plants

N.K. Dhal* and S.C. Sahu

Environment and Sustainability Department,
CSIR-IMMT, Bhubaneswar – 13, Odisha

Introduction

Plants and plant based medicaments have been employed not only to ameliorate human and cattle diseases but also to improve the quality of life of man by meeting his three basic needs *i.e.*, food, shelter and clothing. For the over all development of an area its natural resources need to be utilized judiciously and this is possible only when the resources are identified and evaluated scientifically. For this, detailed floristic investigation need to be undertaken so that necessary plans can be drawn for its integrated development.

Hence, several botanists (Mukherjee 1953, Janaki Ammal 1954, Santapau 1958, Rao and Sreeramulu 1986) are of the opinion that this can be achieved only when smaller areas are explored intensively with critical assessment of its resources for proper utilization. Preparation of People's Biodiversity Registers (PBR) is another

* Corresponding Author: E-mail: nkdhal@immt.res.in, dhalnk@yahoo.com

step which awakened the Intellectual Property Rights of indigenous people following the Earth summit at Rio de Janeiro (Anonymous, 1992). Many have accepted Revenue district as an ideal geographical unit for floristic study since almost all Indian states have been divided into districts with definite jurisdiction for administrative convenience.

Of the 30 districts of Odisha, Nawarangpur lying on the western part of Odisha bordering Chhattisgarh suffer from non-utilization of its forest resources though it has vast potentialities. Nawarangpur has been in news every now and then because of frequent clashes between the natives and the East Bengal (Bangladesh) refugees. Vast stretches of virgin forestlands were cleared to accommodate the refugees under the Dandakaranya project. Continuous influx of more and more settlers in the guise of 'refugees' caused the gradual shrinkage of forest wealth and 'income generation opportunities' of the local inhabitants including tribals. The indigenous people have become minorities and this has adversely affected their cultural heritage and resulted in the form of serious clashes every now and then. The composition of the forest flora of the Eastern Ghats was also seriously affected due to removal of vegetation which contained some typical forest species. Large number of exotics has intruded and a few forest tracks have been reduced to scrub jungles.

The present work therefore was started in the year 2000 with a view to provide systematic account of floristic resources and identify vulnerable species and habitats for conservation either *in situ* or *ex situ*. Moreover, this will help in assessing the impact of the new settlers on the original floristic composition of the district and to evaluate to what extent the exotic weed flora exercised its inhibitive influence over the regeneration of indigenous floristic elements.

Geographical Features of Nawarangpur

Location and Area

Nawarangpur (Nabarangapur) lying between 19°10' 42" and 20°6' 12" N latitudes and 81°51' 30" and 82°52' 36" E longitudes is carved as a separate district from the erstwhile Koraput district of Odisha by an administrative approval and it has started functioning as a distinct unit from 1st April 1993 onwards. It is on the western part of Odisha through which the Eastern Ghats pass southwards. With an area of 5290.1 sq. km., Nawarangpur is a land locked district surrounded on the west and north by Bastar and Raipur districts of Chhattisgarh state, and on the east and south by the Kalahandi and Koraput districts of Odisha. The whole district is more or less an elevated plateau of Eastern Ghats with occasional valleys and peaks ranging from 2,000 ft to 3,000 ft. (Figure 3.1). It is interesting to note that the district is completely devoid of any railway line and the road system is still in its primitive stage. Because of high rain fall and judicious utilization of land and water, the agricultural production has increased many folds and hence it is often referred to as 'granary of Odisha'.

Rivers

Indravati the chief river of the district with its numerous tributaries drain more or less westwards except in the north where it slopes down towards east dropping

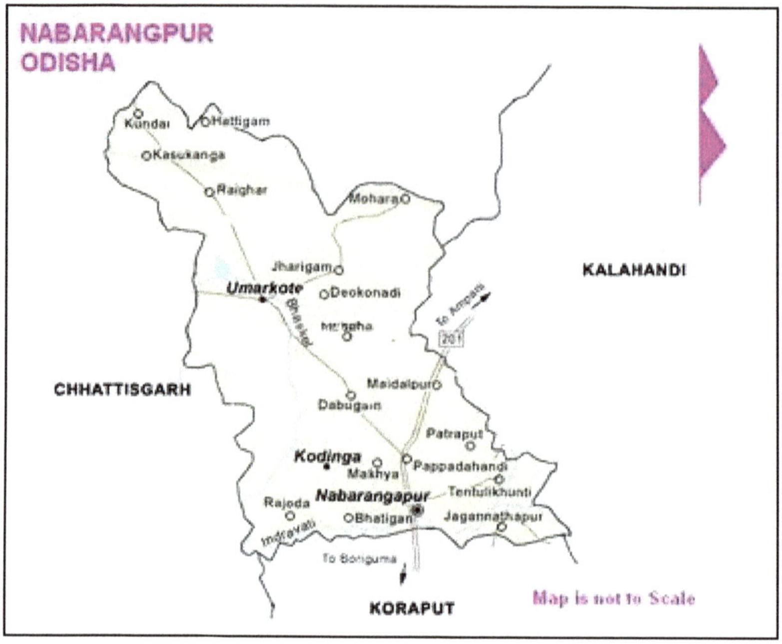

Figure 3.1: Map of Study Area

into the valley of the river Tel, a tributary of Mahanadi. The other river that drained the district is the Baskel. The climatic and edaphic factors coupled with riverine system have given rise to a rich and varied flora. The natural vegetation, which was predominantly arboreal before 1970, became considerably modified due to the settlements of East Bengal (Bangladesh) refugees under the rehabilitation programme of Dandakaranya project.

Geology

Major portion of geological formation is Archaean which includes unclassified crystalline-granites, gneisses and other magnetitic rocks, the stratification of which is uncertain. Towards the western margin bordering Chattisgarh, upper Precambrian consisting of lime-stones, shales and quartzites are noticed in detached patches. Among mineral wealth, iron, china clay and fire clay dominate the occurrence. Kaolin and pottery clay deposits are the frequent sites of the district.

Soils

Soils broadly speaking are the surface layers of earth and are classified by the constituents of rocks underlying them combined with some environmental phenomena. Three types of soils in distinct patches are found in Nawarangpur. These are:

Black Cotton Soil

This is seen towards extreme north-western part of the district with Raigarh as the epicentre. The vegetation is luxuriant dominated by Teak (***Tectona grandis***). The soils crack when exposed to hot sun. This is not suitable for erection of buildings because the foundation sinks. Soils are fertile and per hectare yield is high. Cotton grows well.

Red Soil

These soils are found towards Chandahandi bordering Kalahandi district. This is poor for plant growth but Sal (*Shorea robusta*) in different magnitudes dominates the vegetation. Iron, Manganese and limestone etc. are the principal minerals found in this type.

Lateritic Soils

Major portion of the district through which Eastern Ghats pass through contain lateritic soils. The soils suffer from nutrient deficiency and hence very poor in plant nurturing. It is good for building construction.

Climate

The climate of Nawarangpur is characterised by an equable temperature all through the year.The area has a seasonal rhythm as prevailing in rest of south-west Odisha. The whole year can conveniently be divided into three distinct seasons. The winter extends from November to February and this is followed by summer from March to June. The period from July to September is the monsoon while October and November constitute the post monsoon transition period. May is the hottest month with mean daily maximum temperature of 39.62°C and mean daily minimum of 11.7 °C. The minimum temperature goes as low as 9.7°C during middle of December and the maximum goes up 40.2 °C during middle of May.

Average annual rainfall for the period 2000 to 2003 is 1423 mm. Maximum rainfall is received in the month of August and the average stands at 486.12 mm. The June, July and August months record highest rainfall accounting for more than 74 per cent of the total annual rainfall. Relative humidity varies between 70 and 85 per cent. The R.H. for winter is 75 per cent, for summer 70 per cent, pre monsoon 85 per cent and post monsoon 78 per cent. Winds are moderate and thunderstorms occur in pre-monsoon months and occasionally they become violent.

Previous Botanical Explorations

Nawarangapur, Umerkote, Jharigan, Dabugan, Papadahandi, Kodinga, Raighar, Hathigam, Tentulikhunti and Indrabati form the main skeleton of the district and these areas appear to have been bypassed by the British Botanists even though the Eastern Ghats passes through the district. A careful scan into the literature shows that there is no mention of any species attributable to any place of the district either in Roxburgh's **Flora Indica** (1820-24) or in Sir J. D. Hooker's **Flora of British India** (1872-1897). **Flora of the Presidency of Madras** by J.S.Gamble and C.E.C.Fischer (1915-36) is the only consolidated work, which supposed to have references to the floristic elements of Nawarangpur as it was included in Koraput when the flora was written. Neither Gamble nor his predecessors and contemporaries like Col. R.H. Beddome, William Roxburgh, A.G. and Lady Bourne, A. Meebold, Retzius, F.G. Koenig, J.H. Burkill, E. Barnes, C.E.C. Fischer etc. ever visited the area with a view to assess the floristic composition. This might perhaps be due to the restrictions imposed by the native rulers on their movement coupled with lack of proper tracks in the inaccessible hilly terrain. **Botany of Bihar and Odisha** by H.H. Haines (1921-25) did not contain any references to the floristic elements of this area as it was a part of Madras Presidency by the time the flora was written. Mooney's **Supplement to the Botany of Bihar and Odisha** (1950) also did not speak about the floristic composition of the district because the supplement had the same jurisdiction as that of Haines. **The Flora of Odisha** by H.O. Saxena and M. Brahmam (1994-96) though very comprehensive in approach also lacks adequate representative specimens from the district thereby indicating that the areas was not adequately botanised.

A thorough search by the author for the plant specimens of Nawarangpur in the Central National Herbarium, Calcutta (CAL), Herbarium of the Forest Research Institute, Dehra Dun (DD), Herbarium of the Central Ayurvedic Research institute, Bhubaneswar and the Herbarium of the Regional Research Laboratory, Bhubaneswar has not yielded much except for noting 32 stray collections of Mr. P.C. Sharma and K.S.Murthy (Hemadri *et al.,* 1996) and Shri Trinath Das (Bahadur and Trinath Das, 1969). Some botanists in fact took interest in the flora of erstwhile Koraput but did not cover Nawarangpur because of various reasons with the result this extreme piece of land has remained more or less un-botanised. It is ironical to mention here that the Botanical Survey of India also did not cover this district in their floristic explorations. Though, a few workers showed interest on the ethnobotany of undivided Koraput (Hemadri and Rao 1989, Hemadri *et al.,* 1996, Das and Misra, 1986, 1987, 1988a, b, c, Dash and Misra, 1999, Mishra, 1992, Srivastava and Rout, 1994 etc.), none seems to have investigated Nawarangpur either from floristic or ethnobotanical point of view. The increased trend of human settlements and consequent conversion of prime forestlands into arable fields, the indigenous flora is drastically reduced. Uncontrolled grazing by the cattle of new settlers on woodlands coupled with the age-old shifting cultivation by the indigenous people had further reduced the forest wealth. The invasion by the exotic weed flora of unknown origin is also partly responsible for the forest shrinkage. Realising the state of affairs and the urgent need to have a comprehensive flora including the ethnobotanical aspects, this work was taken up for detailed study between 2000 and 2004.

Present Study and Methodologies Adopted

Survey and Plant Collection

Realising that the area was not adequately botanised earlier, systematic surveys were undertaken. The study involved intensive explorations and critical study of specimens for the last four years. The field trips were organised in such a way so as to cover all the areas of the district at regular intervals in different seasons between 2000 and 2004. As a result, it became possible to record the seasonal variations in the vegetation including distributional patterns and collect most of the plants in different developmental stages of their life cycle. Plant specimens were collected in sets of four both in flowering and fruiting stages taking due care to collect the healthy specimens. Field observations on phenology, habit, habitat, local names, local uses, frequency of occurrence etc. were recorded in the field note books at the time of collection and the collected specimens were tagged with field book numbers.

Preparation of Herbarium Specimens

All the specimens were dipped in 10 per cent $HgCl_2$ in alcohol solution before pressing them in blotting papers. The papers were changed in every alternate day till specimens became completely dry. After perfect drying, one set of the specimens were mounted on hand made Herbarium sheets of 42 x 28 cm size with the help of modern synthetic adhesives like fevicol or mowicol. For additional security, they were stitched at 4-5 places. Each mounted specimen was provided with a label wherein data pertinent to field number, date of collection, botanical name, family to which it belongs, place of collection, brief diagnostic description, uses if any, local name(s) etc.

Fleshy specimens, delicate flowers and large plant parts like tubers, rhizomes, corms, fruits, etc which could not be mounted on herbarium sheets were preserved in museum jars following customary preservative methods.

Critical Study and Identification

Care was taken to identify the specimens in the field itself while the specimens were fresh. Their identity was ascertained in the Herbarium with the help of floras, monographs, revisions and other taxonomic literature. Dried flowers and fruits were dissected by dissolving either in hot water or in Pohl's softening reagent (Pohl, 1965). This reagent is prepared by mixing 1 per cent solid Aerosol OT + 74 per cent of distilled water + 25 per cent methyl alcohol.

For confirmation, the Central National Herbarium (CNH), Howrah and Forest Research Institute Dehradun were consulted. Voucher specimens are preserved in the Herbarium of the CSIR-IMMT, Bhubaneswar, Odisha. Another set is made ready for depositing in the CNH, Howrah.

Plant Collection and Herbarium Studies

The supportive plant specimens of folklore claims were collected, processed, critically studied, identified and preserved in the Herbarium.

Critical Identification

Correct identification of plant was given highest priority. Identification by the local name is non-scientific and redundant. It so happens that a plant may be known by several names and one name may be applied to more than one species. If the plant cannot be identified, all the effort of ethnobotanical knowledge becomes a futile exercise. Hence, every effort was taken to identify the plant in the field itself by extending the stay in the field. Whenever the doubts arose, special effort was taken to separate out such plants for critical studies in the Herbarium. Identification of all the specimens was confirmed with the help of RRL Herbarium and Central National Herbarium, Kolkata.

Plan Followed for Presentation

An artificial dichotomous key is given to the families of flowering plants of the district in the beginning followed by the treatment of individual families. The sequence and arrangement of families is primarily after Bentham and Hooker's Genera Plantarum (1862-83) but an exception has been made for Gymnospermae. Bentham and Hooker had placed Gymnospermae between Dicotyledones and Monocotyledones. Following the embryological evidences as shown by Hofmeister (1851) and morphological and phylogenetical evidences as enumerated by Van Tiegham (1898), the Gymnospermae has been removed from its intermediate position and installed as one of the 2 divisions of Spermatophyta – Gymnosperms and Angiosperms. The concept of delimiting the families is after Hutchinson (1973) but the Genera within families and species within Genera are arranged in alphabetical sequence for easy reference.

Keys

Easily workable dichotomous keys to the families, genera, species and varieties were prepared taking the gross morphological characters into consideration for quick identification.

Arrangement and Nomenclature

The sequence of Genera and Species under each family is arranged in alphabetical order to facilitate easy reference. Nomenclature has been brought up-to-date as far as possible. The valid name is given in bold letters in the first place with the reference of its original publication followed by its basionym if any and more important synonyms in chronological order. References to the flora of Hooker (1872-92), Gamble (1915-1936), Haines (1921-1925), Mooney (1950), Saxena and Brahmam (1994-1996) and recent monographs have been given through out the work.

Local Names

Vernacular names as employed by the local inhabitants in Oriya and tribal languages are given. For some popular species English names are also given if they have relevance.

Plant Description

The description of each species contains important diagnostic features of the taxon as observed in the field and critical study of the specimen in the Herbarium. A note on its distribution denoting 'common', 'occasional', 'frequent', 'rare' 'planted', 'cultivated' etc. with respect to Nabarangpur are mentioned. Other features such as habitat, flowering-fruiting period etc. are given based on personal observations in the field.

Results and Discussion

Floristic and Vegetation

Nawarangapur falls under Deccan region of Hooker's (1904) sketch of the Flora of British India but Gamble (1915-36) placed it in the Sal region of Madras presidency. The existing vegetation as surveyed now is found to fall broadly under the following three types as per the classifications of Revised Forest Types of India (Champion and Seth, 1968). These are:

1. Tropical semi-evergreen forests
2. Tropical Dry Deciduous Forests and
3. Tropical Moist Deciduous Forests.

Besides the above, several sub types and seral types like Bamboo brakes, Riparian Fringing Forest, Scrubs, Grass lands etc. are noticed mostly either developed or resulted due to biotic influence.

Sal Forest

Major portion of Nawarangapur is occupied by sal (*Shorea robusta*) and it occurs in different plant communities in different frequencies. It ascends even to the top of the hills if sufficient soil is available. When sal occurs in continuation with the semi-evergreen type close to the moist valleys, then it is classified as Tropical moist Deciduous Forests (Champion and Seth, 1968). The top storey in this consists of predominantly deciduous elements but the second storey has some evergreens. Predominant associates are *Terminalia bellirica, T.chebula, Lannea coromandelica, Lagerstroemia parviflora, Madhuca indica, Pterocarpus marsupium, Bambusa arundinacea, Bombax malabaricum, Adina cordifolia* etc. The shrubby layer consists of *Daedalacanthus nervosus, Ardisia solancea, Barleria strigosa* etc. The ground flora is rich with several members of Zingiberaceae, Asteraceae, Orchidaceae, Poaceae, *Justicia* spp., *Andrographis* spp., *Dicliptera* spp. etc.

Wherever the conditions are hot and dry, the sal quality is also reduced and this type may be classified as Tropical Dry Deciduous (Champion and Seth, 1968). Both upper and lower canopies are composed of deciduous elements and important associates of sal are: *Terminalia alata, Anogeissus latifolia, Diospyros melanoxylon, Mangifera indica, Adina cordifolia, Dendrocalamus strictus, Butea monosperma, Cassia fistula, Buchanania lanzan, Sterculia urens, Cochlospermum religiosum* etc. The shrubby layer is dominated by *Cleistanthus collinus, Randia dumetorum, Zizyphus rugosus, Gardenia latifolia, Pavetta indica, Ixora parviflora, Helicteres isora* etc. The climbers are not many, but the following are very frequent - *Combretum roxburghianum, Bauhinia vahlii, Leea macrophylla, Ventilago maderaspatana, Ampelocissus tomentosa* etc.

Mixed Forest

Wherever drier conditions prevailed in the district, mixed forests made their presence. Moist valleys, steeper hill slopes, low and open places in plains etc. are the commonest sites for this type of vegetation. Besides a few evergreens, deciduous spp., climbers, epiphytes etc. made their presence felt. The important constituent elements frequently encountered are – *Diospyros peregrina, Mangifera indica, Dillenia pentagyna, Litsea monopetala, Schleichera oleosa, Leea indica, L. macrophylla, Flemingia macrophylla, Pongamia pinnata, Bridelia stipularis, Polyalthia cerasoides, Albizia odoratissima, Cissus repanda, Cayratia auriculata, Gnetum ula, Ampelocissus latifolia, Entada pursaetha, Clematis gouriana, Smilax macrophylla, S. prolifera* etc. The epiphytic flora include *Aerides odorata, Bulbophyllum trista, Cymbidium aloifolium, Dendrbium bicameratum, Luisia trichorhiza, Viscum articulatum* etc. Several ferns and mosses also occur along moist valleys.

On steep slopes and below the ridges Dry mixed forests are generally found. Here sal is usually absent but other associates become dominant. Common constituents are – *Terminalia alata, Sterculia urens, Anogeissus latifoila, Bursera serrata, Cochlospermum religiosum, Oroxylum indicum, Callicarpa arborea, Dalbergia latifolia, Grewia hirsuta, Bridelia retusa, Petalidium barlerioides* etc. Mixed forests at lower elevations are composed *of Diospyros sylvatica, Gelonium multiflorum, Capparis zeylanica, Alstonia venenata, Chloroxylon sweitenia, Strychnos nux-vomica, Randia malabarica, Croton caudatus, Hugonia mystax, Glycosmis arborea* etc.

Scrub Forest

These are found scattered throughout the length and breadth of the district. These are the by-products of heavy biotic interference and may be termed as 'Induced scrubs'. Important species in this type are: *Ziziphus oenoplia, Ochna squarrosa, Hugonia mystax, Flacourtia sepiaria, Toddalia asiatica, Pavetta crassicaulis, Webera corymbosa, Randia malabarica, Capparis zeylanica, Allophylus serratus, Euphorbia neriifolia, Gymnosporia emarginata, Carissa spinarum, Cassytha filiformis, Cipadessa baccifera, Mimosa rubicaulis, M. pudica, Diospyros ferrea* etc.

Grass Lands

The upper slopes and flat summits of hills are mostly grasslands. These are considered to be biotic sub climax or plagioclimax. It is also possible that above 1000 m, their ecological status is of climax vegetation as there is no indication that in course of time forest would again supervene naturally though there is no appreciable biotic interference. The important grass species met with in this type are: *Themeda triandra, T. quadrivalvis, Cymbopogon flexuosus, Heteropogon contortus, Apluda mutica, Chrysopogon verticillatus, Arundinella setosa, A. ciliata, Imperata cylindrica, Bothriochloa glabra, Eragrostis coarctata, Sporobolus fertilis, Cynodon dactylon* etc. Other herbaceous species found scattered among the grasses are: *Crotolaria* spp., *Cassia leschenaultiana, Viola betonicifolia, Glossogyne pinnatifida, Sonchus arvensis, Swertia angustifolia, Rubia cordifolia, Scleria* spp. etc.

Aquatic Flora

Nawarangapur falls under monsoon zone and receives on an average 1423 mm rainfall annually. This has helped in the formation of a number of tanks, pools, lakes, ditches, reservoirs etc. in low elevations that supply water for drinking and irrigation purposes. Additionally, they harbour hydrophytic vegetation and common species in this type are: *Nelumbo nucifera, Nymphaea nouchali, N.stellata, Trapa bispinosa, Ipomoea aquatica, Ludwigia adscendens, Limnophila indica, Hydrilla verticillata, Vallisneria spiralis, Ottelia alismoides, Eichhornea crassipes, Monochoria hastata, Pistia stratiotes, Eriocaulon quinquangulare, Wolfia arrhiza* etc. In the stream edges, *Ipomoea pescaprae, Geniosporum prostratum, Eleocharis dulcis* etc. are seen over sand heaps.

Rheophytes

Rheophytes are those plants which are riverine and found among rocky, gravely, sandy river beds below the flood zone. These species are resistant to sudden overflow of swift running water. Some of these that occur in Nawarangapur are: *Homonoia riparia, Barringtonia acutangula, Polygonum hydropiper, Salix tetrasperma, Ludwigia adscendens,* etc.

Statistical Analysis

Based on the results obtained from the field surveys and critical studies of the specimens it is found that the total number of species of vascular plants in the flora is 1008 including 199 species of exotic and cultivated origin. A statistical analysis (Table 3.1) of the indigenous flora shows the percentage of Dicot and Monocot species as 72.99:27.01 as against 70.7: 29.3 for the flora of Odisha. This shows higher percentage of Dicots in the flora of Nawarangapur. This is mainly due to the presence of Eastern Ghats spreading the most part of the district, which rarely allow predominance of grasses and sedges. Out of 215 species of monocots, 130 species (*i.e.* >60 per cent) belong to 2 families - *Poaceae* and *Cyperaceae* and the remaining 85 species are distributed over 22 families. The families *Zingiberaceae* (11 spp.) and *Commelinaceae* (7 spp.), though poor in number of species, are otherwise fairly abundant and form the dominant feature of the herbaceous vegetation. Pteridophyta is represented by 11 species of Ferns and Gymnosperms by only 2 indigenous species.

Table 3.1: Floristic Synopsis

Class	*Families*	*Genera*	*Indigenous Species*	*Exotic and Cultivated spp.*	*Total Species*
Angiospermae	141	615	796	196	992
Monocotyledonae	24	122	215	24	239
Dicotyledonae	117	493	581	172	753
Gymnospermae	4	4	2	3	5
Pteridophyta	11	11	11	-	11
Total	156	630	809	199	1008

The genus–species ratio for indigenous vascular taxa is 1:1.3 as against 1:7 for Flora of British India (Hook.f, 1904), 1:2.4 for 'Flora of Bihar and Odisha' (Panigrahi *et al.*, 1964) and 1:2.57 for 'Flora of Odisha' (Saxena and Brahmam, 1994). This shows larger number of genera and smaller number of species to genera in this flora. Low Genus-Species ratio confirms to the general rule that the floras of smaller regions like district floras are richer in genera and poorer in Genus-Species ratio.

Exotic Flora

Another striking feature of the flora is the presence of large number (199) of species of exotic and cultivated origin. This accounts for 19.74 per cent of the flora, which may be considered as 'high' by all standards. Normally, 6-7 per cent is expected but nearly 20 per cent is an unusual figure and this is largely due to forest felling for rehabilitation programme. When the existing vegetation is cleared, weeds make their way and occupy the forest floor thus becoming complimentary to the existing vegetation. Moreover, the new refugee settlers have introduced many species for their livelihood. A few exotics like *Eupatorium, Lantana, Bidens, Blumea, Prosopis, Pennisetum, Vernonia* etc. have become very aggressive and replaced the original flora at certain places and some weeds have even become obnoxious. Their fast spread can be attributed to their high

1. Germination index
2. Survival rate
3. Aggressive growth performance
4. Varied adaptability
5. Viable seed output
6. Ability to survive in nutrient deficit substrata

Dominant Families

The ten most dominant families of Nawarangpur, Odisha (Saxena and Brahmam 1994-96), Flora of British India (Hooker 1872-97) are plotted in the order of their dominance with respect to number of species represented in Table 3.2. While more than 70 per cent of the plant families in the floras of the 4 regions are almost the same, yet there are a few glaring sequence disturbances. The family 'Orchidaceae' which occupies 1st position in British India, 7th position in Flora of Bihar and Odisha and 4th position in 'Flora of Odisha' does not figure in the flora of Nawarangpur and this is due to the fact that Nawarangpur is more of a dry area and the forest has become thoroughly disturbed due to the settlement activities of Bangladesh refugees. Any disturbance in the site is detrimental for the development of Orchid flora. Orchids as a rule do not tolerate the direct beam of sunlight and hence they did not flourish much in the district.

Table 3.2: Dominant Families of Nawarangpur, Odisha, Bihar and Odisha and British India

Sl.No.	*Nawarangpur (Dhal, 2004)*	*Odisha State (Saxena and Brahmam 1996)*	*Bihar and Odisha (Haines, 1921-25)*	*British India (Hook.f. 1872-97)*
1	Poaceae	Poaceae	Leguminosae	Orchidaceae
2	Fabaceae	Fabaceae	Graminae	Leguminosae
3	Cyperaceae	Cyperaceae	Cyperaceae	Graminae
4	Asteraceae	Orchidaceae	Compositae	Rubiaceae
5	Euphorbiaceae	Asteraceae	Euphorbiaceae	Euphorbiaceae
6	Rubiaceae	Euphorbiaceae	Acanthaceae	Acanthaceae
7	Acanthaceae	Rubiaceae	Orchidaceae	Compositae
8	Malvaceae	Acanthaceae	Rubiaceae	Cyperaceae
9	Lamiaceae	Scrophulariaceae	Labiatae	Labiatae
10	Convolvulaceae	Lamiaceae	Scrophulariaceae	Urticaceae

The dominant families in the floras of Nawarangpur and Odisha are more or less same but their sequence is disturbed due to site change. Due to the settlement activities of refugees, the indigenous flora is replaced by the more aggressive exotic species especially from new world countries. The trees have lost their arboreal habitat and are reduced to shrubby condition with crooked stems and stunted growth. *Thuja orientalis, Araucaria columnaris* and *Cycas revoluta* are grown in plains especially near dwellings of new settlers. Since the Bangladesh settlers are fond of fish, several low-lying areas hither to harbouring numerous Hydrophytes were converted into fish ponds thus reducing Hydrophytic growth.

Threatened Plants

Rapid destruction of forests causes severe damage to natural forest of the district, thus threatening the very survival of several important plants. Recently, the assessment of threatened medicinal plants of Odisha was carried through Conservation Assessment and Management Priotisation (CAMP, 2007). Floristic assessment of the district resulting 2 critically endangered, 9 endangered and 12 vulnerable species along with their botanical name, family and IUCN status. So it is an urgent necessary to conserve these medicinal plants prior to its extinction. This may be planned through In-Situ or Ex-Situ conservation method, for preserving the biodiversity of the state.

Table 3.3: List of Threatened Plants in Nawarangpur District of Odisha

Sl.No.	*Botanical Name*	*Family*	*IUCN Status*
1.	*Blepharispermum subsessile*	Asteraceae	EN
2.	*Celastrus paniculatus*	Celastraceae	VU
3.	*Caesalpinia digyna*	Caesalpiniaceae	VU
4.	*Crataeva magna*	Capparaceae	VU
5.	*Cycas circinalis*	Cycadaceae	VU
6.	*Embelia tsjeriam-cottam*	Myrsinaceae	VU
7.	*Garcinia xanthochymus*	Clusiaceae	VU
8.	*Gardenia gummifera*	Rubiaceae	VU
9.	*Gloriosa superba*	Liliaceae	EN
10.	*Hedychium coronarium*	Zingiberaceae	VU
11.	*Mesua ferrea*	Clusiaceae	EN
12.	*Oroxylum indicum*	Bignoniaceae	EN
13.	*Paedaria foetida*	Rubiaceae	VU
14.	*Piper longum*	Piperaceae	EN
15.	*Pterocarpus marsupium*	Fabaceae	EN
16.	*Pueraria tuberosa*	Fabaceae	VU
17.	*Rauvolfia serpentina*	Apocynaceae	EN
18.	*Saraca asoca*	Caesalpiniaceae	CR
19.	*Schrebera swietinoides*	Oleaceae	VU
20.	*Stereospermum chelonoides*	Bignoniaceae	EN
21.	*Strychnos potatorum*	Strychnaceae	VU
22.	*Symplocos racemosa*	Symplocaceae	CR
23.	*Uraria picta*	Fabaceae	EN

Species Identified for Industrial Uses

Species for Vaneers and Plywood

- *Bombax ceiba*–Semul
- *Bridelia retusa*–Kasi
- *Dalbergia latifolia*–Sissoo
- *Desmodium oojeinense*–Bandhan
- *Dillenia pentagyna*–Rai
- *Garuga pinnata*–Moi
- *Gmelina arborea*–Gambhari
- *Kydia calycina*–Kapasia
- *Lannea coromandelica*–Moi, Jhingan
- *Mangifera indica*–Ambo
- *Michelia champaca*–Champa
- *Mitragyna parvifolia*–Mundi
- *Pterocarpus marsupium*–Piasal, Bija
- *Syzygium cumini*–Jamun
- *Terminalia alata*–Sahaja, Asan
- *Terminalia bellirica*–Bahada

Species for Paper and Pulp

- *Anogeissus latifolia*–Dhaura
- *Cleistanthus collinus*–Karada
- *Cymbopogon flexuosus*–Dhanvantri
- *Eulaliopsis binata*–Sabai
- *Gargua pinnata*–Moi
- *Haldinia cordifolia*–Haldu, Koim
- *Helicteres isora*–Murkuri, Antomulo
- *Kydia calycina*–Kapasia
- *Lagerstroemia parviflora*–Sidha
- *Lannea coromandelica*–Moi
- *Mitragyna parvifolia*–Mundi
- *Saccharum spontaneum*–Jangli akhu
- *Shorea robusta*–sal
- *Terminalia alata*–Asan, Sahaja

Species for Match-Box and Splinters

- *Alstonia scholaris*–Chhatina (Splinters)
- *Anthocephalus chinensis*–Kadambo (Splinters)
- *Bombax ceiba*–Semul (match-box and splinters)

- ☆ *Hymenodictyon oriexense*–Bhurkunda (Splinters)
- ☆ *Kydia calycina*–Kapasia (match-box)
- ☆ *Mangifera indica*–Ambo (match-box)
- ☆ *Salix tetrasperma*–Panibegunia (Splinters)
- ☆ *Sterculia villosa*–Udal (Splinters)
- ☆ *Symplocos racemosa*–Lodha (Splinters)

Fodder Grasses

- ☆ *Andropogon ascinodes*
- ☆ *Apluda mutica*
- ☆ *Arundinella holcoides*
- ☆ *Bothriochloa bladhii*
- ☆ *Bothriochloa pertusa*
- ☆ *Brachiaria distachya*
- ☆ *Brachiaria ramosa*
- ☆ *Capillipedium assimile*
- ☆ *Chrysopogon aciculatus*
- ☆ *Chrysopogon verticillatus*
- ☆ *Cynodon dactylon*
- ☆ *Dactyloctenium aegyptium*
- ☆ *Desmostachya bipinnata*
- ☆ *Dichanthium caricosum*
- ☆ *Digitaria abludens*
- ☆ *Digitaria bicornis*
- ☆ *Digitaria ciliaris*
- ☆ *Digitaria longiflora*
- ☆ *Digitaria setigera*
- ☆ *Digitaria stricta*
- ☆ *Echinochloa colona*
- ☆ *Eleusine indica*
- ☆ *Eragrostis tenella*
- ☆ *Hackelocloa granularis*
- ☆ *Heteropogon contortus*
- ☆ *Hygroryza aristata*
- ☆ *Isachne globosa*
- ☆ *Iseilema laxum*
- ☆ *Oryza sativa*
- ☆ *Panicum trypheron*
- ☆ *Paspalum scrobiculatum*
- ☆ *Pennisetum purpureum*
- ☆ *Rottboellia cochinchinensis*
- ☆ *Sacciolepis indica*
- ☆ *Setaria intermedia*
- ☆ *Setaria italica*
- ☆ *Sorghum vulgare*
- ☆ *Sporobolus indicus* var. *fertilis*
- ☆ *Themeda quadrivalvis*
- ☆ *Themeda triandra*

Thatching Grasses

- ☆ *Eulaliopsis binata*–Babuli, Sabai
- ☆ *Heteropogon contortus*–Sukla
- ☆ *Imperata cylindrica*–Chhanna ghas
- ☆ *Saccharum spontaneum*–Payal (K.)

Fibre-Yielding Species

- ☆ *Abutilon persicum*–Pedipedika
- ☆ *Bauhinia vahlii*–Siali
- ☆ *Careya arborea*–Kumbhi
- ☆ *Entada rheedii*–Geredi

- ☆ *Eulaliopsis binata*–Babuli, Sabai
- ☆ *Grewia tiliifolia*–Dhaman
- ☆ *Helicteres isora*–Murkuri, Antomulo
- ☆ *Hibiscus cannabinus*–Kanuriya
- ☆ *Ichnocarpus frutescens*–Swanlata
- ☆ *Kydia calycina*–Kapasia
- ☆ *Sesbania bispinosa*–Tentua
- ☆ *Sida rhombifolia*–Bajramuli
- ☆ *Trema orientalis*–Jivani, Khadkas
- ☆ *Urena lobata*–Billo kapas

Species for Brooms

- ☆ *Phoenix acaulis*–Khajuri
- ☆ *Sida acuta*–Bajarmuli
- ☆ *Thysanolaena maxima*–Phuljhadu ghas
- ☆ *Sida rhombifolia*

Species Yielding Flosses

- ☆ *Bombax ceiba*–Semul, Simuli
- ☆ *Cochlospermum religiosum*–Kapasia
- ☆ *Cryptolepis buchananii*–Dudhi
- ☆ *Holarrhena pubescens*–Kurchi

Dye Yielding Species

- ☆ *Acacia farnesiana*–(bark)
- ☆ *Butea monosperma*–Palas (flower)
- ☆ *Mallotus philippensis*–(fruits)
- ☆ *Morinda pubescens*–Achu–(root–bark)
- ☆ *Nyctanthes arbortristis*–Gangaseoil (flower)
- ☆ *Artocarpus heterophyllus* (stem–bark)
- ☆ *Rubia cordifolia*–Rango chireta.(root)
- ☆ *Terminalia alata*–Sahaja, Asan

Potential Tanning Bearing Species

- ☆ *Anogeissus latifolia*–Dhaura (leaf)
- ☆ *Bridelia retusa*–Kasi (bark)
- ☆ *Cassia fistula*–Sonari (bark)
- ☆ *Cleistanthus collinus*–Korada (fruit)
- ☆ *Emblica officinalis*–Amla (fruit, leaf)
- ☆ *Garuga pinnata*–Moi (bark)
- ☆ *Lagerstroemia parviflora*–Sidha (bark)

- *Shorea robusta*–Sal (bark)
- *Terminalia alata*–Sahaj (bark, leaf-galls)
- *Terminalia bellirica*–Bahada (fruit)
- *Terminalia chebula*–Harada (fruit)
- *Ziziphus xylopyrus*–Goto (fruit)

Important Essential Oil–Bearing Species

- *Cymbopogon flexuosus*–Lemongrass
- *Cymbopogon martini*–Palmarosa grass
- *Hyptis suaveolens*–Bantulsi
- *Michelia champaca*–Champa
- *Neocinnamomum caudatum*–tejapatro
- *Ocimum gratissimum* (Janglitulsi)

Plants Yielding Gums and Resins

- *Anogeissus latifolia*–Dhaura gum
- *Bauhinia malabarica*–Amta gum
- *Cochlospermum religiosum*–Kapasis
- *Lannea coromandelica*–Jhingan
- *Pterocarpus marsupium*–Piasal, Bija; Kinogum
- *Shorea robusta*–Sal Jhuna
- *Sterculia urens* and *Sterculia villosa* Udal; Gum karaya, Genduli gum

Conclusion

The species identified for industrial uses may be tested for intensive industrial bio-processing for globally accepted quality products as well as for phytochemical screening and pharmacognosy in search of modern herbal drugs. Based on economic and market survey extensive studies of different aspects of value additions are suggested. As more than 50 per cent of people are tribal in the district, their indigenous knowledge in conservation activity should be encouraged through *in situ* which is beneficial for their livelihood. Floristic analysis results documentation of threatened medicinal plants which needs immediate conservation. *In-situ* or *ex-situ* conservation methods including sustainable collection may be adopted for better germ plasm conservation.

Acknowledgements

The authors are thankful to Prof. B.K. Mishra, Director CSIR-IMMT, Bhubaneswar for providing the facilities to carry out the research work. The financial help by the Department of Biotechnology, Government of India is gratefully acknowledged.

References

Bahadur, K.N. and Das, T. 1969. On the occurrence and gregarious flowering of *Perilepta edgeworthiana* (Nees) Brem. in Odisha. Indian Forester, Vol. 209-211.

Bentham, G. and Hooker, J. D. 1862-1883. Genera Plantanum, London.

Champion, H.G. and Seth, S. K. 1968. *A revised survey of forest types of India*, Manager of Publications, Delhi.

Das, P.K. and Misra, M.K. 1986. Ferns of Koraput district (Odisha). *J. Odisha Bot. Soc.* 8:77.

Das, P.K. and Misra, M.K. 1987. Some medicinal plants used by the tribals of Deomali and adjacent areas areas of Koraput district Odisha. *Indian J. For.* 10:301-303.

Das, P.K. and Misra, M.K. 1988. Ethnobotanical studies of the tribal belt of Koraput (Odisha). *Bull. Medico-Ethnobot. Res.* 9:123-128.

Das, P.K. and Misra, M.K. 1988. Some ethnobotanical plants of Koraput district, Odisha. *Ancient Science of life* 8:60-70.

Das, P.K. and Misra, M.K. 1988. Some medicinal plants among Kondhas around Chandrapur (Koraput). *J. Econ. Tax. Bot.* 12: 103-109.

Dash, S.S. and Misra, M.K. 1999. Taxonomic survey and systematic census of economic plants of Narayanapatna hills of Koraput District, Odisha. *J.Econ. Tax. Bot.* 23(2): 473-498.

Gamble, J.S. and Fischer, C.E.C. 1915-1935. *Flora of Presidency of Madras*. Vols 1-3. Adlard and Son Ltd, London.

Haines, H.H. 1921-25.*The Botany of Bihar and Odisha 6 parts*: Bishen Singh and Mahendra Pal Singh. Dehradun.

Hemadri, K. and Rao, R.S. 1989. Folklore claims of Koraput and Phulbani districts of Odisha state. *Indian Medicine* 1(4): 11-13.

Hemadri, K., Sharma, P.C. and Murty, K.C. 1996. Medico-botanical exploration of Phulbani and Koraput districts of Odisha. Central Council of Research in Ayurveda and Siddha, Ministry of Health and Family Welfare, Government of India.

Hooker, J.D. 1872-97. *The Flora of British India*, London.

Misra, R.C. 1992. Medicinal plants among the tribals of Upper Bonda region, Koraput (Odisha). *J.Econ.Tax.Bot.* Addl. Ser. 10: 275-279.

Mooney, H.F. 1950. *Supplement to the Botany of Bihar and Odisha,* Ranchi, India.

Roxburgh, W. 1820-24. *Flora Indica* (Carey and Wallich editions) 2 Vols. Serampur.

Saxena, H.O. and Brahmam, M. 1996. *The Flora of Odisha. Vols I-IV.* Odisha Forest Development Corporation Ltd, Bhuabneshwar, India.

Srivastav, S.C. and Rout N.C. 1994. Some plants of ethnopaediatric importance in district Koraput. *Bull. Bot. Surv. India* 36(1-4): 166-168.

Ved D. K., Kinhal G. A., Ravikumar K., Vijayasankar R., Sumathi R., Mahapatra A.K. and Panda P.C. 2007. *CAMP Report*: Conservation Assessment and Management Prioritisation for Medicinal plants of Odisha, India. Foundation for Revitalisation of Local Health Traditions, Bangalore.

Chapter 4

Floral Structure and Phenology in Relation to Pollination of some Endangered *Impatiens* of Western Ghats

R. Ramasubbu[1]*, A.K. Sreekala[2], A.G. Pandurangan[2] and A. Chandra Prabha[3]

[1]Department of Biology, Gandhigram Rural University, Gandhigram, Dindigul, Tamil Nadu, India
[2]Plant Systematics and Evolutionary Science Division, Tropical Botanic Garden and Research Institute, Palode, Thiruvananthapuram, Kerala, India
[3]SFR College for Women, Sivakasi, Tamil Nadu, India

ABSTRACT

Flowers of *Impatiens* are basically 5-merous, though various modifications of structures are evident.The basic structure and phenology in relation to the pollination of selected rare endangered and threatened *Impatiens* species growing in the high altitude areas of Southern Western Ghats have been carried out. Five population were studied in selected forest areas of Idukki district of Kerala and Palni hills of Tamil Nadu, India. Morphology of floral parts, Anthesis, phenology, pollinator visitation etc. were examined. *Impatiens* have complex flowers among

* Corresponding Author: E-mail: racprabha@yahoo.com

zygomorphic angiosperms. The floral parts (lip with spur and wing petals) are specially adopted for endomophilous pollination especially by bees and butterflies. The flowering periods of the selected *Impatiens* are almost in same period (July-December) which attract all the pollinators towards pollination by their beautiful colour and mild fragrance. The colour of the flowers attract mostly than the fragrance. The lateral and wing petals act as a platform for landing of honey bees and butterflies while forage. Position of anthers also prevent self pollination and promote cross pollination. Large numbers of butterflies and honey bees visited the flowers and transferred the pollen grains to stigma. Whenever cross pollination failed, there was no fruit set. The successsful pollination by the agents only determine the survival of the species in the wild condition.

Keywords: *Impatiens, Phenology, Pollination, Western Ghats.*

Introduction

Angiosperm flowers are incredibly diverse in size, shape, colour, scent and features. These floral traits may be augmented by non-floral organs such as bracts. Functionally, a flower is a compound structure in which the constructional complexities presumably are adapted to sexual reproduction. Thus floral structure, phenology and the evolutionary ecology of pollination partnership are intimately interwoven-pollination biology makes sense only through evolutionary adaptations for reproductive success, through non-adaptive influences, such as exaptations and phylogenetic constraints must be allowed. Floral structure and phenology have functional significance for both pollination and breeding systems, simultaneous study of both will lead to a better understanding of the process that drive floral traits diversification (Holsinger, 1996; Barrett, 2003). Plants in isolated lands may be self breeding, particularly during colonization and establishment or when pollinator abundances are low (Baker, 1995). However, reduced genetic diversity is found associated with selfing in some insular plant species. Pollination specialization has long been considered as an important process underlying the evolution of floral diversity. Consequently, floral traits have been viewed as adaptation to attract specific pollinators and to enhance the efficiency of pollen transfer and out crossing (Darwin, 1862; Stebbin, 1970). More knowledge on the floral biology of plants in tropical regions is necessary to achieve a better understanding of the overall ecological and evolutionary patterns of specialization.

Habitat fragmentation, degradation and overexploitation are the more apparent causal factors for population reduction of many endemic species. In certain cases, the failure of reproductive process due to environmental changes is often the fundamental reason for species loss. Sexual reproduction is the only natural process that incorporates variability and ensures survival of a species under adverse condition (Monika and Bhatnagar, 2007). Sexual reproduction is based on the phenomenon of syngamy and double fertilization. Successful fertilization is dependent on effective pollination. Therefore, pollination studies can provide a lot of information about the population reduction of many species, because it is the fundamental step in plant reproduction. Successful pollination is an essential prerequisite for survival of plants in natural communities and is dependent on many biotic and abiotic factors. Plants have co-evolved with their pollinators and any ecological changes can decouple the process of flowering and breeding cycles.

The family Balsaminaceae comprises annual as well as perennial herbs that show conspicuous and striking floral morphology. It comprises two widespread genera, *viz.*, *Hydrocera* Wight and Arn., a monotypic genus and *Impatiens* L., a genus with a large number of species. *Hydrocera* can be distinguished by its five free petals and capsular berry while *Impatiens* has four petals that are fused to form two lateral petals and a 5-valved capsule. *Hydrocera triflora* (L.) Wight and Arn. is a semi-aquatic herb distributed in the Indomalesian region. *Impatiens* with more than 1,000 species (Janssens *et al.*, 2006, 2009) is distributed in the tropical and subtropical regions of the Old World as well as in the northern temperate regions (GreyWilson, 1980). The genus is known to have 5 distinct diversity hotspots, *viz.*, Tropical Africa, Madagascar, Southern India and Sri Lanka, the Eastern Himalaya and Southeast Asia (Yuan *et al.*, 2004). *Impatiens* are beautiful plants bearing curious and variously coloured flowers with peculiar floral structure. Majority of the *Impatiens* have great horticultural potential. Many species are cultivated as ornamental and some are used in medicine and cosmetics. Species belonging to this genus are commonly referred to as 'balsams' or 'jewel weeds'. The genus is represented by *c.* 210 species in India with two centres of diversity; the Eastern Himalaya and the Western Ghats. Both regions show a high degree of endemism and hence recognized as two amongst the 34 biodiversity hotspot regions in the world (Mittermeier *et al.*, 2004). Species occurring in India can be recognized under two groups: the first group with short, swollen spindle-shaped pods occurring in the peninsular region and the second group with long and narrow pods distributed in the Himalayan region (Hooker, 1874).

The survival of the selected candidate species in the Western Ghats is questionable due to various factors including its reproductive constrains (Ramasubbu, 2009). The selected candidate species are in critically endangered (CR) and Endangered (EN) category. So the survivability mainly depends on pollination by the visitors. The high altitude *Impatiens* species are pollinated mostly by honey bees, *Trigona*, butterflies etc. The visitation rate of pollinators depends upon its phenology, colour, size, shape and nectar availability. With this background a study has been framed to analyze the relationship between floral structure and phenology in relation to the pollination of *Impatiens* in the Western Ghats.

Materials and Methods

Study Area

The Western Ghats stretches along the western coast of Peninsular India from Tapti valleys in southern Gujarat to southwest Tamil Nadu (8-20° N), covering an area of 1,60, 000 sq. km and ranging from sea level to 2695m. The northern parts and extreme southern tip are less moist, while the central and southern Western Ghats receive heavy annual precipitation of up to 6000mm, which supports dense diverse tropical forests rich in species diversity and endemism. The mountainous configuration, high altitude, high rain fall, humid climate etc. make this tract one of the botanically richest areas of India and therefore aptly designated as one among 34 globally recognized biodiversity hot-spots in the world. The hills of southern Western Ghats have been attributed to this status due to the high levels of endemic forms of biodiversity distributed in genera, species and races as well as the severe threats to such biodiversity posed by humans. It is generally accepted based on several explorations conducted by various researchers that nearly 2000 species of higher plants are endemic to the Western Ghats (Ahmedulla and Nayar, 1987). But the existing number of endemic species among lower plants and microorganisms unfortunately remains unknown even today.

The Western Ghats and its slopes were explored during 2005-2008 for locating the viable populations of candidate species selected for reproductive biological study. These exploration trips suggested that the selected balsams are exclusively endemic to southern Western Ghats especially in Kerala high ranges and Palni hills of Tamil Nadu (Plate 4.1). During the field exploration, sufficient populations of *Impatiens coelotropis* and *I. platyadena* were located at three sites each from Kerala high ranges. *Impatiens phoenicea* was identified at four sites of which three from high ranges and one from upper Palni hills bordering Kerala. Further exploration trips could not locate these three *Impatiens* other than the sites mentioned above in the entire Western Ghats. Therefore, the detailed studies on reproductive biology of *Impatiens coelotropis, I. henslowiana* and *I. platyadena,* were carried out in Neyamakkad gap, Rajamalai, and Pettumudi in Idukki district, Kerala. The reproductive behaviour of *I. phoenicea* was studied in Top station–Shola National Park, Neyamakkad gap, Rajamalai, Mannavan shola from Kerala and Kodaikkanal (Palni hills) of Tamil Nadu. These studies were conducted in both natural habitats and in *ex-situ* gene bank developed in Tropical Botanical Garden and Research Institute, Palode, Thiruvananthapuram, Kerala. The details of geographic and climatic data of the study area are given in the Table 4.1.

Western Ghats: Kerala High Ranges

The study area comprises mainly on the high altitudes of Western Ghats of Kerala with an altitude ranging from 1600-2400m asl. It has a wide range of temperature, especially in winter months. In this region, the climate is that of temperate montane and the average temperature has been recorded to about 24°C during the hottest months (March-May) and less than 4°C during the cold season of December-February. From November to February heavy frost occurs in all the study sites and the surroundings. The warmest months are from February to May. The area is situated on the edge of the escarpment which faces the Arabian Sea. Hence, during the months of June to September, the area receives the full force of south west monsoons that lead to precipitate to an extent of 2000-4000mm (60 per cent of total rainfall). The north east retreating monsoon also brings rainfall up to 30 per cent during October-November and remaining 10 per cent during summer showers in the month of April-May. There are two perennial rivers flowing through the Rajamalai forest in addition to numerous streamlets crisscrossing the study area. The soil is acidic and made up of sandy clay loam over the crest and along the slopes of the precipitous hills. The soils composition forms 30.5-37.5 per cent sand fractions and 11.3-17.4 per cent silt. The percentage of clay and organic matter available in this area is 47.9-49.3 per cent and 2.8-3.03 per cent respectively. The pH range is being 5.4-5.7.

Plate 4.1: Study Area: Western Ghats–Kerala High Ranges and Palni Hills

Table 4.1: Geographic and Climatic Data of the Study Area in Western Ghats

Sl.No.	*Details of the Forest Area*	*Kerala High Ranges*				*Upper Palni Hills*
		Neymakkad	*Top Station*	*Rajamala, Pettumudi*	*Mannavan Shola*	*Kodaikkanal*
1.	Forest range	Munnar	Devikulam	Devikulam	Marayur	Kodaikkanal
2.	Latitude	10' 8' E	10° 05'–10° 20' N	10° 05'–10° 20' N	10° 10' N–10° 12' N	10° 6' N–10° 21' N
3.	Longitude	77° 9' E	77° 0'–77° 10' E	77° 0'–77° 10' E	77° 90'E–77° 12' E	77°16'E–77°42' E
4.	Altitude (asl)	1800–2197	1900–2100	2000–2400	1600–2100	1800–2195
5.	Status	Disturbed	Disturbed	Disturbed	Undisturbed	Disturbed
6.	Vegetation	Southern wet temperate forest and shola	Wet temperate forest and shola	Southern wet temperate forest and shola	Southern sub tropical forest and shola	Tropical ever green and shola
7.	Temperature (°C)	Max. 25 min. 11	Max. 23 min. 6	Max. 21 min. 7	Max. 24 min. 16	Max. 24 min. 4
8.	Rain fall (mm)	4000–4500	4500–4700	4050–4500	2000–3000	1650–1800
9.	Relative humidity (per cent)	85–90	80–90	85–90	80–95	85–95

The vegetation of the Western Ghats of the Kerala high ranges varies very much in character from place to place due to steep mountain ranges, fertile valleys, perennial rivers and streams. Further, there is a remarkable variation between the eastern and western slopes of the Western Ghats falling in the Kerala high ranges. The difference in altitude, climate, rainfall, the prevalence of wind of varying nature, the geological and edaphic factors are mainly responsible for bringing about such unusual variety and diversity in the vegetation pattern of this region. Tropical dry deciduous, tropical moist deciduous, tropical wet evergreen, montane wet temperate sholas and grasslands are the main types of forests in this area (Champion, 1936; Shetty and Vivekananthan, 1971). The vegetation on the eastern slopes of this high mountain is rich and following distinct types could be noted at various altitudes and valleys.

The evergreen forests are perhaps the richest in terms of the occurrence of number of species in a given area. But they have been subjected to interference continuously by man and because of that the existing forests in certain area are in the process of attaining resilience with change of species composition to some extent. However, some of the forest patches have still survived in the sheltered valleys, protected slopes, certain pockets of remote and less accessible ravines contain original species composition where the candidate species exist at present. These evergreen forests consist of trees such as *Elaeocarpus variabilis, Garcinia gummi-gutta, Garcinia indica,*

Hydnocarpus alpina, Neolitsea fischeri, Oreocnide integrifolia, Persea macrantha, Phoebe wightii, Photinia integrifolia, Trichilia connaroides etc. Shrubs like *Allophylus cobbe, Chassalia ophioxyloides, Ixora nigricans, Leea indica, Pavetta breviflora* var *subcoriacea, Pogostemon pubescens, Rauvolfia verticillata, Sarcandra chloranthoides, Sarcococca coriacea, Solanum giganteum* and *Tarenna asiatica* are generally encountered along with the candidate species. The herbaceous plants such as *Adenostemma lavenia, Phyllocephalum rangacharii, Droguetia iners* ssp. *urticoides, Elatostema lineolatum, Conyza japonica, Euphorbia rothiana, Hydrocotyle javanica, Physalis peruviana, Pouzolzia wightii* var. *scabra,* and *Wedelia urticaefolia* var. *wightii* formed the ground flora. *Clematis gouriana, Embelia ribes, Jasminum rottlerianum, Passiflora leschenaultii, Piper mulleusa, P. trioicum, Senecio corymbosus, Shuteria vestita, Smilax zeylanica, Stephania japonica, S. wightii* etc. are some of the prominent climbers of this area. Orchids of both terrestrial and epiphytes namely *Bulbophyllum fischeri, Cleisostoma tenuifolium, Dendrobium herbaceum, Diplocentrum recurvum, Eria reticosa, Oberonia mucronata* var. *denticulata* etc. grow luxuriously.

The high ranges of Neymakkad, Top station and Mannavan shola (1800m asl), are inhabited with southern montane wet temperate forests (sholas) interspersed with vast stretches of grasslands. Some of the common tress in these forests are *Daphniphyllum neilgherrense, Eurya japonica, Meliosma pinnata* ssp. *barbulata, M. simplicifolia* ssp. *pungens, Michelia nilagirica, Pittosporum tetraspermum, Archidendron clypearia* var. *subcoriaceum, Symplocos cochinchinensis* ssp. *laurina, Syzygium arnottianum, Ternstroemia gymnanthera, Turpinia nepalensis, Vaccinium leschenaultii, Vernonia arborea, Viburnum coriaceum* etc. It is of interest to note that the tree fern *Cyathea spinulosa* is very common in these regions. *Gaultheria fragrantissima, Leucas lanceaefolia, Moonia heterophylla, Hedyotis albo-nervia, Psychotria glandulosa, Rhodomyrtus tomentosus, Rubus niveus* etc. are some of the common shrubs and under shrubs. The ground flora represents herbs like *Phyllanthus rheedii, Scutellaria violacea, Spilanthes calva, Viola pilosa* etc. The epiphytes, other than orchids noted in these areas are *Aeschynanthus perrottetii, Impatiens jerdoniae, Medinella malabarica, Peperomia tetraphylla* etc. Scandent shrubs and climbers are *Clematis gouriana, Elaeagnus kologa, Lonicera leschenaultii, Senecio scandens, Tetrastigma nilagiricum* etc.

In places like Munnar, Rajamala and Pettumudi, the grasslands are dominated with species of *Brachiaria reptans, Chrysopogon hackelii, Eragrostis unioloides, Ischaemum mangaluricum, Tripogon bromoides* etc. Occasionally gregarious patches of *Didymocarpus lanuginosa* and *Rhododendron arboreum* ssp. *nilagiricum* are seen in the midst of these grasslands. The vegetation of this area consists of montane sholas and sub-tropical evergreen forests interspreads with vast expanse of grasslands. The trees comprising the sholas are all evergreen and mostly short boled. Shrubs like *Gaultheria fragrantissima* and *Hypericum mysurense* are also extensively found in this region. In the grassland, non grass species like *Anaphalis subdecurrens, Bupleurum distichophyllum, Chlorophytum tuberosum, Justicia japonica, Leucus ternifolia, Osbeckia aspera, Pimpinella candolleana, Ranunculus peninsularis, Wahlenbergia marginata* etc. were also reported. In moist places during wetter months, ephemerals like *Drosera peltata, Eriocaulon brownianum, E. odoratum, Lysimachia procumbens, Utricularia roseo-purpurea, U. smithiana, U. striatula* were also found growing luxuriantly.

The southern Western Ghats is the home to many endemic balsams. Extensive field trips conducted by Pandurangan (1996) reported nearly 60 species of *Impatiens* from the Kerala high ranges. During the study period, the following balsams namely *Impatiens anaimudica, I. campanulata, I. chinensis* var. *brevicornis, I. coelotropis, I. cordata, I. cuspidata, I. dasysperma, I. elegans, I. flaccida, I. floribunda, I. goughii, I. henslowiana, I. herbicola, I. grandis, I. latifolia, I. leschenaulti, I. maculata, I. minor, I. munnarensis, I. pandata, I. phoenicea, I. platyadena, I. tangachee, I. tomentosa, I. trichocarpa, I. verticillata,* and *I. wightiana* were recorded. Epiphytic balsams like *Impatiens acualis, I. jerdoniae, I. parasitica, I. viridiflora, I. violacea, I. kulamavensis* and *I. scapiflora* were also collected for the documentation.

Western Ghats: Upper Palni Hills

Palni hills are an eastward spur of the Western Ghats with a maximum length of 65 Km and an area of 2068 Km^2. These hills fall into two geographically distinct zones, the upper and lower Palni hills along a ravine running from Palni in the north to Periakulam in the south along the Parappar- Thevankarai valley. The upper Palni hills are the western block with an altitude ranging from 1800-2654 m. Like Anaimalai and Cardamom hills, Palni hills also consist of arachaic plutonic rocks. The soil of this region may vary to laterite, sandy loam, humid, organic soil etc., with high level of aluminum oxide. The pH range is being 5.8-5.9. The temperature and humidity ranges from 2.9-26.1°C and 85-90 per cent respectively. The bulk of the annual rainfall is from north-east monsoon (October-December), with a lesser amount during the south-west monsoon (June-September).

Most of the plateaus covered with coarse grass, the continuity is occasionally broken by sheltered valleys with some streams supporting isolated woods locally called sholas. The dominant tree species distributed in this area belong to the genus *Ilex, Meliosma, Elaeocarpus, Michelia, Phoebe, Rhododendron* etc. Some common shrubs and herbs are species of *Crotalaria, Knoxia, Indigofera, Leucas, Polygala, Ranunculus, Strobilanthes, Valeriana, Cyanotis, Sopubia, Gentiana, Juncus* etc. The grassland is dominated by *Chrysopogon, Aristida, Panicum* etc. 11 *Impatiens* species such as *I. campanulata, I. henslowiana, I. viscida, I. latifolia, I. leschenaultii, I. tangachee, I. tomentosa, I. balsamina, I. chinensis* and *I. goughii* were recorded from this area including *I. phoenicea*. Most of the forest land is now being used to grow commercial plantations of Pines, Conifers, Wattles, *Eucalyptus, Cinchona* and other fruit trees and their expansion will affect the natural forest in these areas.

Materials

Five populations of each species were selected in the natural habitat *i.e., Impatiens coelotropis* (in and around Anaimudi, Idukki district), *I. henslowiana* (Neymakkad gap of Idukki district Kerala, and Palni Hills, Tamil Nadu), *I. phoenicea* (Top station, Idukki district of Kerala and Palni hills, Tamil Nadu) *I. platyadena* (Pettumudi and Neymakkad gap of Idukki district Kerala). Voucher specimens of each species were deposited in the Tropical Botanic Garden Travancore Herbarium (TBGT), Thiruvananthapuram, Kerala, India.

Methods

Field work was carried out throughout the year especially on the monsoon and flowering period of the individual species. Fifty healthy plants were selected on each site and observations were made on day to day basis on flushing and flowering phenology which includes flowering season, flower initiation and development, anthesis, anther dehiscence, etc. Duration of flowering was recorded as per the method suggested by Dafni *et al.* (2005). Flower growth of ten selected individuals was monitored in the vaqueros population and the buds were marked twice a day until they opened. Floral morphology was studied using dissection microscope. The measurements of the floral parts were taken with the help of scale and vernier calipers. Twenty randomly selected flowers were analyzed from eight to ten different plants in each population. The fresh colour of the flower (corolla and calyx) was also recorded. The number of pollen grains and ovules per flower, pollen type, stigma type, flower-fruit ratio were calculated as per the method suggested by Dafni *et al.* (2005). Pollen-ovule ratio was also calculated as per the method suggested by Cruden (1977).

Observations on Visitors

Observations were made on floral visitors throughout the day. The number of floral visitors, percentage of floral visit and stigma touch by the insects were recorded. Foraging period and nature of the individual visitors were observed. After each visit of the pollinators, the stigmas were observed by hand lens and confirmed the transfer of viable pollens by their visit to the virgin stigma. Insects were trapped using butter paper bags while they were foraging the flowers and were fixed using ethyl acetate vapours. They were inspected individually in small screw cap vials (10ml) containing 3ml ethanol. The vials were vigorously shaken for 2min to remove the pollen grains from the body of the insects. The insects were taken out from the vial and the suspension was allowed to evaporate. After evaporation, pollen grains were mounted in a few drops of acetocarmine-glycerin stain and observed under a microscope. The number of stained and unstained pollen grains was counted. The insects were identified with the help of entomologists at the Department of Entomology, Kerala Agricultural University, Vellayani, Thiruvananthapuram and Kerala Forest Research Institute, Peechi, Thrissur, Kerala, India. The foraging period and the type of food collected by different pollinators/visitors on daily basis were recorded by close observation. To measure the amount of nectar availability per flower at different time intervals, a total of fifty flowers were collected for each experiment. The nectar present in the lip and spur of the flower was pipetted out by calibrated microcapillary tube. The length of nectar column of capillary tube was measured (for volume calculation) and also measured the concentration with hand refractometer (0-35 per cent). The results were used to correlate the number of visitors and their frequency in a given period.

Results

Morphology of the Flower

Plants in the Balsaminaceae have complex flowers. The sepals are modified into spur with lip and the petals are modified into standard, the uppermost petal and lower two wing petals. But every species of *Impatiens* have its specific type, colour and shape of floral organs. The typical *Impatiens* flower has a zygomorphic symmetry, but there is a group of non-related taxa, *i.e.,* Orchids have some morphological similarities with *Impatiens.*

Calyx

Sepals five; lateral two sepals are green (*I. henslowiana*) or coloured. The lower three sepals are modified into lip with spur. Size and colour of the sepals may vary from species to species. In *I.coelotropis,* the lip is large, deeply succate and deep green in colour. The lip is boat shaped and the spur is long and coloured in all the remaining three species (Plates 4.2 and 4.3).

Corolla

The corolla consists of five petals; standard- a large and often partially sepaloid. One upper portion (abaxial in origin) and four lateral ones which are connate in pairs on the left and right side of the flower. The lateral petals are bright red in colour (purple colour and white colour observed in Kerala high ranges and Palni hills respectively). The arrangement of wing petals (deeply/partially dissected) adopted for landing of bees and butterflies during pollination (Plates 4.2 and 4.3).

Androecium

The androecium has five stamens with short, broad filaments possessing partly fused and inwardly growing appendages at the top. The dithecal tetrasporangiate anthers are connivent and open either apically or laterally by means of pores or slits.

Gynoecium

The anthers lie as a cap above the gynoecium which has a five locular, syncarpus ovary and five stigmas. The style is very short or absent. Only after the pollen has been released and androecia shed will the star shaped stigmas ripen. The coherent stigma commonly spread and star shaped receptive surface is exposed. Hence, the flowers are protandrous favouring cross pollination. The ovules are anatropous, bitegmic and tenuinucellate and they are arranged in one or two vertical rows on the axile placenta. The micropyle is directed upward and inward (Table 4.2).

Phenology

Young seedlings of selected *Impatiens* begin to appear after the commencement of monsoon. Vegetative buds and leaf initiation occurs simultaneously at the second and third week of June in all the selected species. During the development of floral buds, initially the free sepals and petals increase their zygomorphic structure with increase in the size of the lower sepal and upper petal causing a subsequent decrease in size and displacement of lateral sepals. The greater enlargement of the lower sepal and the upper petal causing the disappearance of the upper pair of lateral sepal and subsequent displacement and uniting of the lateral petals into pairs. The flowering period of the individual species may vary from species to species. But the period had completed within December–January. The plants were in flower for a total of 180 days in a year. Fruit development was completed between 18-20 days after pollination. Fruit maturation and dehiscence were commenced before drying the capsule wall. The detailed phenological data are presented in the Table 4.3.

Plate 4.2

A) *Impatiens coelotropis* flower at anthesis: lateral view; B) *Impatiens coelotropis* full bloomed flower: front view; C) *Impatiens phoenicea* flowers at anthesis; lateral view; D) *Impatiens phoenicea:* Scarlet flower at anthesis: front view.

LS: Lateral sepal; L: Lip; S: Spur; WP: Wing petal; AN: Androecium; O: Ovary

Plate 4.3

A) *Impatiens henslowiana:* full bloom flower at Idukki, Kerala; B) *Impatiens henslowiana* full bloom flower at Kodaikkanal, Tamil Nadu; C) *Impatiens platyadena* flowers at anthesis: front view; D) *Impatiens platyadena:* Scarlet flower at anthesis: lateral view

LS: Lateral sepal; L: Lip; S: Spur; WP: Wing petal; AN: Androecium; O: Ovary

Table 4.2: Morphological Characters of *Impatiens*

Sl.No.	*Characters*	*I. coelotropis*	*I. henslowiana*	*I. phoenicea*	*I. platyadena*
1.	Habit	Shrub	Shrub	Herb	Shrub
2.	Habitat	Under the trees/shrubs	Sloppy streams	Sholas/shady areas	Under the trees/shrubs
3.	Height of the plant (M)	1.25±0.25	1.5±0.78	0.84±0.18	0.97±0.23
4.	Branching pattern	Less branched	Highly branched	Branched	Rarely branched
5.	Type of Inflorescence	Axillary racemes	Solitary/fascicles	Axillary racemes	Axillary racemes
6.	No. of inflorescence/plant	18.12±2.47	14.0±2.7	13.7±3.2	13.4±4.5
7.	No. of flowers per inflorescence	9.3±1.83	04.4±0.8	10.84±1.68	13.9±3.8
8.	Flower type	Zygomorphic	Zygomorphic	Zygomorphic	Zygomorphic
9.	Colour of the flower (petals and lip)	Scarlet with deep green	Purple/white	Brilliant scarlet with yellow tinge	Scarlet with pale purple
10.	Size of the flower (cm)	2.7±0.57	3.08±0.43	2.43±0.49	1.38±0.17
11.	Size of the standard petal (cm)	1.5±0.42	2.43±0.27	1.87±0.29	0.73±0.09
12.	Size of the wing petal (cm)	1.5±0.27	2.08±0.28	1.68±0.43	1.48±0.48
13.	Size of the lip (cm)	1.8±0.13	0.62±0.16	0.84±0.09	0.76±0.11
14.	Size of the spur (cm)	0.73±0.07	4.5±0.21	3.98±0.57	2.76±0.27
15.	Size of androecium (cm)	0.92±0.02	1.42±0.11	0.58±0.20	0.47±0.33
16.	Size of ovary (cm)	1.08±0.18	1.78±0.41	0.64±0.13	0.56±0.17

Table 4.3: Floral Phenology of Selected *Impatiens*

Sl.No.	*Phenological Characters*	*I. coelotropis*	*I. henslowiana*	*I. phoenicea*	*I. platyadena*
1.	Period of flowering	August–January	July–December	July–December	September–December
2.	Odour	Mild fragrance	Good fragrance	Mild fragrance	Mild Fragrance
3.	Resupination time	2 days before anthesis	3 days before anthesis	3 days before anthesis	2 days before anthesis
4.	Nectar (μl/flower)	5.7±1.87	6.4±1.43	4.1±1.08	8.3±2.81
5.	Anthesis time	2130-2330h	2130-0530h	0330-0740 h.	0600 - 0930 h
6.	Anther dehiscence time	One day before anthesis (protandrous)	One day before anthesis (protandrous)	One day before anthesis (protandrous)	One day before anthesis (protandrous)
7.	Mean no. of anthers/flower	5	5	5	5
8.	Mean no. of pollen grains/flower	42520±23.08	38250±29.13	28860±21.14	37530±24.06
9.	Mean no. of ovules/flower	24±2.05	22±2.04	14±1.09	16±2.13
10.	Pollen-ovule ratio	1772:1	1738:1	1990:1	2346:1
11.	Pollen type	3-colpate	3-colpate	4-colpate	4-colpate
12.	Pollen size	38.3±2.67 (oval)	41.31±3.08 (oval)	31.06±3.37 × 24.7±4.32μm (rectangular)	42±2.8 μm × 36.4±3.5μm (rectangular)
13.	Stigma type	Wet and non-papillate	Wet and non-papillate	Wet and non-papillate	Wet and non-papillate
14.	Fruit type	Capsules	Capsules	Capsules	Capsules
15.	Flower-Fruit ratio	12.5:1	7.6:1	10:1	10:3.8

Pollinators Activity

The pollination is effected by honeybees (*Apis cerana indica* and *Apis dorsata*), Rock bee (*Trigona iredipennis*), Hawk moths (*Macroglossom corythus* and *M. variegatum*) moths and butterflies. The pollinators are attracted by the bright colour and mild fragrance of the flowers of all the selected species. Honeybees forage for both nectar and pollen grains during the daytime. They regularly visited the flowers and spend 2-6 seconds in each visit. Among honeybees, *Apis dorsata* is the most abundant visitor, visiting more flowers than any other pollinators for pollen and nectar collections (Table 4.3). Honeybees are entering into the lip region and collecting the nectar. During foraging, they transfer the viable pollen grains from one flower to another of the same plant or flowers of different plants within and between the populations. *Trigona iridipennis* commonly known as "pollen robbers" forage during day time for pollen. Pollen loads (with viable pollen grains) were found on the hind limbs and heads and they pollinated the flowers by transferring the pollen grains to the stigma. During the night time, 2 species of hawk moths such as *Macroglossom corythus* and *Macroglossom variegatam* foraged the flowers of *I. coelotropis* and *I. henslowiana* They help the flowers in the pollination by their rapid and frequent visit to the flowers for nectar collections. Fertile pollen grains found on the head and long proboscis of hawk moths showed its effective pollen transfer to the flower and confirmed as a pollinator. The hawk moths transferred the pollen grains from one flower to another of the same plant or flowers of different plants and thus promoting geitonogamy and xenogamy (Table 4.4).

Butterflies are important visitors of selected *Impatiens.* There was a strong relationship found between the weather and butterflies activity in the study area. Temperature and relative humidity were identified as important environmental correlates controlling honeybee and butterflies foraging activities (Bisht and Pant, 1968). Butterflies were active during daytime and in fine weather, they actively visited the flowers and spending 4-8 seconds in each flower. However, during the rainy season, butterflies were completely inactive. Butterflies land on the lower two wing petals, slightly bend the body and insert the proboscis in the long and deep lip region for collecting the nectar from the flower. While harvesting nectar, the head portion of the butterflies having pollen grains from the previous visits was transferred to the stigma of another flower. The visitation continued day time and thereby effective pollination takes place. This specialized floral structure promoting the effective pollen transfer from one flower to another.

Xylocopa sp. was rarely visiting the flower and damaged the ovary by rough entering into the lip without pollination. All the insects are pollen carriers and their frequent inter plant movement facilitates cross-pollination. However, the visits of all the insects are not sufficient to pollinate all the flowers in the populations. It was calculated that 47 insects are visiting nearly 1000 flowers per hour at the peak foraging period. This may also lead to the poor fruit set in the natural condition. The main reason of this problem was due to the poor climatic condition prevailing during monsoon season in the selected area. In addition, the candidate species are also growing under shady areas and under bushes, therefore the pollinators have difficulty to identify the flowers under bushes/low light density. By the bright colour and fragrance, the pollinators are identifying the flowers of the selected species.

Table 4.4: Pollinators and their Foraging Behaviour of High Altitude *Impatiens*

Sl.No.	*Name of the Pollinator*	*Family*	*Foraging Nature*	**M.N.I.V*	*Foraging Period*	*Stigma Touch*
1.	*Apis cerana indica* [c,h,p,q]	Day	Nectar and pollen	12.34±1.08	0730-1600	+++
2.	*Apis dorsata* [c,h,p,q]	Day	Nectar and pollen	11.78±0.98	0900-1300	++
3.	*Trigona iredipennis* [c,q]	Day	Nectar and pollen	9.52±0.58	0630-1000	++
4.	*Danaus chrysippus* [c,q]	Day	Nectar	4.21±0.32	0600-1300	++
5.	*Danaus genutia* [c,q]	Day	Nectar	7.03±1.08	0700-0900	++
6.	*Tirumala limniace*[q]	Day	Nectar	5.24±1.02	0600-1100	++
7.	*Parantica aglea*[q]	Day	Nectar	2.43±1.13	0700-1600	++
8.	*Caprona ransonnett*[h]	Day	Nectar	6.03±1.04	0800-1200	++
9.	*Mycalesis mineus*[p]	Day	Nectar	6.54±0.97	0730-1100	++
10.	*Rpthima baldus*[p]	Day	Nectar	5.42±0.54	0600-1400	++
11.	*Hasora chromus*[p]	Night	Nectar	4.15±0.99	0800-1200	++
12.	*Badamia exclamationis*[p]	Night	Nectar	2.38±0.28	0900-1600	++
13.	*Delias eucharis*[p]	Day	Nectar	1.68±0.54	0700-1200	++
14.	*Macroglossum corythus* [c,h]	Night	Nectar	2.40±0.93	2200-2400	++
15.	*Macroglossum variegatum* [c,h]	Night	Nectar	2.70±0.82	2230-0330	++
16.	*Papilio demoleus* [c]	Day	Nectar	3.10±0.91	0600-1600	++
17.	*Pachliopta aristolochiae* [c]	Day	Nectar	2.50±0.72	0700-1600	++
18.	*Pachliopta hector* [c]	Day	Nectar	2.70±0.83	0600-1700	++
19.	*Delias eucharis* [c]	Day	Nectar	3.30±0.75	0600-1600	++
20.	*Pereronia valeria* [c]	Day	Nectar	1.00±0.69	0900-1600	++
21.	*Pelopids conjucta* [c]	Night	Nectar	0.80±0.13	0700-1200	++
22.	*Xylocopa* sp. [c]	Day	Nectar and pollen	0.62±0.17	0800-1100	++

(Visitors of c: *I. coelotropis*, h: *I. henslowiana; p. I. phoenicea; q. I. platyadena)* (No. of pollen grains transfer of to the virgin stigma, +: Less than 30 pollens; ++ : 30-50 pollens; +++: more than 50 pollens) M.N.I.V.: Mean no. of individual pollinator visited/population/hr.

Discussion

Roper (1830) was the first author who stated that, the perianth of *Impatiens* consists of a calyx and a corolla that both are pentamerous. The flowers are always zygomorphic and this had led to an enormous diversity in development of floral structures. In *Impatiens*, the floral structure, particularly the perianth parts, vary not only in colours, but also in shapes characters such as the anterior petals, the lateral petals and especially the spurred sepal are extremely variable even within the same species (Hooker and Thomson, 1859). The rich variation may be linked to co-adaptation with pollinators. Several authors have reported that there was a relationship between nectar spur curvature and set of pollinators (Grey Wilson 1980; Travers *et al.,* 2003). Grey Wilson (1985) also suggested that selection of flowers for more successful visitation by pollinators is based on their behaviour and it might be much more important with respect to evolution of floral characters than adaptation to improve the mechanical fit between pollinators and flowers. In *Impatiens,* the flowers are zygomorphic and usually having 3 or occasionally 5 sepals, of which the lower one is modified into lip and much larger and often coloured. Generally, the lip is fused with nectary tipped spur. The spur is lined with nectar producing secretary cells (Ramasubbu, 2009). The corolla consist of 5 petals of which the standard one often partly sepaloid especially in the upper position. The 4 lateral ones are connate in pairs on the left and the right side of the flower. The androecium has 5 stamens which are partly fused and lie as cap over gynoecium. The ovary is syncarpus and has five locular and 5 stigmas. The style is short and almost fused with ovary. The stigma become receptive only after pollens released from stamens. Hence, the flowers of *Impatiens* are protandrous and favouring cross-pollination (Pieter *et al.,* 2006).

Floral Structure and Resupination

Selection for successful visitation based on pollination behaviour might be much more important with respect to evolution of floral structure than adaptation to improve the mechanical fit between pollinators and flower. The zygomorphic structure and the elaboration of the lower petals in *Impatiens* have three fold function such as, to attract the pollinators, to act as a platform for the pollinators to land and to provide an access point to the lower sepal (lip) and spur where nectar is stored (Ramasubbu *et al.,* 2011). The labellum (standard/upper petal) also provides guide to visiting pollinators in the flowers (most of the species). The extra floral nectarines found in the lower/adjacent to the reproductive part will helpful to the pollination and further mechanism. The nectarines tissues carries important role in the stigma touches of the pollinators during forage time. The nectarines were observed to secrete both cane and grape sugars which act as an attractant to pollinators.

In the majority of balsams, buds are positioned with the standard uppermost and the lip below them. However, some flowers are born with lip above and the standard petals which are lower most. The reversal of position of flower buds occurs as a result of a process called resupination which takes place as before 2-3 days of anthesis. In most of the species, the buds turn only to the extent necessary to place standard lower most which is usually 180°, but depending on the position of inflorescence the turning can be more or less than, some species do not resupinate at all and their flowers are often described as being borne upside down (Grey Wilson 1980; Joseph Arditti, 2003). Darwin suggested that, resupination facilitates pollination because the labellum assumes the position of a lower petal, so that insects can easily visit the flower. Resupination usually occurs 2-3 days prior to anthesis. Once the flowers are fully open, they can no longer resupinate. However, the flowers of some species of Orchids de-resupinate following pollination. Another interesting characteristic of resupination is that in some species like *Impatiens grandis, I. maculata* and *I. flaccida* and hybrids, the buds alternate in resupinating clock (CL) and counter clock (CL) wise (Ramasubbu, 2009). Resupination may also provide flowers with space to open because in many balsams, the buds arranged tightly on the developing inflorescence. In Orchids, the resupination may also be influenced by the weight of the flower on a slender inflorescence. In this context, the resupination is a key factor in the evolution of lower sepals and spur otherwise, it would be difficult to visualize the development of such a wide range of bulky structure if resupination did not occur. Therefore, the resupination is closely associated with pollinator/pollination evolution. Zygomorphy and the accompanying resupination in *Impatiens* is clearly an evolutionary progress towards highly specialized pollinations, a trend that often seem to lead a high degree of speciation. Possible selection process on such systems permits rapid speciation, often in small areas geographically and may help to account for the high degree of endemism (Grey Wilson, 1980).

The flower size has a special functional significance in plant forager association and their evolutionary shift from out-crossing to selfing is mediated through decreased flower size and alterations in floral morphology (Ornduff, 1969). *Xylocopa* bees prefer medium sized flowers (Cruden, 1977). The relationship between bees and flowers is mutualistic where the former one is searching for food and the latter for pollination which is a key aspect in the reproduction of plants, enabled by appropriate structural and functional characteristics. Generally insects have the ability to recognize and distinguish colours. It was found out that, insects visit more yellow and white flowers whereas bees favour blue or purple flowers and rarely visiting red flowers which reflect ultraviolet light (Solmon Raju and Purnachandra Rao, 2006). Floral odours are diverse and have evolved to attract potential pollinators.

Phenological Adaptation

Phenology is the study of phenoevents which are critical for the survival and reproduction of plant species (Rathcke and Lacey, 1985). Therefore, the detailed information regarding the phenology is a prerequisite for the studies on the breeding systems and silvicultural practices. The most important and intensely studied phenoevents in majority of the cases are on flowering only (Widen, 1990). The timing of flowering influence the competition for pollinations and thus availability of pollinators (Bawa, 1983). Phenological pattern of flowering can be analyzed at various levels such as individuals, populations and communities (Bawa *et al.,* 1990). Community based phenological patterns have been already explored by several workers whereas population based phenological studies are rather limited inspite of the fact that such events have direct bearing on the survival and establishment (Primack, 1980; Bawa *et al.,* 1990).

The genus *Impatiens* L. is representing the true balsam and their members have most fascinated with peculiar flower structures and spectacular colours. Knowledge on phenology and floral morphology is essential for conducting studies on pollination biology and breeding systems. About 62 per cent of *Impatiens* species growing in the Western Ghats are flowering in July-December, 16 per cent during April-June and 15 per cent between January-March. Interestingly, 18 per cent of the *Impatiens* flowers throughout the year if conditions are favourable (Pandurangan, 1996; Pandurangan and Sreekala, 2008; Rajalal *et al.*, 1996). In the present study, the candidate species starts flowering in the months of July-September, extended up to January and reaches a peak during September-October. *Impatiens coelotropis* and *I. henslowiana* bloom in the night between 2130-2330h and 2130-0530h but in the case of *I. phoenicea* and *I. platyadena*, anthesis commenced in the early morning between 0330-0740h and 0600-930h respectively. Bhaskar and Razi (1974b) had already reported that, majority of the wild *Impatiens* growing in the high altitude areas are night blooming and have wide range of timing with regard to pollen germination. In all cases, the anther dehiscence takes place one day before flower opening which in turn confirmed the protandrous condition of the flowers. This observation was similar to that of *Impatiens reptans* in China (Tian *et al.*, 2004).

Floral Structure and Phenology in Relation to Pollination

The balsams are highly evolved members among the order Geraniales as evident from the marked zygomorphic flowers and nectiferous spur. The arrangement of stamens and its protandry nature, pistil and spur are markedly adopted for cross pollination (Bhaskar and Razi, 1974b) and hence most of the species of *Impatiens* reproduce by means of cross pollination including *I. coelotropis, I. henslowiana, I. phoenicea* and *I. platyadena*. It is well known that the flowers of *Impatiens* have enormous diversity in structure, shape, colour etc., which attract different pollinators. In the present study, the selected species of *Impatiens* are pollinated by honey bees, hawk moths, *Trigona iredipennis*, moths, butterflies etc. In different climatic regions, species of pollinators are also varying. In subtropical regions of Africa, the *Impatiens* species are pollinated by humming birds as well as insects. In temperate zones, pollinators are bumble bees and humming birds (Rust, 1977, 1979; Heinrich, 1979; Kato *et al.*, 1989). In Sumatra and Japan, hawk moths and honey bees are major pollinators of *Impatiens* (Kato, 1988; Kato *et al.*, 1991). In Sumatra, *Impatiens platypetala* was visited by crepuscular hawk moths, *Macroglossum corythus* (Kato *et al.*, 1991). Kato (1988) reported that some species of hawk moths were also visited the flowers of *Impatiens* in Japan. In *I. coelotropis*, honey bees, *Trigona iredipennis* and butterflies were also foraged during day time. Among the honey bees, *A. dorsata* visits the flowers in the day time but actively foraged in the early morning. All the insects are pollen carriers and their frequent and interplant movement facilitates cross pollination. Butterflies were actively foraged only in day time. In fine weather, they actively visited the flower of *Impatiens coelotropis*. However, during the rainy season, butterflies were completely inactive. Similar observation was noticed in *Impatiens reptans* in China (Tian *et al.*, 2004). Floral damage caused by the larvae and caterpillar are detrimental to the reproduction of *I. coelotropis*. The pollinators are not interested to visit the infected flowers which may lead to the poor fruit set in *I. coelotropis*.

The geographical distribution of *Impatiens* is very localized and endemic (Tian, *et al.*, 2004). In this study, the flowers in all the cases are protandrous and favouring cross-pollination. The insects land at the front of the flower and passes through corolla to the spur behind the sepal to collect nectar. The insects while collecting the nectar from the spur, their back comes in contact with the androecium. In *Impatiens*, the 5 stamens are connected together to form a staminal cone (formed from the filaments) and release sticky pollen grains. The staminal cone completely encloses the pistil. The insect comes in contact a second time, the androecium falls off and the ovary with stigma becomes exposed. The arms of the stigma then open gradually and become receptive, so that when the next insect visits the flower for nectar, its back comes into contact with the stigma and cross pollination takes place. Finally, the perianth falls off and the insects visits have come to cease. These observations were supported by the findings of several authors in *Impatiens pallida* and *I. capensis* (Rust, 1977), *I. reptans* (Tian *et al.,* 2004) and *I. henslowiana* (Sreekala *et al.,* 2007).

Daumann (1967) showed experimentally that the insects were attracted visually to the flowers of *Impatiens,* rather than by odour. Pollinators preference at some point of time is determined principally by the quantity and quality of the floral reward (Morquin, 1971; Heinrich, 1979). The dominant factors affecting bee visits appear to be nectar abundance, sugar concentration and chemical attractants in pollens (Martin and Mc Gregor, 1973). The quantity of nectar present in the flowers of *I. coelotropis* is 5.7±1.87µl whereas in *I. phoenicea* and *I. platyadena* are 4.1±1.43µl and 8.3±2.81µl respectively. However, the concentration of nectar in the selected species is ranging from 28-35 per cent. Many species of flowering plants have evolved structures or exudates that act as reward, ensuring that the animal pollinator will consistently return to flowers to transport pollens. The most common floral reward is nectar which is secreted from specialized organ in flowers called nectaries. In *Impatiens,* nectar is present in the spur because the inner surface of the spur is lined with nectar secreting cells.

Conclusion

The flowers of selected species of *Impatiens* are self-incompatible, the modification and adaptations of floral parts attract pollinators towards successful pollination. Whenever cross pollination fails, no fruit set which affect the distribution of the species. The failure of fruit set which leads to the poor distribution of these species in the Western Ghats.

References

Baker, H.G. 1995. Self compatibility and establishment after long distance dispersal. Evolution **9:**347-349.

Barrett, S.C.H. 2003. Mating strategies in flowering plants: the outcrossing-selfing pardigm and beyond. Philosophical Transaction of the Royal Society B, Biological Sciences **358:** 991-1004.

Bawa, K.S. 1983. Patterns of flowering in tropical plants. Hand book of experimental pollination Biology, Johnes, C.E., Little, R.J. (eds.), Von Nostrand Reinhold Company, New York. pp. 95- 410.

Bawa, K. S., Ashton, P.S. and Nor, S.M.1990. Reproductive ecology of tropical forest plants: management issues. Reproductive ecology of tropical forest plants, Bawa, K.S., Hadley, M. (eds.), MAB series UNESCO, Paris, **7:** 3-12.

Bhaskar, V. and Razi, B.A. 1974a. Nocturnal pollen germination in *Impatiens* (Balsaminaceae). Curr. Sci. **43(19):** 626-628.

Bhaskar, V. and Razi, B.A. 1974b. New chromosome counts in South Indian *Impatiens* localities (Balasaminaceae). Proc. Ind. Sci. Congr. Assoc. **61(III B)**: 35.

Bisht, D.S. and Pant, N.C. 1968. Indian J. Entomol. **30**: 163-168.

Champion, H. G. 1936. A preliminary survey of the forest types of India and Burma. Indian For. Res. N.S. **1(1):** 1-286.

Cruden, R.W. 1977. Pollen-ovule ratios: A conservative indicator of breeding systems in flowering plants. Evolution **31**: 32-36.

Dafni, A., Kevan, P.G. and Husband, B.C., 2005. Practical Pollination biology, Enviroquest Ltd, Cambridge, Otntario, Canada.

Darwin, C. 1862. On the various contivances by which British Orchids and foreign orchids are fetilized by insects, and on the good effect of outcrossing. London, Murray.

Daumann, E. 1967. Zur Bestaubungs-Und Verbreitungsokologie dreier *Impatiens*-Arten. Presalia **39**: 43- 58.

GreyWilson, C. 1980. *Impatiens* of Africa. Balkema, A. A., Rotterdam, Netherlands.

GreyWilson, C. 1985. Balsaminaceae. A revised hand book to the flora of Ceylon. vol 5. Dassanayaka, M. D. and Fosberg, F. R. (eds.), Amerind Publishing Co., New Delhi.

Heinrich, B. 1979. Resource heterogeneity and patterns of movements in foraging bumble bees. Oecologia **40**: 235-245.

Holsinger, K.E. 1996. Pollination biology and the evolution of mating systems in floweirng plants. Evolutionary Biology **29**: 107-149.

Hooker, J. D. and Thomson, T. 1859. Precursores ad floraum indicam: Balsamineae. J. Linn. Soc., **4**: 106-157.

Joseph Arditti, 2003. Resupination. Lankesteriana **7**: 95-96.

Kato, M., 1988. Bumble bee visits to *Impatiens* sp. Pattern and efficiency. Oecologia **76**: 364-370.

Kato, M., Itino, I., Hotta, M., Abbas, I. and Okada, H. 1989. Flower visitors of 32 plant species in West Sumatra. Occasional papers of the Kagoshima University Research Centre for South Pachific **16**: 15-31.

Kato, M., Itino, I., Hotta, M. and Inoue, T., 1991. Pollination of four Sumatran *Impatiens* species by hawk moths and bees. Tropics **1**: 59-73.

Martin, E. C. and Mc Gregor, S.E. 1973. Changing trends in insect pollination of commercial crops. Amer. Rev. Entomol. **18**: 207-226.

Monika, K. M. and Bhatnagar, A. K. 2007. Plant reproductive biology studies crucial for conservation. Curr. Sci., **92(9)**: 1207.

Morquin, T., 1971. Competition for pollinators as a stimulus for the evolution of flowering time. Okios **22**: 398-402.

Orunduff, R., 1969. Reproductive biology in relation to Systematics. Taxon **18**: 121-133.

Pandurangan, A.G. 1996. Systematic studies on Balsaminaceae in Peninsular India, Final Project report submitted to Kerala Forest Department, Thiruvananthapuram, Kerala.

Pandurangan, A.G. and Sreekala, A.K. 2008. Studies on reproductive biology of selected rare, endemic and horticulturally promising balsams from the Western Ghats, Final Project report submitted to Department of Science and Technology (SERC), New Delhi, India.

Pieter L.C., Koen, P.G., Steven, B.J. and Erik, F.S. 2006. Floral development in three species of *Impatiens* (Balsaminaceae). Amer. J. Bot. **93(1):** 1-14.

Primack, R.D. 1980. Phenological varigation within natural population flowering in New Zealand Montane shrub. J. Ecol. **68**: 849-862.

Rajalal, R., Pandurangan, A.G. and Pushpangadan, P. 1996. Systematic studies on Balsaminaceae in Peninsular India. J. Swamy. Bot. Cl. **13**: 59-62.

Ramasubbu, R. 2009. Reproductive biology of three endemic and endangerd balsams (*Impatiens coelotropis* Fischer, *I.phoenicea* Bedd. and *I. platyadena* Fischer) of Western Ghats. Ph.D. thesis submitted to the University of Kerala, Thiruvananthapuram, Kerala, India.

Ramasubbu, R., Sreekala A.K. and Pandurangan A.G. 2009. Resupination and flower structure: is it the adaptation for Pollination in *Impatiens* L.? Trends in Biosciences **2(2):**89-90

Ramasubbu, R., Sreekala A.K., Pandurangan A.G. and Kulloli, S.K. 2011. Reproductive Ecology of *Impatiens platyadena* Fischer- A critically endangered balsam of Western Ghats. Curr. Sci. **100(10)**: 1551-1554.

Rathcke, B. and Lacey, E. P. 1985. Phenological patterns of terrestrial plants. Ann. Rev. Ecol. Syst. **16:** 179 – 214.

Roper, J.A.C. 1830. De floribus et affinitatibus Balsaminearum, Basel, Switzerland.

Rust, R. W. 1977. Pollination in *Impatiens capensis* and *I. pallida* (Balsaminaceae). Bull. Torrey Bot. Club. **104:** 361-367.

Rust, R.W. 1979. Pollination in *Impatiens capensis*: Pollinators and nectar robbers. J. Kansas. Entom. Soc. **52**: 297-308.

Shetty, B. V. and Vivekananthan, K. 1971. Studies on the vascular flora of Anaimudi and the surrounding regions, Kottayam District, Kerala. Bull. Bot. Surv. India, **13**: 16-42.

Solomon Raju, A.J. and Purnachandra Rao, S. 2006. Nesting habits, floral resources and foraging ecology of large carpenter bees (*Xylocopa latipes* and *X. pubescens*) in India. Curr. Sci. **90(9)**: 1210-1217.

Sreekala, A.K., Pandurangan, A.G., Ramasubbu, R. and Kulloli, S.K. 2008. Reproductive biology of *Impatiens coelotropis*.- a critically endangered balsam from the Southern Western Ghats, Curr. Sci., **95(3)**: 386-388.

Stebbin, G.L. 1970. Adaptive radiation of reproductive characteristics in angiosperms: pollination mechanisms. Annual review of Ecology and Systematics **1:** 307-326.

Tian, J. P., Liu, K. and Hu, G. 2004. Pollination ecology and pollination system of *Impatiens reptans* (Balsaminaceae) endemic to China. Ann. Bot. **93:** 167-175.

Travers, S. E., Temeles, E. J. and Pan, I. 2003. The relationship between nectar spur curvature in Jewel weed (*Impatiens capensis*) and pollen removal by humming bird pollinators. Can. J. Bot. **81:** 164-170.

Yuan, Y. M., Song, K., Geuten, E., Rahelivololona, S., Wohlhauser, E., Fischer, E. and Smets, Kufer, P., 2004. Phylogeny and Biogeography of Balsaminaceae inferred from ITS sequence data. Taxon **53:** 391-403.

Widen, B.1990. Environmental and genetic influences on phenology and plant size in a perennial herb, *Senecio integrifolius*. Can. J. Bot. **69:** 209-217.

Chapter 5

Flora of North Andaman Islands, India

C. Sudhakar Reddy*

Forestry and Ecology Division, National Remote Sensing Centre, Balanagar, Hyderabad – 500 625, India

ABSTRACT

The presence of 1025 indigenous and 191 non-indigenous species of higher plants (incl. 31 species of fern flora and 3 gymnospermous species) within a land area 1710 sq km is a significant feature of North Andaman Islands (representing 629 genera under 145 families). Among the indigenous Dicotyledons, the Polypetalae is represented by 64 families, 220 genera and 378 species, the Gamopetalae by 30 families, 150 genera and 240 species, where as Monochlamydeae by 23 families, 75 genera and 141 species. Life form analysis shows that Phanerophytes were dominant due to tropical humid bioclimate. At family level, *Poaceae* (78), *Euphorbiaceae* (67), *Papilionaceae* (61), *Cyperaceae* (52) and *Rubiaceae* (50) were dominant familes. Of the genera, *Cyperus* is the largest represented by 18 species, followed by *Ficus* (14), *Fimbristylis* (12), *Ixora* (10). At the species level, 101 species are endemic and at generic level endemism is rather less with only two genera *viz. Pubistylis* Thoth. (Rubiaceae) with one species, *Sphyranthera* T. Chakrab. et Vasud. (Euphorbiaceae) with two species. Among the non-endemic angiosperm species about 40 per cent are not found in mainland India but extend their distribution to Myanmar and Thailand.

* Corresponding Author: E-mail: drsudhakarreddy@gmail.com

Introduction

The Andaman and Nicobar Islands are a group of about 572 islands, Islets rocks and keys situated off the eastern coast of India in a junction box with Bay of Bengal and Indian Ocean on one side and South China Sea and the pacific on the other (between 10° 30' and 13° 42' North latitude and 92°14' and 94°16' East longitude). Covered with lush green tropical rain forests, about 1200 km away from the mainland of the Indian subcontinent, these islands are also called "the Bay Islands". Out of a total geographical area of 8239 sq km., 7464 km are covered under reserve and protected forests. 36 per cent of the reserve forests have been earmarked for primitive aboriginals, 13 per cent of the total area all along the coast has to be protected against soil erosion and for mangroves and about 12 per cent of the area have been occupied under revenue settlements. Andaman islands, where inhabitation is found, are comprised of *North Andaman*, Middle Andaman, and South Andaman. However, the same in Nicobar Islands includes Car Nicobar, Nancowry and Great Nicobar Islands.

Study Area

North Andaman Islands

Topography in general ranges from hilly undulating running north to south to highly rolling, narrow valleys escorted by the spurs emerged from the main hill ranges which are moderate. Flat terrain is extremely limited. The slope of the hills are mostly rugged and moderate to very steep. The larger islands are deeply indented by long meandering saline creeks. The highest peak in the Andaman and Nicobar islands is Saddle peak in north Andaman with an altitude of 737m (MSL). Long narrow sandy beaches are salient features of the topography here. The major portions are undulating to hilly with maximum altitude rising to 180 m. There are no rivers, however, a few perennial streams remain alive due to contribution of catchment areas under forest cover. North Andamans characterized by the presence of the volcanoes of Narcondum and Barren Islands. The major part of Andaman and Nicobar Islands is covered by rocks belonging to the Port Blair group, which lithologically consist of a vast thickness of alternating sequences of grey sandstone and black shells with subordinate grits and conglomerates. Being recent in geological origin, the soils of Andamans are immature and fragile, poor in drainage with low moisture retention capacity. By nature mostly acidic, medium textured on the surface (sandy and clay loam) formation and conglomerates below. Owing to its proximity to the equator and surrounded by the sea, Andaman group of Islands are endowed with a true maritime climate year round with least variation between maximum (30.1°C) and minimum (23.1°C) temperatures. The maximum temperatures are experienced during the dry season when evapo-transpiration losses are found to be highest. Since these islands are under the influence of both the south-west and north-east monsoons, they receive rain from April to December. The mean annual precipitation is around 3100 mm unevenly distributed throughout the year. Average relative humidity ranges from 68 to 86 per cent. The islands receive North eastern winds between the months November and March and South Western between the months from May to October. Presently, North Andamans are inhabited by settlers from Bangladesh (refugee

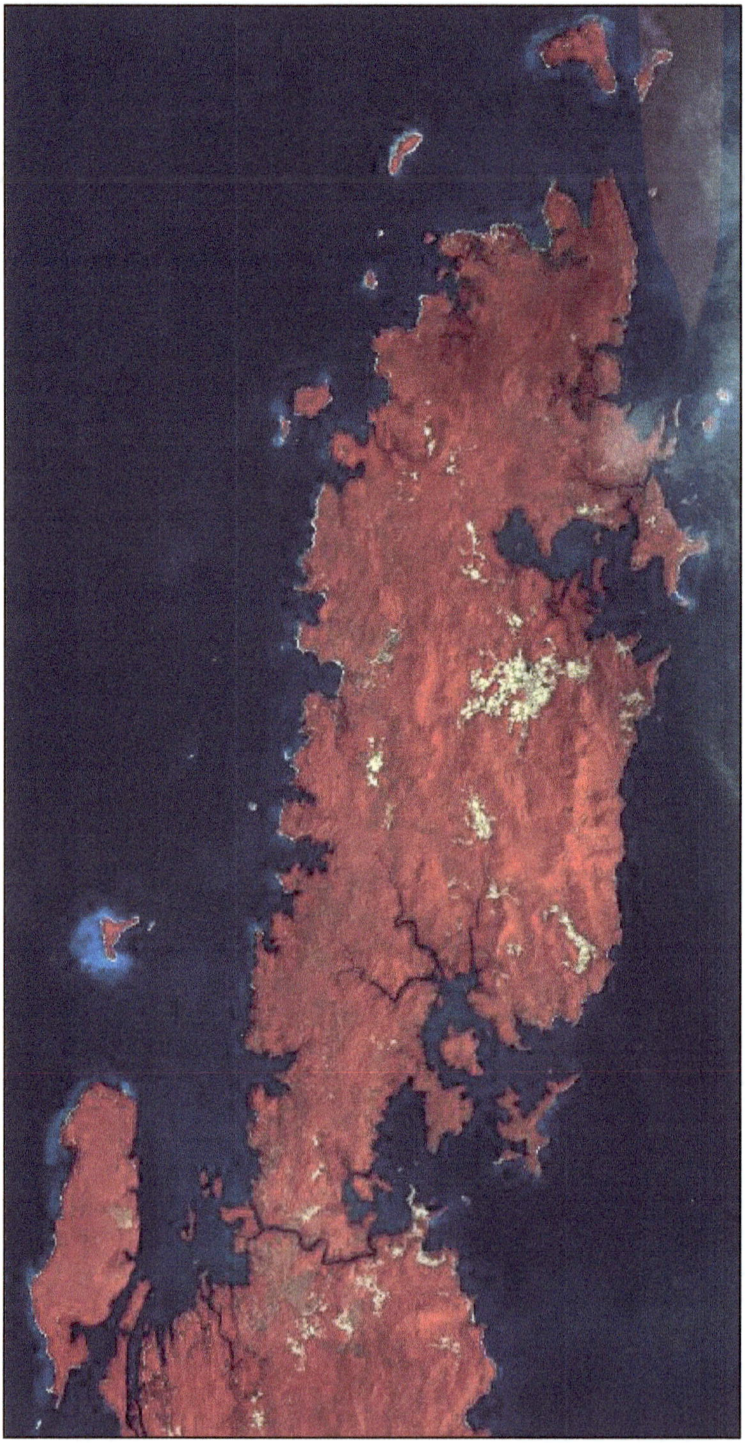

Figure 5.1: IRS P6 LISS III FCC Image of North Andaman

population), Tamil Nadu and Andhra Pradesh. But, majority of population is represented by Bengalis (now Bangladesh)

Out of total 1400.85 sq.kms geographical area of North Andamans, about 8.47 per cent have been cleared for agriculture and related activities. In Andamans 15578 ha is the net cropped area and 4117 ha is the fallow land. Agriculture in the Bay Islands is about 140 years old and obviously at the expense of forests lands. To ensure food security the Government allotted 2 ha of cultivable paddy land and 3 ha of hilly land to all settlers. As these settlements were basically planned for rice cultivation, 2 ha of hilly land have hardly been put to any productive use. In spite of Government grant and input assistance production has fallen short of demand necessitating shipment of food grains, vegetables and fruits from the mainland.

Islands of North Andaman Group

Table 5.1: Protected Forests

Sl.No.	*Protected Forests*	*Area (ha.)*
1	North Andaman	137599
2	Smith Island	2470
3	Aves Island	16
	Geographical area	140085

Table 5.2: Wildlife Sanctuaries:
The Following Islands have 100 per cent Forest (No human inhabitation)

Sl.No.	*Island*	*Area (ha.)*	*Sl.No.*	*Island*	*Area (ha.)*
1.	Landfall	2948	2.	East	611
3.	West	640	4.	Peacock	62
5.	White Cliff	47	6.	Thorn Hill	81
7.	Reef	174	8.	Mayo	10
9.	Paget	736	10.	Point	307
11.	Shearme	785	12.	Ox	13
13.	Sugar loaf	NA	14.	South Button	3
15.	North	49	16.	Chatham	247
17.	Jub-Jub	NA	18.	Rowe	1
19.	Snark	60	20.	Kwatung	57
21.	North Reef	348	22.	Latouche	96
23.	Interview	13387	24.	Jungle	52
25.	Trilby	96	26.	Table (Excelsior)	169
27.	Table (Delgarne)	229	28.	Temple	104
29.	Turtle	39	30.	Ross	101
31.	Brush	1	32.	Craggy	62

Contd...

Table 5.2–*Contd...*

Sl.No.	*Island*	*Area (ha.)*	*Sl.No.*	*Island*	*Area (ha.)*
33.	Bamboo	5	34.	Blister	26
35.	Dot	18	36.	Curlew	3
37.	Stewart	723	38.	Olevor	16
39.	Oyster	21	40.	Sound	2246
41.	Orchid	10	42.	Curlew (B.P.)	16
43.	Egg	5	44.	Swamp	409
45.	Dottrel	13	46.	Gurjan	16
47.	Sea serpent	78	48.	Snake	73
49.	Bondeville	255	50.	Buchanan	933
51.	Surat	31	52.	Entrance	96
53.	Benett	346	54.	Roper	146
55.	South Reef	117	56.	Mask	78
57.	Anderson	1973	58.	Tuft	29
59.	Humb	47	60.	Cander	5
61.	Goose	1	62.	Flat	936
63.	Speke	42	64.	Wharf	11
65.	Tree	3	66.	Channal	13
67.	Narcondum	738			
	Reserved and notified forest			30993	
	Geographical area			30993	

Total geographical area of North Andaman (incl. all Islands) is about 171078 ha. (1710.78 km^2)

National Parks

1. Saddle Peak National Park : 36 km^2.

Vegetation

There are three important natural ecosystems in the north Andaman islands as like other Andaman and Nicobar Islands – the forest ecosystem, the marine ecosystem, and the interphase between the two, the mangrove ecosystem. The forest cover in the island is a rich and unique land resource, basic to the survival of the island ecosystems. At the same time it is fragile and can easily be destroyed (Saldanha, 1989). Although in many islands forest vegetation is greatly damaged or under severe pressure. Champion and Seth (1968) have classified the Andaman and Nicobar forests into the following types:

- ☆ 1A/C1: Giant evergreen forest
- ☆ 1A/C2: Andamans Tropical evergreen forest
- ☆ 1A/C3: Southern hilltop tropical evergreen forest

- 1/E1: Cane brakes
- 1/E2: Wet bamboo brakes
- 2A/C1: Andamans semievergreen forest
- 3A/C1: Andamans moist deciduous forest
- 3A/C1/2S1: Andamans secondary moist deciduous forest
- 4A/L1: Littoral forest
- 4B/Ts: Tidal swamp forest
- 4C/FS2: Submontane hill valley swamp forest

Of these evergreen type (Giant Andaman Evergreens, Andaman Tropical Evergreens, Andaman Hilltop evergreens), Andaman semievergreen, Andaman Moist Deciduous Forests, Littoral Forests and Mangrove Forests are the major and distinguishable. Bamboo and Cane brakes form local variations.

Andaman Evergreen Forests

Giant Andaman Evergreens are regarded as climax formation in lowlands and occur in areas where there is a rainfall of over 300 cm. It once grew on most of the alluvial soil near banks of streams with deep soil (Upadhyay, 2000) and in the lower valleys. Magnificant buttressed trees festooned with climbers and epiphytes and interspersed with canes and bamboos rose to 30 m and more. Tall Gurjan (*Dipterocarpus alatus*) was common in these forests. However, these forests are almost extinct due to large-scale clearance of flat terrain for the agriculture. Andaman Tropical Evergreens are multi-storied, closed canopy forests which grow on hilly terrain. They are less luxuriant than the Giant Evergreen with lesser height and lesser density.

In evergreen forests of Andaman Islands, three or more strata of vegetation can be found and forest floor does not receive much of the direct sunlight. According to our study in North Andamans the species composition in evergreen forests as follows:

- Canopy trees: *Dipterocarpus alatus, Dipterocarpus griffithii, D. turbinatus, Sideroxylon longipetiolatum, Hopea odorata, Endospermum malaccense, Planchonia andamanica.*
- The second storey trees: *Baccaurea sapida, Myristica glaucescens, M. andamanica, Buchanania splendens, Randia pulcherrima, Pometia pinnata, Fagrea racemosa, Micromelum minutum* and *Xanthophyllum andamanicum.*
- Shrubs: *Clerodendrum viscosum, Leea asiatica, Leea indica, L. guineensis, Dinochloa andamanica, Maesa andmanica, Morinda umbellata, Breynia rhamnoides, Grewia disperma.*
- Woody climbers: *Calamus longisetus, Calamus viminalis, Daemnorops manii, Entada rheedei, Mezoneuron cucullatum, Illigera appendiculata, Thunbergia laurifolia.*
- Epiphytic plants: *Asplenuim nidus, Dendrobium aphyllum, D. secundum, Drymoglossum piloselloides, Drynaria quercifolia* and *Lycopodium phlegmaria.*

Southern Hill-Top Evergreen Forest

Andaman Hill-top Evergreens are stunted, wind-resistant forests. Their physiognomy is greatly influenced by strong winds. Saddle Peak (737 m) in the north Andaman present good example of such vegetation. Frequently occurs on the slopes and top of hills and sometimes on the steep slopes lower down. A more or less inferior to typical wet evergreen because of exposure to wind and less favorable soil conditions. Rainfall and humidity are usually high.

Dominant tree species are *Eugenia kurzii, Myristica glaucescens, Parishia insignis, Anacolosa frutescens, Dipterocarpus gracilis, Canarium manii, Dipterocarpus costatus. Caryotis mitis, Chionanthus parkinsonii* form shrub layer and *Amorphophallus cornosus, Colocasia virosa, Hedyotis andamanica* forming herb layer. Associated climbers are *Anamirta cocculus, Anodendron paniculatum, Calamus palustris, C. viminalis,* etc.

Andaman Semi-Evergreen Forest

A luxuriant type of forest, occuring mostly in the Valleys with many giant trees of both deciduous and evergreen species often intimately mixed in groups. Climbers are often heavy.

Predominant tree species are *Pterocarpus dalbergioides, Celtis wightii, Dipteocarpus gracilis, Pterygota alata, Pterocymbium tinctorium* etc. Climbers associated are *Abrus pulchellus, Ancistrocladus extensus, Acacia pennata, Dalbergia confertifolia, Cayratia pedata* with canes like *Calamus longisetus, C. pseudoviralis, C. palustris.* Shrubs include *Licuala peltata, Anaxagoreas luzioniensis, Carytois mitis* etc. Herbs found are *Colocasia virosa, Costus speciosus, Curcuma latifolia* and *Strobilanthes glandulosus.*

Andaman Moist Deciduous Forests

This forest type mostly occurs on lower slopes and is characterized by open canopy and seasonal leaf fall. The deciduous trees often occupy the area between the tidal zone and the evergreen patches. Generally the dominant trees in these forests have very large girth with buttresses and grow more than 20 m height. Moist deciduous forests are widely distributed covering approximately 50 per cent area of Andamans. The forests are important as they yield timber of more economic value. The white Chuglum (*Terminalia bialata*), Padauk (*Pterocarpus dalbergioides*) and Pema (*Lagerstroemia hypoleuca*) are important constituents of the moist deciduous forest.

In the islands secondary moist deciduous forests are also found which have grown on worked out areas as semievergreen type. They are more or less even-aged. Depending on the work site the regenerated plants in deciduous forests have affinity with semi-evergreen and moist deciduous forests (Ellis, 1989).

Species composition of the deciduous forest is as follows :

- ✰ Canopy trees: *Pterocarpus dalbergioides, Termidralia bialata, T. manii, T. alata, Canarium euphyllum, Ailanthes kurzii, Parishia insignis, Diploknema butyracea, Albizia lebbeck, Terrameles nudiflora* and *Pterocymbium tinctorium.*
- ✰ Second storey trees: *Sageraea elliptica, Sterculia villosa, Miliusa tectona, Semecarpus kurzii, Diospyros pyrhocarpa, Garuga pinnata, Cratoxylon formosum, Pterospermum aceroides, Lannea coromandelica* etc.

- ✰ Shrubs: *Actephila excelsa, Ixora grandifolia, Licuala peltata, Bridelia griffithii, Mallotus acuminatus, Rinorea benghalensis* etc.
- ✰ Climbers: *Calamus viminalis, Ventilago madraspatana, Sphenodesma unguiculata, Pothos scandens, Tetracera sarmentosa, Dinochloa andamanica* etc.
- ✰ Herbs and grasses: *Paspalum conjugatum, Echinochloa crus-galli, Dichanthium annulatum, Chrysopogon aciculatus, Imperata cylindrica, Alysicarpus vaginalis, Peperomia pellucida, Pteris quadriaurita* (Fern) etc.

Littoral Forests

Found as a narrow belt on sandy beaches above high water level, littoral forests are open formations of tall, sturdy trees with limited undergrowth. They are capable of withstanding the fury of monsoons and cyclones and act as wind breaks protecting the island shores. Bullet wood (*Manilkara littoralis*) and Sundri (*Heritiera littoralis*) are common in these forests.

Trees - *Barringtonia asiatica, Hernandia peltata, Thesepsia populnea, Pandanus, Mimusops littoralis, Pongamia pinnata, Instia bijuga, Glochidion calocarpum. Cerbera odollam, Heritiera littoralis, Barrigtonia racemosa, Ficus retusa, Scyphiphora hydrophyllacea, Excoecaria agallocha, Lumnitzera rocemosa, Xylocarpus granatum, Sonneratia* spp.; shrubs- *Clerodendrum inerme, Desmodium umbellatum*; climbers - *Vigna marina, Caesalpinia crista; Derris scandens, D. heterophylla, Flagellaria indica, Dischidia bengalensis*; epiphytes - *Dendrobium, Bulbophyllum, Eria* etc.; creepers - *Ipomoea pes-caprae, Phyla nodiflora,* ferns – *Drymglossum pilosellioides, Asplenium nidus.*

Mangrove Forests

As dense formations well adapted to salinity, tidal fluctuations and marshy soil, Mangrove Forests interphase between land and marine ecosystems.

Mangroves in the North Andaman occupy about 14 per cent of geographical area. The coast line of the Andaman Islands is irregular and deeply indented thereby giving rise to a number of tidal creeks. These creeks are densely populated by mangrove plant community. Among plant species recorded 25 were true mangroves and 15 were other common associated species. Community of these islands is greatly dominated by the family *Rhizophoraceae* with 9 species and the seaward line is always occupied by *Rhizophora* spp.

In mangrove forests of Andaman Islands three distinct vegetation zones has been described by Mall *et al.* (1985): (i) Proximal zone dominated by *Rhizophora mucronata* and *R. apiculata*, (ii) Middle zone with characteristic species as *Ceriops tagal, Bruguiera gymnorhiza, B. cylindrica, B. parviflora, Rhizophora stylosa, Lumnitzera stylosa, L. littorea, Sonneratia alba, S. apetala, Exoecaria agallocha, Schyphiphora hydrophyllacea* etc. and (iii) distal zone represented by *Hibiscus tiliaceus, Phoenix paludosa, Nypa fruticans, Heritiera littoralis,* etc. However, in disturbed mangrove forest, proximal zone is not very clear and its species composition shifts to *Avicennia marina, Bruguiera gymnorhiza, Ceriops tagal*, etc.

Threats

Physical isolation between groups of islands and also from the neighbouring mainlands through millions of years has resulted in the evolution of a rare and distinct flora. Though related to mainland Indian flora, vegetation shows much closer similarity with Burmese, Malaysian and Indonesian floras. These islands also contain species, which are endemic and not found anywhere else. About 40 per cent of the flora of these islands are absent in mainland India (Balakrishnan, 1989).

Taking the endemics and non-Indian southeast Asian elements of the flora, about 50 per cent of the flora are threatened and deserving immediate conservation before they totally disappear from Indian territory. Increasing human settlement results in clearance of large tracts of virgin mangrove forests for conversion to agriculture, domestic and industrial development, and construction of harbours and channels, etc. The demand for wood and wood products during pre-1950 period was low. Earlier good quality logs of Padauk (*Pterocarpus dalbergioides*), Gurjan (*Dipterocarpus alatus*) and White Chuglam (*Terminalia bialata*) found a ready market in the mainland. Post-1950 period was characterized by an increasing demand for timber within the islands. Due to deforestation and over-exploitation of forests, the rapid run-off during the rainy season, and erosion on steep slopes and degradation of soil are evident in many parts of the Andamans. Invasion of alien species is appearing to be a major threat to flora and fauna of Andaman and Nicobar Islands. The invasion of exotic weeds, *Mikania scandens* and *Chromolaena odorata* and their rapid spread is posing problem to native forest vegetation. The growth of the weeds are very fast and they may invade the whole of the worked forest. The coverage of the species on the forest floor is so intense that any seedling or sapling growth is difficult.

Botanical History

A knowledge of the botanical history of any geographic area is of paramount importance for understanding its phytogeography. The botanical history of the Andaman and Nicobar Islands dates back to 1791 when Colonel Kyd visited the islands and introduced some Andaman plants in the Royal Botanic Gardens, Calcutta, which were later described in Roxburgh's Flora of India.

In 1839, Helfer, Russian geologist visited the islands with a view to ascertaining mineral resources and also made extensive botanical collections, but these after his murder by aborigines on the North Andaman, were unfortunately mixed up with his Tenasserim plants and all were labelled as "Tenasserim and Andaman" which resulted in great deal of confusion.

Subsequently, botanists like Kurz, Parish, Prain, King and Rogers made some botanical studies between 1866 and 1903, which together and with his own efforts enabled Parkinson (1923) to bring out a "Flora of the Andaman Islands" dealing with the woody elements (650 species).

Since then, Indian explorers, Thothathri (1959, 1974), Ellis and Ramamurthy (1964, 1987), N.P. Balakrishnan (1973, 1976-77), N.G. Nair (1973, 1976-77), Bhargawa (1974), R.K. Premnath (1982), P. Basu (1978), B.K. Sinha (1982, 1992), T. Chakraborty and Kindo (1986), M.K. Vasudeva Rao (1982), L.N. Ray (1990), Debnath (1991),

Sanjappa (1989), P.L. Narasimhan (1990-92), P.S.N. Rao (1999), Sumathi (2000), P.V. Sreekumar (2000) made botanical collections from North Andamans. Their collections representing about 425 species were available at PBL, BSI, Port Blair (up to 2003).

A preliminary report of angiosperms of Andaman and Nicobar Islands by M.K.V. Rao (1986), supplementary lists by Lakhsminarasimhan and P.S.N. Rao (1996), S.P. Mathew (1998) have greatly contributed towards understanding the floristic composition of the archipelago. So far, a total of 2000 indigenous and 500 non-indigenous species of flowering plants are recorded in Andaman and Nicobar Islands.

Significant works on endemic flora are of Balakrishnan and M.K.V. Rao (1983) who has given an account 110 endemic taxa, which are rare and endangered. M.K.V. Rao (1986) in the preliminary report of angiosperms mentioned about 221 endemic taxa. The most exhaustive work is that of Dagar (1989) who enlisted 289 endemic taxa. Since publication of it so many new taxa are described from Andaman and Nicobar Islands. Reddy *et al.* (2004) recorded 353 endemic taxa, which was based on the existing literature, and herbarium studies. Reddy and Dutt (2005) and Reddy and Raju (2005) recorded 32 species as additions to flora of Andaman and Nicobar Islands from North Andaman.

Floristic Diversity

The presence of 1025 indigenous and 191 non-indigenous species of flowering plants within a land area 1710 sq km is a significant feature of North Andaman Islands (representing 629 genera under 145 families and 101 endemic species). Though related to the mainland Indian flora, the flora of the North Andaman Islands shows much affinity with the Myanmar flora and included under Indo-Burma hotspot. North Andamans flora is diverse when compared to Flora of Great Nicobar (the latter is represented with 648 species).

On the whole, 1216 species of higher plants were recorded from North Andaman group of islands (included 31 species of fern flora and 3 gymnospermous species). Among the indigenous Dicotyledons, the Polypetalae is represented by 64 families, 220 genera and 378 species, the Gamopetalae by 30 families, 150 genera and 240 species, where as Monochlamydeae by 23 families, 75 genera and 141 species (Tables 5.3–5.7).

Table 5.3: Statistical Analysis of the Flora (Wild)

Group	*No. of Families*	*No. of Genera*	*No. of Species*
Dicotyledons	114	445	759
Monocotyledons	23	131	232
Total	135	576	991
Gymnosperms	3	3	3
Pteridophyta	17	24	31
Grand Total	155	603	1025

Of the wild species herbs represents 425 species followed by trees with 300 species (Table 5.4). Life-form analysis shows that Phanerophytes were dominant due to tropical humid bioclimate (Table 5.5). At family level, *Poaceae* (78), *Euphorbiaceae* (67), *Papilionaceae* (61), *Cyperaceae* (52), *Rubiaceae* (50) were dominant familes (Table 5.6). Of the genera, *Cyperus* is the largest represented by 18 species, followed by *Ficus* (14), *Fimbristylis* (12) and *Ixora* (10) (Table 5.7).

Table 5.4: Habit-wise Distribution of Species

Sl.No.	*Habitat*	*No. of Species*	*Per cent of Species*
1.	Trees	300	29
2.	Shrubs	111	11
3.	Climbers	189	18
4.	Herbs	425	41
		1025	100

Table 5.5: Life-form Wise Distribution of Species

Sl.No.	*Habitat*	*No. of Species*	*Per cent of Species*
1.	Phanerophytes	619	60
2.	Chamaephytes	68	7
3.	Cryptophytes	64	6
4.	Hemicryptophytes	84	8
5.	Therophytes	190	19
		1025	100

* High percentage of phanerophytes indicates Tropical humid bioclimate

Table 5.6: A Comparison between Ten Dominant Families of North Andamans with those of Flora of A.&N. Islands and Flora of Great Nicobar Island

Sl.No.	*Flora of North Andaman*	*Flora of Great Nicobar*	*Flora of A.& N. Islands*
1.	Poaceae (78)	Euphorbiaceae (39)	Poaceae (150)
2.	Euphorbiaceae (67)	Rubiaceae (37)	Rubiaceae (115)
3.	Papilionaceae (61)	Orchidaceae (31)	Euphorbiaceae (110)
4.	Cyperaceae (52)	Cyperaceae (28)	Orchidaceae (98)
5.	Rubiaceae (50)	Poaceae (26)	Papilionaceae (89)
6.	Orchidaceae (33)	Annonaceae (16)	Cyperaceae (80)
7.	Asteraceae (25)	Papilionaceae (14)	Moraceae (52)
8.	Annonaceae (24)	Moraceae (13)	Annonaceae (50)
9.	Convolvulaceae (23)	Menispermaceae (12)	Arecaceae (37)
10.	Moraceae (20)	Meliaceae (12)	Meliaceae (31)

At the species level, 101 species are endemic and at generic level endemism is rather less with only two genera *viz. Pubistylis* Thoth. (Rubiaceae) with one species, *Sphyranthera* T. Chakrab. and Vasud. (Euphorbiaceae) with two species. Among the non-endemic angiosperm species about 40 per cent are not found in mainland India but extend their distribution to Myanmar (Burma) and Thailand. The phytogeographical distinction between the Andaman islands and Nicobar islands is evident from the fact that only 28 per cent of the angiosperm plants reportedly have a common distribution on both.

Table 5.7: Genera with Maximum Diversity (more than 7 species)

Sl.No.	Genera	No. of Species	Sl.No.	Genera	No. of Species
1.	*Cyperus*	18	2.	*Ficus*	14
3.	*Fimbristylis*	12	4.	*Ixora*	10
5.	*Cassia*	8	6.	*Ipomoea*	8
7.	*Phyllanthus*	8	8.	*Blumea*	7
9.	*Leea*	7	10.	*Desmodium*	7

Plan of Analysis

The families were arranged according to Bentham and Hooker's system of classification. Taxa were arranged alphabetically under families of Magnoliopsida (*Dicotyledonae*) and Liliopsida (*Monocotyledonae*). Information on habit, vegetation type, distribution and reference number (place and date of collection, species collection number by various collectors) for each taxon was furnished. Plant collections by BSI Scientists and others available at PBL, BSI, Port Blair are provided. The species which are recorded during phytosociological studies as part of biodiversity characterization project are also presented here (CSR). Local names of species are provided in paranthesis after botanical name. Species of endemic nature are marked as Endemic after collection number and distributional records to Andaman and Nicobar islands are given in bold letters after collection number as **New Record**. Nomenclature is up dated.

SYSTEMATIC ENUMERATION

MAGNOLIOPSIDA

POLYPETALAE

RANUNCULACEAE

Clematis smilacifolia Wall. subsp. **andamanica** Kapoor

Climber. Rare in evergreen forests. Saddle peak, 2-12-1976 Balak and Nair 4784 (700 m); 23-11-2001. CSR 2240.

Endemic.

DILLENIACEAE

Dillenia andamanica Parkinson (Chalta, Karkad)

Tree. Common in moist decidous and semievergreen forests. Saddle peak foot hill (50 m): 23-11-2001, CSR 2221.

Endemic.

Dillenia aurea J.E. Smith (Karkad)

Tree. Rare, in semievergreen forests. Shyamnagar: 11-11-2001, CSR 2213.

Dillenia indica L.(Karkad)

Tree. Occasional in open moist decidous forests. Ramnagar, 4-12-2001. CSR 2450.

Dillenia pentagyna Roxb. (Sambu, Sambidium)

Deciduous trees to 15 m. Frequent in Moist deciduous forests. Radhanagar: 7-11-2001, CSR 2021. Saddle peak foot hill: (50 m): 24-11-2001, CSR 2251.

Tetracera sarmentosa (L.) Vahl subsp. **andamanica** (Hoogl.) Hoogl. (Chalta lata, Kujili bel)

Climber. Common in moist decidous and semievergreen forests. Saddle peak (75 m), 20-9-1982, RKP 9203; 23-11-2001, CSR 2222. Mohanpur: 7-12-2001, CSR 2494.

Endemic.

MAGNOLIACEAE

Magnolia andamanica (King) Raju et Nayar (*Talauma andamanica*)

Large trees. Rare, in evergreen forests. Mohanpur: 7-12-2001. CSR 2491.

Endemic.

Michelia champaca L. is planted in temple premises.

ANNONACEAE

Anaxogoria javanica Bl. (Tip Phal).

Shrubs. Common in moist deciduous and semievergreen forests. Diglipur-4, 19-11-1976. Nair 4812; Smith Island: 14-12-2001, CSR 2571.

Anaxogoria luzonensis A.Gray (Tip Phal).

Shrubs. Common in moist deciduous and semievergreen forests. Paget Island: 5-11-2001. CSR 2001.

Artabotrys speciosus Kurz ex Hook.f. et Thoms.

Woody climbers. Occasional in moist deciduous forests. Interview Island: 5-3-1990, PLN 15122; Saddle peak foot hill: 23-11-2001, CSR 2223.

Cyathostemma micranthum (DC.) Sinclair

Large climber. Growing in inland forests. Way to Saddle peak (150 m): 20-9-1982, RKP 9203; 23-11-2001, CSR 2224.

Endemic.

Cyathostemmma viridiflorum Griffith

Scandent shrubs. Occasional in moist deciduous forests. Keralapuram: 13-2-1991, Debnath 4856; Radhanagar: 7-11-2002, 2022. Saddle peak (150 m): 22-12-2001, CSR 2662.

Desmos dasymachalus (Blume) Safford

Shrub. Occasional in forests. Laxmipur: 23-11-1976, NAIR 4877. Sitanagar: 15-5-1982, MKV 9026. Interview Island: 16-11-1990, PLN 15630; Karmatong-9: 7-12-1992, BKS 16290. Way to Saddle Peak, 20-9-1982 (150 m), RKP 9201.

Friesodielsia fornicata (Roxb.) Das (Jalpai lata)

Climber. Occasional in forests. Radhanagar: 7-11-2001, CSR 2025.

Goniothalamus macranthus (Kurz) Boerl. (Batamballi).

Small tree to 6 m. Occasional in moist deciduous forests. Saddle peak (50 m): 23-11-2001 CSR 2226.

Endemic.

Goniothalamus malayanus Hook.f. et Thoms.

Trees to 10 m high. Occasional in moist deciduous forests. Interview Island: 17-11-1990, PLN 15637; 22-2-1991, PLN 15634.

Miliusa andamanica (King) Finet et Gagnep.

Small tree to 5 m. Rare, in evergreen forests. Saddle peak (100 m): 22-11-2001, CSR 2225.

Endemic.

Miliusa globosa (DC.) Panigr. et Mishra

Small tree. Occasional, in evergreen forests. Saddle peak Foot hill, 11-5-1982, (150 m) MKV 8967; Interview Island: 6-3-1990, PLN 15135.

Miliusa tectona Hutch. ex Parkinson (Jangli-sagwan, Jangli-teak).

Trees to 15 m high. Occasional, in moist deciduous forests. Lamia Bay: 22-11-2001. CSR 2202.

Orophea hexandra Blume

Small tree. Occasional in evergreen forests. Interview Island: 16-2-1959, Thoth. 9224.

Orophea katschallica Kurz

Small tree. Occasional in semievergreen forests. Interview Island: 17-11-1990, PLN 15636; Paget Island: 5-11-2001, CSR 2003; Point Island: 15-11-2001. CSR 2121.

Endemic.

Polyalthia jenkinsii (Hook.f. et Thoms.) Hook.f. et Thoms.

Small tree. Rare, in evergreen forests. Interview Island: 1974, Thoth. and Nair 1074; 1978, P. Basu 6844; 6-3-1990, PLN 15132.

Polyalthia lateriflora (Blume) Kurz

Shrubs to 5 m. Rare in inland forests. Interview Island: 6-3-1990, PLN 15140.

Polyalthia rufescens Hook.f. et Thoms.

Shrub to 3 m high. Interview Island: 7-3-1990, PLN 15168.

Pseuduvaria prainii (King) Merr.

Small tree. Occasional in evergreen and semievergreen forests. Saddle peak, 28-3-1977 (50 m), Balak 5193; 11-5-1982 (50 m), MKV 8979.

Sageraea elliptica (DC.) Hook.f. et Thoms. (Chooi)

Trees, 6-15 m. Occasional in moist deciduous and Semievergreen forests. Saddle peak, 23-11-2001 (100 m), CSR 2663.

Sageraea listeri var. andamanica Chatterjee et Mukherjee

Small tree. Rare, in semievergreen forests. Paget Island: 5-11-2001, CSR 2002. Endemic.

Trivalvaria argentea (Hook.f. et Thoms.) Sinclair

Trees. Rare, in evergreen forests. Anderson Island: 9-11-1990, PLN and L.N. Ray 15253A.

Trivalvaria dubia (Kurz) Sinclair

Shrubs to 2 m. Rare in inland forests. Interview Island: 8-3-1990, PLN 15190.

Uvaria andamanica King

Trees. Rare, in evergreen forests. Kalpong area: 25-12-1986, GCB and GSK 12389.

Endemic.

Uvaria cordata (Dunal) Alston

Trees. Rare, in evergreen forests. Austin-9: 3-12-1992, BKS 16224.

The following species are cultivated for economic purposes.

Annona muricata L. (Laxmana phal).

Annona reticulata L. (Rama phal, Bullocks heart).

Annona squamosa L. (Sita phal, Custard apple)

Artabotrys hexapetalus (L.f.) Bhandari (Champak).

Polyalthia longifolia (Sonner.) Thw. (Ashok).

MENISPERMACEAE

Anamirta cocculus (L.) Wight et Arn.

Climber. Rare in coastal forests. Interview Island: 23-2-1991, PLN 15574; Buchanan Island: 10-12-2001, CSR 2502.

Cissampelos pariera L.

Climber. Rare, in inland forests. Mohanpur: 7-12-2001, CSR 2495.

Cyclea peltata (Lam.) Hook.f. et Thoms.

Climber. Rare, in inland forests. Sadddle peak: 24-11-2001 (150 m), CSR 2450.

Cyclea pendulina Miers

Climber. Rare, in inland forests. Durgapur: 20-11-1976. Nair 4835.

Endemic.

Diploclisia glaucescens (Blume) Diels

Climber. Scattered in hill forests. Sadddle peak: (600 m), 31-9-1977, Balak 5422; 10-5-1982, MKV 8958. Saddle peak (700 m): 25-11-2001, CSR 2263.

Hypserpa nitida Miers

Climber. Rare in inland forests. Keralapuram: 21-11-1976, Nair 4852.

Parabaena sagittata Miers ex Hook.f. et Thoms.

Climber. Rare in inland forests. Lamia Bay to Kalipur: 1-4-1977, Balak, 5430.

Pycnarrhena lucida (Teijsm.& Binn.) Miq.

Climbers. Rare in inland forests. Interview Island: 15-2-1959, Thoth. 9221.

Stephania japonica var. discolor (Blume) Forman

Climber. Rare in inland forests. Mohanpur: 7-12-2001, CSR 2496.

Tinospora cordifolia (Willd.) Miers ex Hook.f. et Thoms.

Climber. Common in inland forests. Point Island: 15-11-2001. CSR 2122.

NYMPHAEACEAE

Nymphaea nouchali Burm.f.

Aquatic herbs. Rare, in stagnant waters. Ganesh Nagar:17-11-2001, CSR 2148.

PAPAVERACEAE

Argemone mexicana L.

Erect herb to 50 cm. Weed of driedup fields. Mohanpur: 7-12-2001, CSR 2497.

BRASSICACEAE (*nom. alter.* CRUCIFERAE)

Brassica juncea (L.) Czern. et Coss. is cultivated by settlers. Runs wild.

CAPPARACEAE

Capparis floribunda Wight

Climber. Rare in moist deciduous forests. Lamia Bay to Ramnagar, 29-3-1977, Balak 5322.

Capparis micrantha DC.

Climber. Rare in moist deciduous forests. Ray hills: 27-4-1964. Ellis and Ramamurthy 1896.; Interview Island: 8-3-1990, PLN 15192.

Capparis sepiaria L.

Climber. Occasional in inland forests. Sadddle peak: 20-9-1982 (150 m), 20-9-1982, RKP 9205. Saddle peak (100 m): 24-11-2001, CSR 2251.

Capparis zeylanica L.

Climber. Occasional in moist deciduous forests. Mohanpur: 7-12-2001, CSR 2498.

Crateva adansonii DC. ssp. **odora** (Buch.-Ham.) is planted in gardens.

CLEOMACEAE

Cleome aspera J. Koen. ex DC.

Herbs. Rare in sandy places. Mohanpur: 7-12-2001, CSR 2499.

New record.

Cleome gynandra L.

Herbs. Frequent on heaps and on dumping grounds. Mayabandar: 6-12-2001, CSR 2486.

Cleome viscosa L.

Herbs. Frequent weed of wastelands. Radhanagar: 10-11-2001, CSR 2065.

VIOLACEAE

Hybanthus enneaspermus (L.) F.V Muell.

Herbs. Rare in sandy localities. Kiranbasti: 20-11-2001, CSR 2174. **New record.**

Rinorea bengalensis (Wall.) O.K.

Shrubs. Occasional in moist deciduous forests. Sadddle peak: 10-9-1982 (150 m). MKV 8940; Sitanagar: 15-5-1982, MKV 9030. Interview Island: 25-4-1964, Ellis and Ramam. 18939.

Endemic.

Rinorea heteroclita (Roxb.) Craib

Shrubs. Occasional in inland forests. Laxmipur: 23-11-1976, Nair 4883.

BIXACEAE

Bixa orellana L. is planted in gardens.

FLACOURTIACEAE

(incl. SAMYDACEAE)

Casearia andamanica King

Large shrub. Rare in evergreen forests. Sadddle peak: 10-5-1982 (300 m). MKV 8955, Mohanpur: 7-12-2001, CSR 2500.

Endemic.

Casearia insularis Vasud. et T. Chakrab.

Large shrub. Rare in evergreen forests. Reported from Sadddle peak by Vasudeva rao and Chakraborty (500 m).

Endemic.

Casearia elliptica Willd.

Large shrub. Rare in Moist deciduous and Semievergreen forests. Mohanpur: 8-12-2001, CSR 2501.

Flacourtia indica (Burm.f.) Merr.

Large shrub. Rare in Moist deciduous and Semievergreen forests. Mayabandar: 6-12-2001, CSR 2485.

Ryparosa javanica (Bl.) Kurz

Trees to 10 m high. Rare in inland forests. Interview Island: 1-3-1991, PLN 20724.

PITTOSPORACEAE

Pittosporum ferrugineum Ait.

Tree. Rare in evergreen forests. Sadddle peak: 1-2-1976 (400 m), Balak. and Nair 4733; 27-3-1977 (250 m), Balak 5154.

POLYGALACEAE

Polygala arvensis Willd.

Herbs. Rare in driedup fields. Mohanpur: 8-12-2001, CSR 2502. **New record.**

PORTULACACEAE

Portulaca oleracea L.

Herbs. Frequent near by cultivation and moist places. Mayabandar: 6-12-2001, CSR 2484.

Portulaca grandiflora Hook., and **Talinum triangulare** (Jacq.) Willd. are planted in gardens.

XANTHOPHYLLACEAE

Xanthophyllum andamanicum King

Trees to 15 m tall. Occasional in evergreen forests. Interview Island: 25-2-1990, PLN 15694.

Endemic.

Xanthophyllum vitellinum (Bl.) Dietr.

Trees to 15 m tall. Rare in evergreen forests. Interview Island: 23-2-1990, PLN 15683.

HYPERICACEAE

Cratoxylum formosum (Jack) Dyer

Shrub. Frequent in inland forests. Saddle peak: 10-5-1982, MKV 8949.

CLUSIACEAE (*nom. alter.* GUTTIFERAE)

Calophyllum austro-indicum Kosterm.ex P. Stevens.

Trees 4-5 m tall. Rare, in stunted evergreen forests at higher elevations. Saddle peak-2nd Tikri (650 m): 3-2-2000, Sumathi 17815.

Calophyllum calaba L. var. **bracteatum** (Wight) Stevens.

Tree. Rare in evergreen forests. Mohanpur: 9-12-2001, CSR 2520.

Calophyllum inophyllum L.

Tree. Common in beach forests. Point Island: 15-11-2001. CSR 2123.

Garcinia andamanica King

Trees. Rare in evergreen forests. Mohanpur: 8-12-2001, CSR 2503.

Endemic.

Garcinia cowa Roxb.ex DC.

Trees. Occasional in inland forests. Laxmipur: 22-11-1976. Nair 4864; Saddle peak: 11-5-1982, MKV 8977.

Garcinia xanthochymus Hook.f.

Trees. Occasional in inland forests. Lamia Bay: 28-3-1977, Balak 5198.

Mesua ferrea L.

Trees. Rare in evergreen forests. Mohanpur: 8-12-2001, CSR 2504.

THEACEAE

Ternstroemia wallichiana (Griff.) Ridley

Trees to 10 m high. Rare in evergreen forests. Laxmipur: 22-11-1976. Nair 4870.

DIPTEROCARPACEAE

Dipterocarpus alatus Roxb. (Garjan, Bada Garjan)

Very tall trees. Occasional in evergreen forests. Interview Island: 13-2-1959, Thothathri 9217. Mohanpur: 8-12-2001, CSR 2505.

Endemic.

Dipterocarpus costatus (Garjan, Chota-patti Garjan)

Very tall trees. Occasional in evergreen forests. Saddle peak: 31-3-1977, Balak 5382; 2-12-1976, Balak and Nair 4780. Saddle peak (500 m): 24-11-2001, CSR 2262.

Dipterocarpus gracilis Blume (*D. turbinatus* Buch.-Ham.) (Garjan)

Very tall trees. Occasional in evergreen forests. Mohanpur: 8-12-2001, CSR 2520.

Dipterocarpus grandiflorus (Blanco) Blanco (*D. griffithii* Miq.) (Garjan)

Very tall trees. Occasional in evergreen forests. Swamp Island: 9-12-2001, CSR 2521.

Hopea helferi (Dyer) Brandis

Trees. Very rare, in forests of Saddle peak. Saddle peak (400 m): 12-1976, Balak and Nair 4739; 27-3-1977, Balak 5168.

Hopea odorata Roxb. (Tingan)

Trees. Common, in inland forests. Swamp Island: 9-12-2001, CSR 2522.

ANCISTROCLADACEAE

Ancistrocladus attenuatus Dyer (Khoon bel)

Climbers. Frequent in inland forests. Mohanpur: 8-12-2001, CSR 2507.

Ancistrocladus tectorius (Lour.) Merr. (*A. extensus* Planch.) (Khoon bel)

Climbers. Frequent in inland forests. Ray hills: 27-4-1964, Ellis and Ramamurthy 18960; Interview Island: 7-3-1990, PLN 181.

MALVACEAE

Abutilon indicum (L.) Sweet

Undershrubs. Occasional in waste lands. Radhanagar: 10-11-2001, CSR 2063.

Hibiscus lobatus (Murray) Kuntze

Undershrubs. Occasional in waste lands. Kishorinagar: 5-12-2001, CSR 2451.

Hibiscus scandens Roxb.

Climbers. Frequent in inland forests. Diglipur, 24-3-1973, Balak 72.

Hibiscus tiliaceus L.

Trees. Common in coastal forests. Point Island: 15-11-2001. CSR 2124.

Malvastrum coromandelianum (L.) Garcke

Undershrubs. Grown in waste lands. Mohanpur: 8-12-2001, CSR 2508.

Sida acuta Burm.f.

Undershrubs. Common in waste lands. Radhanagar: 10-11-2001, CSR 2062.

Sida cordata (Burm.f.) Borssum

Rambling undershrubs. Occasional in waste lands. Kishorinagar: 5-12-2001, CSR 2452.

Sida cordifolia L.

Undershrubs. Common in near by human inhabitation. Mohanpur: 8-12-2001, CSR 2519.

Sida mysorensis Wight et Arn.

Undershrubs. Occasional in near by human inhabitation. Mohanpur: 8-12-2001, CSR 2521.

Sida rhombifolia L.

Undershrubs. Occasional in near by human inhabitation. Kishorinagar: 5-12-2001, CSR 2453.

Sida spinosa L.

Undershrubs. Common in waste lands. Kishorinagar: 5-12-2001, CSR 2458.

Thespesia lampas (Cav.) Dalz. ex Dalz. et Gibs.

Shrubs. Occasional in open forests. Kishorinagar: 5-12-2001, CSR 2456. **New record.**

Thespesia populnea (L.) Sol. ex Corr.

Trees. Common in coastal forests. Point Island: 15-11-2001. CSR 2125.

Urena lobata L. ssp. **sinuata** (L.) Borssum

Shrubs. Occasional in open forests. Diglipur-4: 5-12-1976, Balak and Nair 4964. Kishorinagar: 5-12-2001, CSR 2454.

The following species are cultivated for economic / ornamental purposes.

Abelmoschus esculentus (L.) Moench

Abelmoschus moschatus Medic

Hibiscus hirtus L.

Hibsicus rosa-sinensis L.

Hibiscus schizopetalus (Masters) Hook.f.

BOMBACACEAE

Bombax ceiba L. (Didu)

Trees. Rare in inland forests. Mohanpur: 8-12-2001, CSR 2518.

Bombax insigne Wall. (Didu)

Trees. Occasional in inland forests. Mohanpur: 8-12-2001, CSR 2509.

Ceiba pentandra (L.) Gaertn. (Didu) is planted along avenues of Diglipur town.

STERCULIACEAE

Byttneria andamanensis Kurz

Climbers. Occasional in inland forests. Durgapur: 20-11-1976, Nair 4842. Radhanagar: 24-12-2001, CSR 2681; Swamp Island: 9-12-2001, CSR 2522.

Endemic.

Byttneria grandifolia DC. (*B. aspera* Colebr. ex Wall.)

Shrubs. Occasional in inland forests. Diglipur-4: 19-11-1976, Nair 4819.

Firmiana colorata (Roxb.) R.Br. (*Sterculia colorata* Roxb.)

Trees. Occasional in inland forests. Kishorinagar: 5-12-2001, CSR 2456.

Helicteres isora L.

Small trees. Rare in moist deciduous forests. Mohanpur: 8-12-2001, CSR 2511.

Heritiera littoralis Dryand.

Small trees. Occasional in littoral forests. Sibpur: 28-11-1976, Balak 4718.

Melochia corchorifolia L.

Herbs. Rare in moist places near by settlements. Radhanagar: 10-11-2001, CSR 2064.

Melochia umbellata (Houtt.) Stapf.

Herbs. Common in moist places near by settlements and roadsides. Radhanagar: 10-11-2001, CSR 2061. Mayabandar: 6-12-2001, CSR 2487.

Pterocymbium tinctorium (Blanco) Merril (Papita)

Trees. Common in inland forests. Paget Island: 6-11-2001, CSR 2018.

Pterospermum acerifolium (L.) Willd. (Makchun, Masanand)

Trees. Common in inland forests. Ariel Bay: 25-11-1976, Nair 4901;

Keralapuram: 5-4-1977, Balak 5474.

Pterygota alata (Roxb.) R. Br. (*Sterculia alata* Roxb.) (Lakkov)

Trees. Common in inland forests. Point Island: 15-11-2001, CSR 2126.

Sterculia guttata Roxb.

Trees. Rare in deciduous forests. Kishorinagar: 5-12-2001, CSR 2457.

Sterculia macrophylla Vent.

Trees. Frequent in semievergreen and moist deciduous forests. Mohanpur: 8-12-2001, CSR 2512.

Sterculia parviflora Roxb.

Trees. Rare in evergreen forests. Radhanagar: 24-12-2001, CSR 2682.

Sterculia rubiginosa Vent.

Trees. Rare in evergreen forests. Ray hills: 14-11-1990, PLN 15628.

Sterculia villosa Roxb. (Chilka)

Trees. Common in semievergreen and moist deciduous forests. Paget Island: 6-11-2001, CSR 2019.

Guazuma ulmifolia Lam. is planted in gardens.

TILIACEAE

Berrya cordifolia (Willd.) Burrett (*B. ammonilla* Roxb.)

Small tree. Occasional in beach and littoral forests. Sibpur: 28-11-1976, Balak. and Nair 4709.

Brownlowia tersa (L.) Kosterm.

Shrubs to 3m tall. Rarely found along the creeks in mangrove areas. Swamp Island: 9-12-2001, CSR 2524.

Corchorus aestuans L.

Herbs. Occasional in moist places. Radhanagar: 7-11-2001, CSR 2030.

New Record.

Corchorus capsularis L.

Undershrubs. Occasional in moist places. Kishorinagar: 5-12-2001, CSR 2459.

Corchorus fascicularis Lam.

Herbs. Occasional in open moist places. Radhanagar: 24-11-2001, CSR 2684.

Corchorus olitorius L.

Herbs. Occasional in open moist places. Mayabandar: 6-12-2001, CSR 2488.

Grewia acuminata A.L. Juss.

Shrubs. Occasional in open moist deciduous forests. Radhanagar: 9-12-2001, CSR 2685.

Grewia calophylla Kurz ex Masters

Shrubs. Occasional in moist deciduous forests. Diglipur-4: 19-11-1976, Nair 4810; Interview Island: 17-11-1991, PLN 15640.

Endemic.

Grewia heterotricha Masters

Shrubs. Occasional in moist deciduous forests. Ganesh nagar: 17-11-2001, CSR 2144.

Grewia indandaminca Ellis et Ray

Shrubs. Occasional in moist deciduous forests. Saddle peak: Balak and Nair 4797.

Endemic.

Grewia serrulata DC.

Shrubs. Occasional in moist deciduous forests. Lakshmipur: 23-11-1976, Nair 4886. Interview Island: 8-3-1990, PLN 15186.

Triumfetta repens (Blume) Merr. et Rolfe

Woody creepers. Occasional along streams of forests. Kishorinagar: 5-12-2001, CSR 2460.

Triumfetta rhomboidea Jacq.

Undershrubs. Common in waste lands. Diglipur-4, 5-12-1976: Balak and Nair 4982. Kishorinagar: 5-12-2001, CSR 2461.

ELAEOCARPACEAE

Elaeocarpus tectorius (Lour.) Poiret

Trees. Rare in evergreen forests. Kishorinagar: 5-12-2001, CSR 2463.

Elaeocarpus tuberculatus Roxb.

Trees. Rare in evergreen forests. Radhanagar: 7-11-2001, CSR 2031.

Muntingia calabura L. is planted in gardens.

ERYTHROXYLACEAE

Erythroxylum cuneatum (Miq.) Kurz

Medium sized trees to 15 m. Rare in inland forests. Kishorinagar: 5-12-2001, CSR 2464.

MALPIGHIACEAE

Aspidopterys tomentosa (Blume) Jussieu

Climbers. Rare in inland forests. Mayabandar: 6-12-2001, CSR 2484.

Hiptage thothathrii Balak. et Srivastava

Climbers. Rare in stunted hilltop evergreen forests. Saddle peak (700 m): 27-3-1977, Balak 5172.

Endemic.

OXALIDACEAE

Oxalis corniculata L.

Herbs. Rare in shady localities in disturbed places. Introduced weed. Radhanagar: 24-12-2001, CSR 2687.

BALSAMINACEAE (nom. fam. conserv.)

Impatiens balsamina L. is planted in gardens.

RUTACEAE

Atalantia monophylla (Roxb.) DC.

Small trees. Rare in moist deciduous forests. Mohanpur: 8-12-2001, CSR 2514.

Atalantia simplicifolia (Roxb.) Tanaka

Small trees. Rare in moist deciduous forests. Radhanagar: 24-12-2001, CSR 2688.

Euodia lunu-ankenda (Gaertn.) Merril

Trees 12-20 m tall. Rare in inland forests. Radhanagar: 24-12-2001, CSR 2690.

Glycosmis mauritiana (Lam.) Tanaka var. **andamanensis** (Naray. et Tanaka)

Shrubs. Frequent in moist deciduous forests. Aerial Bay: 23-12-2001, CSR 2675.

Endemic.

Glycosmis pentaphylla (Retz.) DC.

Shrubs. Rare in moist deciduous forests. Radhanagar: 10-11-2001, CSR 2066.

Glycosmis pilosa Narayanaswami

Shrubs. Rare in semievergreen forests. Interview Island: 6-3-1990, PLN 15138; 23-2-1991, PLN 15677.

Endemic.

Micromelum minutum (Forst.f.) Wt. et Arn.

Shrubs. Frequent in inland forests. Diglipur-4: 19-11-1976, Nair 4806; Durgapur: 5-4-1977, Balak 5469.

Murraya paniculata (L.) Jack. (*Murraya exotica* L.)

Small trees. Common in moist deciduous forests. Radhanagar: 24-12-2001, CSR 2690.

Paramignya armata (Thw.) Bedd. ex Oliv. var. **andamanica** (King) Tanaka

Armed climbers. Common in inland forests. Sitanagar: 15-5-1982, MKV 9040. Radhanagar: 7-11-2001, CSR 2033.

Endemic.

Pleiospermium alatum (Wt. et Arn.) Swingle

Trees 5-8 m tall. Common in inland forests. Paget Island: 6-11-2001, CSR 2020.

Toddalia asiatica (L.) Lam.

Armed climbers. Rare in forests of Saddle peak at higher elevations. Saddle peak (700 m): 28-11-2001, CSR 2335.

Zanthoxylum rhetsa (Roxb.) DC. (*Z. budrunga* (Roxb.) DC.)

Trees. Occasional in inland forests. Mohanpur: 8-12-2001, CSR 2515.

Aegle marmelos (L.) Correa. **Citrus limon** (L.) Burm. f., **Citrus aurantium** L. (Santra), **Citrus aurantifolia** (Nimma), **Limonia acidissima** L. are cultivated for fruits. **Murraya koenigii** (L.) Spreng. is planted in house compounds for leaves.

SIMAROUBACEAE

Ailanthus excelsa Roxb.

Trees. Grown near by bunds of cultivated fields. Radhanagar: 24-12-2001, CSR 2691.

Ailanthus triphysa (Dennst.) Alston

Trees. Rare in evergreen forests. Mohanpur: 12-12-2001, CSR 2551.

Harrisonia brownii Juss.

Armed shrubs. Rare in inland forests. Saddle peak (200 m): 28-11-2001, CSR 2339.

Harrisonia perforata (Blanco) Merr.

Armed climbers. Common in moist deciduous forests. Aerial Bay: 5-4-1977, Balak.5453. Interview Island: 25-4-1964, Ellis and Ramam. 18949; 5-3-1990, PLN 15112; Mohanpur: 8-12-2001, CSR 2516.

Picrasma javanica Blume

Trees. Occasional in evergreen forests. Saddle peak (50 m): 28-3-1977, Balak. 5194.

Samadera indica Gaertn.

Small trees. Rare in evergreen forests. Mohanpur: 8-12-2001, CSR 2517.

Suriana maritima L.

Small trees. Rarely found along sea coasts and littoral forests. Austin no.1: 12-12-2001, CSR 2552.

OCHNACEAE

Ochna integerrima (Lour.) Merr.

Small trees. Found in moist deciduous and semievergreen forests. Saddle peak: 10-5-1982, MKV 8957; Sitanagar: 15-5-1982, MKV 9031.

BURSERACEAE

Canarium euphyllum Kurz

Trees. Occasional in inland forests. Radhanagar: 7-11-2001, CSR 2033.

Canarium manii King

Trees. Rare in evergreen forests. Saddle peak (500 m): 27-3-1977, Balak 5167; 1-12-1976, Balak and Nair 4752.

Endemic.

Garuga pinnata Roxb.

Trees. Scattered in deciduous forests. North Island:17-12-2001, CSR 2616.

MELIACEAE

Aglaia argentea Blume

Trees. Scattered in inland forests. South Island: 17-12-2001, 2600.

Aglaia lawii (Wight) Sald. et Ramam.(*A. andamanica* Hiern**)** (Laktok)

Trees. Occasional in inland forests. Point Island: 15-11-2001, CSR 2127.

Endemic.

Aglaia oligophylla Miq.

Trees. Occasional in inland forests. Radhanagar: 7-11-2001, CSR 2034.

Aglaia silvestris (Roem.) Merr.

Trees. Occasional in inland forests. Mohanpur: 8-12-2001, CSR 2516.

Aphanamixis polystachya (Wall.) Parker **(***Amoora rohituka* (Roxb.) Wt. et Arn.)

Trees. Occasional in inland forests. Mohanpur: 8-12-2001, CSR 2519.

Chukrasia tubularis Jussieu

Trees. Occasional in inland forests. Kishorinagar: 5-12-2001, CSR 2453.

Trichilia connaroides (Wt. et Arn.) Bentre.

Small trees. Rare in inland forests. Kafeedera: 18-11-2001, CSR 2155.

Walsura pinnata Hassk.

Small trees. Rare in inland forests. Lamia bay: 22-11-2001, CSR 2225.

Xylocarpus gangeticus (Prain) Parkinson

Small trees. Grown in tidal forests. Buchanan Island: 12-12-2001, CSR 2553.

Xylocarpus granatum Koen.

Small trees. Grown in tidal forests. Shearme Island: 13-11-2001, CSR 2098.

Xylocarpus moluccensis (Lam.) Roem.

Small trees. Grown in tidal forests. Sound Island: 18-12-2001, CSR 2626.

Azadirachta indica A. Juss. is planted along avenues.

DICHAPETALACEAE

Dichapetalum gelonioides (Roxb.) Engl. **ssp. andamanicum** (King) Leenh.

Large shrubs or small trees. Common in inland forests. Keralapuram: 21-11-1976, Nair 4849; Sitanagar: 15-5-1982, MKV 9019. Endemic.

OLACACEAE

Anacolosa frutescens (Bl.) Bl. (Haddiped)

Small trees. Common in inland forests. Radhanagar: 24-12-2001, CSR 2695.

Olax imbricata Roxb.

Climber. Occasional in forests. Kafeedera: 18-11-2001, CSR 2154.

Ximenia americana L.

Small trees. Rare in moist deciduous forests. Trilby Island: 16-12-2001, CSR 2596.

ICACINACEAE

Codiocarpus andamanica (Kurz) Howard

Trees. Rare in semievergreen forests. Interview Island: 12-2-1959, Thoth. 9205; Saddle peak foot hill: 11-5-1982, MKV 8972.

Endemic.

Sarcostigma kleinii Wt. et Arn.

Climbers. Lamia Bay: 30-3-1977, Balak 5363; Karmatong-10: 6-12-1992, B.K. Sinha 16273. Kafeedera: 18-11-2001, CSR 2153.

OPILIACEAE

Cansjera rheedii Gmel.

Lianas. Occasional in moist deciduous forests. Kafeedera: 18-11-2001, CSR 2152.

Champeria manilaliana (Bl.) Merr.

Shrubs to 4 m. Occasional in semievergreen and moist deciduous forests. Laxmipur: 6-4-1977, Balak. 5495. Saddle peak foot hill: 9-5-1982, MKV 8922.

CELASTRACEAE

Celastrus paniculatus Willd.

Scandent shrubs. Scattered in inland forests. Radhanagar: 24-12-2001, CSR 2696.

Glyptopetalum calocarpus (Kurz) Prain

Shrubs to 3 m. Rare in hill forests. Saddle peak (600 m): 28-11-2001, CSR 2333.

Endemic.

Euonymus javanicus Bl.

Trees. Rare in inland forests. Radhanagar: 24-12-2001, CSR 2697.

Hyppoctratea grahamii Wight

Scandent shrubs. Scattered in inland forests. Lamia bay: 22-11-2001, CSR 2226.

Salacia chinensis L.

Climbers. Occasional in moist deciduous forests. Radhanagar: 10-11-2001, CSR 2067.

Salacia exsculpta Korth.

Climbers. Occasional in moist deciduous forests. Saddle peak (300 m): 28-11-2001, CSR 2334.

Salacia tortuosa Griff.

Climbers. Occasional in semievergreen and moist deciduous forests. Interview Island: 22-2-1991, PLN 15664; 7-3-1990, PLN 15161.

Siphonodon celastrineus Griff.

Climbers. Occasional in semievergreen and moist deciduous forests. Karmatong-10: BK Sinha 16305.

RHAMNACEAE

Colubrina asiatica (L.) Brongn.

Climbers. Occasional in littoral and beach forests. Smith Island: 14-12-2001, CSR 2571.

Gouania andamanica King

Climbers. Rare in semievergreen forests. Sibpur: 28-11-1976, Balak and Nair 4722.

Endemic.

Gouania leptostachya DC. (*G. tiliaefolia* Roxb.) (Saboon bel)

Climbers. Frequent in semievergreen forests. Aerial Bay: 19-11-2001, Nair 4801; Radhanagar to Swarajgram: 24-12-2001, CSR 2698.

Sageretia parviflora G.Don

Shrubs. Saddle peak (350 m): 16-10-1987, Ellis 12713; 28-11-2001, CSR 2332.

Ventilago denticulata Willd.

Climbers. Rare in moist deciduous forests. Lamia bay: 22-11-2001, CSR 2227.

Ventilago maderaspatana Gaertner

Climbers. Rare in moist deciduous forests. Swarajgram: 24-12-2001, CSR 2699.

Ziziphus andamanica Bhandari et Bhansali (*Z. glabra* King)

Scandent shrubs. Rare in inland forests. Smith island: 14-12-2001, CSR 2572.

Ziziphus horfieldii Miq.

Scandent shrubs. Occassional in inland forests. Swarajgram: 24-12-2001, CSR 2700.

Ziziphus nummularia (Burm.f.) Wt. et Arn.

Shrubs. Rare in disturbed forests. Kalipur: 22-12-2001, CSR 2649.

Ziziphus oenoplia (L.) Miller var. **oenoplia**

Scandent shrubs. Occasional in degraded forests. Radhanagar: 24-12-2001, 2701.

Ziziphus oenoplia (L.) Miller var. **pallens** Bhandari et Bhansali

Climbers. Occasional in moist deciduous forests. Lamia bay: 22-11-2001, CSR 2228.

Ziziphus rugosa Lam.

Climbers. Occasional in moist deciduous forests. Radhanagar: 24-12-2001, 2702.

Ziziphus mauritiana Lam. (Ber) is planted in house compounds.

VITACEAE

Ampelocissus barbata (Wall.) Planch.

Climbers. Occasional in semievergreen forests. Saddle peak (400 m): 28-11-2001, CSR 2331.

Cayratia japonica (Thunb.) Gagnep.

Climbers. Occasional in moist deciduous forests. Lamia bay: 22-11-2001, CSR 2229.

Cayratia pedata (Lam.) A.L. Juss. ex Gagnep.

Climbers. Common in forests. Smith Island: 25-11-1976, Nair 4909.

Cayratia trifolia (L.) Domin.

Climbers. Occasional in semievergreen forests. Radhanagar: 24-12-2001, 2703.

Cissus aristata Bl.

Climbers. Occasional in semievergreen forests. Saddle peak (400 m): 28-11-2001, CSR 2330.

Cissus elongata Roxb.

Climber. Rare in forests. Laxmipur: 24-11-1976, Nair 4897.

Cissus quadrangularis L.

Climbers. Grown on fences. Ramakrishnagram: 22-11-2001, CSR 2182.

Cissus pentagona Roxb.

Climber. Common in open moist deciduous and littoral forests. Buchanan Island: 8-11-1990, PLN 15231.

Cissus repanda Vahl

Climber. Rare in forests. Radhanagar: 24-12-2001, 2704.

Cissus repens Lam.

Climbers. Occasional in moist deciduous forests. Lamia bay: 22-11-2001, 2210.

Tetrastigma andamanicum (King) Suesseng.

Climber. Occasional in forests. Saddle peak: 2-12-1976, Balak and Nair 4782. Austin-9: 3-12-1992, BK Sinha 16220; Kalpong no-10: 22-12-1986, Chakrab. and Kindo 12329.

Endemic.

LEEACEAE

Leea adwivedica Kumar

Shrubs. Rare along streams of forests. Sinduri nallah, near Lamia Bay: 18-6-1998, Krishnakumar 0019.

Endemic.

Leea aeguata L.

Shrubs. Occasional in inland forests. Radhanagar: 24-12-2001, 2705.

Leea angulata Korth. ex Miq.

Shrubs, 7 m high, armed. Interview Island: 5-3-1990, PLN 15104.

Leea asiatica (L.) Rids. (Koya Kaccha)

Shrubs. Common in inland forests. Radhanagar: 7-11-2001, CSR 2038.

Leea guineensis G.Don

Shrubs to 5 m high. Frequent in moist deciduous forests. Lamia bay to Saddle hill: 22-11-2001, CSR 2219.

Leea indica (Burm. f.) Merr. (Koya Kaccha)

Shrubs. Common in inland forests. Durgapur: 28-9-1976, Balak and Nair 4707; Saddle peak (200 m): 28-3-1977, Balak 5408.

Leea macrophylla Roxb. ex Hornem.

Shrubs to 5 m high. Frquent in moist deciduous forests. Lamia bay: 22-11-2001, CSR 2211.

SAPINDACEAE

Allophylus cobbe (L.) Raeusch.

Small trees to 8 mm tall. Rare in moist deciduous forests. Interview Island: 16-2-199, Thoth. 9266. Saddle peak (100 m): 23-11-2001, CSR 2228.

Arytera littoralis Bl. (*A. xerocarpa* (Bl.) Adelb.)

Shrubs. Rare in littoral forests. Interview Island: 26-2-1991, PLN 20702.

Cardiospermum halicacabum L.

Climbing herbs. Diglipur: 24-12-2001, 2706.

Dimocarpus longan Lour.

Trees, common in inland forests. Arial Bay: 5-4-1977, Balak 5454; Interview Island: 5-3-1990, PLN 15117.

Dodonaea viscosa (L.) Jacq.

Evergreen shrubs or small trees. Brush Island: 16-12-2001, CSR 2591.

Ganophyllum falcatum Bl. (Jangli Neem)

Trees. Scattrered in littoral forests. Sound Island: 18-12-2001, CSR 2628.

Harpullia cupanioides Roxb.

Trees. Scattrered in inland forests. Lamia bay: 22-11-2001, 2213.

Lepisanthes rubiginosa (Roxb.) Leenh. (*Erioglossum rubiginosum* Bl.)

Trees. Occasional in semievergreen forests. Lamia Bay: 12-5-1982, MKV 9011.; Saddle peak foot hill: 9-5-1982, MKV 8917. Interview Island: 5-3-1990, PLN 15115. Saddle peak (100 m): 23-11-2001, CSR 2229.

Lepisanthes senegalensis (Juss. ex Poir.) Leenh.

Trees. Occasional in evergreen forests. Tarlait bay: 8-2-1959. Durgapur: 5-4-1977, Balak 5465. Interview Island: 6-3-1990. PLN 15130.

Mischocarpus sundaicus Bl. (Kakda).

Trees to 10 m high. Frequent in evergreen forests. Saddle peak (700 m): 2-12-1976, Balak and Nair 4794. 28-11-2001, CSR 2329.

Pometia pinnata J.R. et G. Forst. (Chidgandu, Lichu ped)

Trees to 25 m high. Frequent in evergreen forests. Interview Island: 2-3-1990, PLN 20728.

Sapindus emarginatus Vahl is planted for soap nuts.

ANACARDIACEAE

Bouea oppositifolia (Roxb.) Meisn.

Trees to 10 m high. Scarce in evergreen forests. Saddle peak (150 m): 23-11-2001, CSR 2230.

Buchanania splendens Miq. (*B. platyneura* Kurz) (Hathpanza)

Trees. Rare in inland forests. Kalpong: 24-12-1986, Chakrab. and Kindo 12371.

Endemic.

Dracontomelon dao (Blanco) Merr. et Rolfo (*D. mangifera* (Bl.) Bl.)

Trees. Frequent in inland forests. Smith Island: 25-4-1976, Nair 4911.

Lannea coromandelica (Houtt.) Merr. (*Odina wodier* Roxb.)

Large deciduous tree. Occasional in moist deciduous forests. Radhanagar: 7-11-2001, CSR 2039.

Mangifera andamanica King

An evergreen tree. Saddle peak (200 m): 28-11-2001, CSR 2328.

Endemic.

Mangifera sylvatica Roxb.

An evergreen tree. Occasional in inland forests. Saddle peak (100 m): 24-11-2001, CSR 2241.

Parishia insignis Hook.f. (Lal Dhup)

Trees. Common in inland forests. Durgapur: 5-4-1977, Balak 5468.

Semecarpus kurzii Engl. (Jangli Kaju)

Tree to 15 m. Common in inland forests. Saddle peak (150 m): 24-11-2001, CSR 2242. Endemic.

Semecarpus prainii King

Trees. Occasional in inland forests. Laxmipur: 22-11-1976, Nair 4861; Karmatong-10: 8-12-1992, BK Sinha 16296. Interview Island: 28-2-1991, PLN 20720.

Spondias pinnata (L.f.) Kurz (*S. mangifera* Willd.) (Amra)

Trees. Occasional in inland forests. Lamia bay: 22-11-2001, CSR 2214.

Anacardium occidentale L. (Cashew nut) is planted in gardens. **Mangifera indica** L. (Mango) is cultivated for edible fruits.

MORINGACEAE

Moringa oleifera Lam. (**M. pterygosperma** Gaertner) (Drum stick)

Small trees. Planted in house compounds; runswild. Mayabandar: 6-12-2001, CSR 2481.

CONNARACEAE

Rourea minor (Gaertn.) Leenh.

Lianas. Scattered in inland forests. Radhanagar: 24-12-2001, 2706.

PAPILIONACEAE

Abrus precatorius L.

Twiners. Occasional in open forests. Durgapur: 20-11-1976, Nair 4831.

Abrus pulchellus Wall. ex Thw.

Twiners. Occasional in open forests. Diglipur: 30-11-2001, CSR 2341.

Aeschynomene indica L.

Marshy herbs. Occasional near perennial marshy localities. Lamia bay: 22-11-2001, CSR 2217.

Alysicarpus vaginalis DC.

Prostrate herbs. Occasional in sandy places. Mayabandar: 6-12-2001, CSR 2480.

Atylosia scarabaeoides (L.) Benth. (*Cajanus scarabaeoides* (L.) Dupetit-Thour.)

Prostrate herbs. Rare in grasslands. Radhanagar: 24-12-2001, CSR 2703.

Canavalia africana Dunn (*C. virosa* (Roxb.) Wight et Arn.)

Climbers. Occasional in the inland forests. Mayabandar: 6-12-2001, CSR 2480.

Canavalia cathartica Thou.

Climbers. Occasional in the inland forests. Sound Island: 18-12-2001, CSR 2629.

Canavalia microcarpa (DC.) Piper

Climbers. Occasional in coastal forests. Durgapur: 26-11-1976, Nair 4990.

Clitoria ternatea L.

Vine. Naturalized along fences and roadsides. Mayabandar: 6-12-2001, CSR 2478.

Crotalaria calycina Schrank

Annual herbs. Rare in the fileds and open forests. Lamia bay: 22-11-2001, CSR 2218.

Crotalaria linifolia L.f.

Annual herbs. Rare along roadsides. Radhanagar: 24-12-2001, CSR 2706.

Crotalaria pallida Ail.

Erect undershrubs. Scattered along roadsides. Diglipur-4: 5-12-1976, Balak and Nair 4946. 30-11-2001, CSR 2342.

Crotalaria retusa L.

Erect undershrubs. Scattered along roadsides. Mayabandar: 6-12-2001, CSR 2479.

Crotalaria spectabilis Roth

Erect undershrubs. Scattered along roadsides. Diglipur: 30-11-2001, CSR 2343.

Crotalaria wildenowiana DC. ssp. **glabrifoliolata** Ellis

Suffruticose herbs. Rare in dried, sunny places in the crevices of boulders. Saddle peak (100 m): 25-11-2001, CSR 2261.

Dalbergia candenatensis (Dennst.) Prain

Climbers. Rare in inland forests. Buchanan Island: 12-12-2001, CSR 2555.

Dalbergia confertiflora Benth.

Climbers. Rare in coastal forests. Entrance Island: 11-12-2001, CSR 2541.

Dalbergia latifolia Roxb.

Deciduous trees to 10 m. Rare in moist deciduous forests. Kafeedera: 16-12-2001, CSR 2600.

Dalbergia pinnata (Lour) Prain

Small trees. Rare in inland forests. Kishorinagar: 5-12-2001, CSR 2465.

Dalbergia volubilis Roxb.

Large shrubby climbers. Interview Island: 22-2-1991, PLN 15660.

Derris elegans Benth. f. var. **elegans**

Climbers. Interview Island: 5-3-1990, PLN 15105.

Derris elegans Benth. f. var. **andamanensis** Thoth.

Climbers. Rare in the inland forests. 22-2-1991, PLN 15652.

Endemic.

Derris scandens (Roxb.) Bentham

Climbers. Occasional in littoral forests. Laxmipur: 23-11-1976, Nair 4875. Shearme Island: 13-11-2001, CSR 2095.

Derris trifoliata Lour. (*D. uliginosa* (Willd.) Benth.)

Scandent shrubs. Common in littoral forests. Interview Island: 8-3-1990, PLN 15178. Shearme Island: 13-11-2001, CSR 2096.

Desmodium gangeticum (L.) DC.

Erect, ascending under shrubs. Kishorinagar: 5-12-2001, CSR 2465.

Desmodium heterocarpon (L) DC. var. **heterocarpon**

Scandent pubescent herbs or under shrubs up to 60 cm high. Occasional in forest openings and along roadsides. Mohanpur: 8-12-2001, CSR 2519.

Desmodium heterocarpon (L) DC. var. **strigosum** Meeuwen

Under shrubs, up to 1 m high. Occasional in forest openings. Diglipur: 29-11-2001, CSR 2344.

Desmodium heterophyllum (Willd.) DC.

Under shrubs, up to 1 m high. Rare in disturbed places of forests. Radhanagar: 25-12-2001, CSR 2710.

Desmodium laxiflorum DC.

Under shrubs, up to 1.5 m high. Diglipur: 30-11-2001, CSR 2344.

Desmodium triflorum (L) DC.

Wiry, spreading herbs. Radhanagar: 10-11-2001, CSR 2068.

Desmodium umbellatum (L.) DC. (*Dendrolobium umbellatum* (L.) Benth.

Shrubs or trees to 6 m tall. Interview Island: 24-2-1990, PLN 15690.

Dioclea hexandra (Ralph) Mabberly (*D. reflexa* Hook.f.)

Lianas, 5-8 m long. Scattered in forests. Included under the authority of Lakhshminarasimhan (1999).

Erythrina fusca Lour.

Deciduous small trees. Lamia Bay: 12-5-1982, MKV 9006.

Erythrina variegata L.

Deciduous small trees. Scattered in littoral forests and planted along roadsides. Trilby Island: 16-12-2001, CSR 2598.

Flemingia macrophylla (Willd.) Ktze

Shrubs to 1.5 m high. Common in stream beds. Radhanagar: 25-12-2001, CSR 2711.

Flemingia strobilifera (L) R.Br. ex Ait.

Erect undershrubs to 1-5 m tall. Common in stream beds and along margins of forests. Lakshmipur: 6-4-1977, Balak. 5496.

Geissaspis cristata Wt. et Arn.

Trailing herbs. Scattered in disturbed places, mainly along roadsides. Diglipur: 30-11-2001, CSR 2350.

Gliricidia sepium (Jacq.) Kunth ex Walp.

Deciduous small trees. Planted along roadsides; runswild. Radhanagar: 10-11-2001, CSR 2069.

Indigofera caerulea Roxb.

Undershrubs. Rare in open forests. Radhanagar: 25-11-2001, CSR 2712.

Indigofera tinctoria L.

Undershrubs. Rare in forest clearings. Diglipur: 30-11-2001, CSR 2345.

Mucuna gigantea (Willd.) DC.

Climbers. Occasional in open forests. Interview Island: 7-3-1990, PLN 15165. Mohanpur: 4-12-2001, CSR

Mucuna monosperma DC.

Climbers. Occasional in open forests. Diglipur: 30-11-2001, CSR 2346.

Mucuna pruriens (L) DC.

Climbers. Occasional in open forests. Shyamnagar: 25-12-2001, CSR 2713.

Ormosia travancorica Bedd.

Trees to 25 m. rare in the protected hill forests of Kalpong at an altitude of ca. 100 m above mean sea level included under the authority of Lakhshminarasimhan (1999).

Pongamia pinnata (L.) Pierre

Densely foliaceous tree to 15 m. common in coastal forests. Interview Island: 25-4-1964, Ellis and Ramamurthy 18940; Lamia Bay: 29-3-1977, Balak. 5352.

Pterocarpus dalbergioides Roxb. ex DC.

Large trees. Common in moist deciduous forests. Lamia Bay: 12-5-1982, MKV 9008; Lakshmipur: 23-11-1976, Nair 4873.

Endemic.

Sesbania bispinosa (Jacq.) Wight

Shrubs to 3 m high. Grown in marshy places. Diglipur: 30-11-2001, CSR 2347.

Sesbania seban (L.) Merr.

Shrubs to 4 m high. Planted in home gardens; runswild. Diglipur: 29-11-2001, CSR 2340.

Smithia sensitiva Ait.

Prostrate herbs. Common along road-sides and forest openings. Sibpur: 28-11-1976, Balak and Nair 4715. Shyamnagar: 11-11-2001, CSR 2080.

Sophora tomentosa L.

Trees to 10 m high. Frequent in inland forests. Kishorinagar: 5-12-2001, CSR 2466.

Strongylodon lucidus (Forst.) Seemann (*S. ruber* Vogel)

Climber. Common in inland forests. Diglipur: 19-11-1976, Nair 4822. Interview Island: 17-11-1990, PLN 15647. Saddle peak (400 m): 28-11-2001, CSR 2327.

Tadehagi triquetrum (L.) Ohashi **ssp. auriculatum** (DC.) Ohashi

Erect undershrubs to 1m. Rare in stunted evergreen forests among the rocks. Saddle peak (600m): 25-11-2001, CSR 2279.

Tadehagi triquetrum (L.) Ohashi **ssp. triquetrum**

Erect undershrubs to 1m. Scattered in the inland forests. Mohanpur: 11-12-2001, CSR 2550.

Uraria picta (N. Jacquin) Desv. ex DC.

Erect herbs. Scattered in the shady places of inland forests. Radhanagar: 25-11-2001, CSR 2714.

Uraria lagopodioides (L.) Desv.

Erect herbs. Scattered in the shady places of inland forests. Sound Island: 18-12-2001, CSR 2629.

Vigna adenantha (Meyer) Marechal et Stainier

Trailing herbs. Occasioanl in evergreen forests. Chatham Island: 15-12-2001, CSR 2581.

Vigna marina (Burm.f.) Merr.

Climbing herbs. Common in littoral forests. Shearme Island: 13-11-2001, CSR 2094.

Vigna pilosa (Willd.) Baker

Twining herbs. Occasional along fences near by settlements. Radhanagar: 7-11-2001, CSR 2040.

Vigna trilobata (L.) Verdc.

Twining herbs. Grows in waste places as weed. Diglipur: 30-11-2001, CSR 2348.

Vigna unguiculata (L.) Walp. ssp. **cylindrica** (L.) Eselt.

Twining herbs. Usually cultivated; also found scattered as an escape along forest fringes. Mayabandar: 6-12-2001, CSR 2489.

Zornia gibbosa Span

Prostrate herbs. Rarely found near by paddy fields. Diglipur: 30-11-2001, CSR 2349.

Canavalia gladiata (Jacq.) DC., **Cyamopsis tetragonoloba** (L.) Taub., **Lablab purpureus** (L.) Sweet (beans) **Sesbania grandiflora** (L.) Poir. are cultivated for vegetables.

CAESALPINIACEAE

Caesalpinia bonduc (L.) Roxb.

Scandent armed shrubs. Common in the littoral forests. Chatham Island: 15-12-2001, CSR 2582.

Caesalpinia crista L.

Climbers. Common in littoral and mangrove forests. Smith Island: 25-11-1976, Nair 4914. Shearme Island: 13-11-2001, CSR 2093.

Caesalpinia cucullata Roxb.

Climbers. Common in inland forests. Interview Island: 22-2-1991, PLN 15668.

Caesalpinia enneaphylla Roxb.

Climbers. Rare in open forests. Mohanpur: 7-12-2001, CSR 2492.

Caesalpinia hymenocarpa (Prain) Hattink

Climbers. Frequent in moist deciduous forests. Lakshmipur: 24-11-1976, Nair 4896. Radhanagar: 10-11-2001, CSR 2070.

Cassia absus L.

Herbs. Rare, in waste places. Shyamnagar: 25-11-2001, CSR 2714.

New record.

Cassia alata L.

Shrubs, up to 4 m high. Scattered along way sides and forest edges. Diglipur: 30-11-2001, CSR 2350.

Cassia fistula L.

Deciduous trees. Occurs in moist deciduous forests; also planted. Kishorinagar: 5-12-2001, CSR 2467.

Cassia obtusifolia L.

Annual herbs or under shrubs. Scattered as a weed in waste places, similar to *C. tora*. Radhanagar: 10-11-2001, CSR 2071.

New record.

Cassia occidentalis L.

Under shrubs. Scattered as a weed in waste places. Radhanagar: 25-12-2001, CSR 2715.

Cassia siamea Lam.

Trees. Planted along avenues. Diglipur-4: 5-12-1976, Balak and Nair 4967.

Cassia sophera L.

Shrubs. Rare in waste lands. Kishorinagar: 5-12-2001, CSR 2468.

Cassia tora L.

Annual herbs or under shrubs. Common along roadsides. Diglipur-4: Balak and Nair 4967.

Cynometra iripa Kostel (*C. ramiflora* L.) (Singara, Singada)

Small trees. Common in coastal forests. Smith Island: 25-11-1976, Nair 4912; Interview Island: 8-3-1990, PLN 15182.

Peltophorum pterocarpum (DC.) Back.ex Heyne

Trees. Scattered in littoral forests; also planted along roadsides; runswild. Kishorinagar: 5-12-2001, CSR 2462.

Pterolobium macropterum Kurz

Armed climbers. Scarce in the inland and coastal forests in sandy and rocky loam. North Island: 17-12-2001, CSR 2620.

Bauhinia acuminata L., **Bauhinia purpurea** L., **Bauhinia variegata** L. **Caesalpinia decapetala** (Roth.) Alston, **Caesalpinia pulcherrina** (L.) Sw., **Cassia javanica** L., **Cassia surattensis** N. Burman., **Delonix regia** (Boj. ex Hook) Raf., **Peltophorum pterocarpum** (DC.) Back.ex Heyne, **Tamarindus indica** L.(Tamarind), **Saraca asoca** (Roxb.) de Wilde (Ashok) are cultivated along avenues and in temples and gardens.

MIMOSACEAE

Acacia pennata (L.) Willd.

Prickles, glabrescent climbing shrubs or trees. Laksmipur: 23-11-1976, Nair 4878. Saddle peak (300m): 25-11-2001, CSR 2276.

Acacia sinuata (Lour.) Merrill (*A. concinna* (Willd.) DC.)

Scandent shrubs. Scattered in the forests. Kishorinagar: 5-12-2001, CSR 2469.

Adenanthera microsperma Teijsm. et Binn. (*A. pavonina* L. var. *microsperma* (Teijsm. et Binn.) Niels. et Guinet (Ooji)

Trees. Common in moist deciduous forests. Lakshmipur: 23-11-1976, Nair 4876. Saddle peak (100 m): 24-11-2001, CSR 2250.

Albizia chinensis (Osb.) Merr. (*A. stipulata* (Roxb.) Boiv.)

Trees. Occasional along forest edges. Swarajgram: 25-12-2001, CSR 2718.

Albizia lebbeck (L.) Bentham (Koko)

Tree. Common in the moist deciduous forests. Smith Island: 14-12-2001, CSR 2574.

Albizia procera (Roxb.) Bentham (Karai)

Trees. Frequent in semievergreen forests. Swarajgram: 25-12-2001, CSR 2720.

Entada rheedei Spr. (*E. pusaetha* DC., *E. scandens* Benth.)

Lianas. Occasional in evergreen forests. Ray hills: 27-4-1964, Ellis and Ramam. 18963; Ramnagar: 29-3-1977, Balak 5333.

Mimosa invisa Mart. ex Colla

Scandent shrubs. Occasional in waste lands and way sides. Introduced. Mayabandar: 6-12-2001, CSR 2490.

Mimosa pudica L. (touch-me-not)

Prickly, suftrutescent herbs. Occasional in waste lands and way sides. Introduced. Kishorinagar: 5-12-2001, CSR 2470.

Acacia auriculiformis A. Cunn. ex Benth, **Acacia nilotica** (L.) Delile **subsp. indica** (Benth.) Brenan, **Leucaena leucocephala** (Lam.) de Wit, **Prosopis juliflora** (Sw.) DC., **Pithecellobium dulce** (Roxb.) Benth., **Samanea saman** (Jacq.) Merr., are cultivated.

ROSACEAE

Prunus javanica (T.&B.) Miq. (*P. martabanica* Kurz)

Trees. Rare in evergreen forests. Mohanpur: 11-12-2001, CSR 2548.

Rubus moluccanus L.

Scandent shrub. Rare in inland forests. Diglipur: 19-11-1976, Nair 4815.

Rosa indica L.(Rose) is planted as ornamental.

CRASSULACEAE

Kalanchoe pinnata (Lam.) Pers.

Semisucculent herbs up to 80 cm tall. Rare in stunted forests at higher elevations, among rocks. Saddle peak (700 m): 28-11-2001, CSR 2326.

RHIZOPHORACEAE

Bruguiera cylindrica (L.) Bl.

Trees to 15 m high. Occasional in mangrove forests. Smith Island: 14-12-2001, CSR 2577.

Bruguiera gymnorrhiza (L.) Lam.

Trees to 18 m high. Abundant in mangrove forests. Interview Island: 8-3-1990, PLN 15179. Shearme Island: 13-11-2001, CSR 2092.

Ceriops tagal (Perr.) Roxb.

Bushy shrub to tree. Occasional in mangrove forests. Brush Island: 16-12-2001, CSR 2593.

Rhizophora apiculata Bl.

Trees to 25 m high. Abundant in mangrove forests. Interview Island: 6-3-1990, PLN 15145.

Rhizophora mucronata Lam.

Trees to 20 m high. Abundant in mangrove forests. Shearme Island: 13-11-2001, CSR 2091.

Rhizophora stylosa Griff.

Trees to 5 m high. Rare in mangrove creeks. Interview Island: 6-3-1990, PLN 15144.

COMBRETACEAE

Calycopteris floribunda (Roxb.) Lam.

Lianas. Rare in inland forests. Swarajgram: 25-12-2001, CSR 2719.

Combretum albidum G.Don

Lianas. Occasional in inland forests. Interview Island: 6-3-1990, PLN 15133.

Combretum latifolium Bl.

Large, scandent glabrous straggling shrubs. Smith Island: 25-11-1976, Nair 4913. Keralapuram: 5-4-1977, Balak 5476. Interview Island: 5-3-1990, PLN 15123.

Combretum punctatum Bl. (*C. squamosum* Roxb. ex Don)

Lianas. Occasional in moist deciduous forests. Radhanagar: 25-12-2001. CSR 2720; North Island: 17-12-2001, CSR 2623.

Combretum roxburghii Spreng. (*C. decandrum* Roxb.)

Deciduous stragglers. Occasional in moist deciduous forests. Chatam Island: 15-12-2001, CSR 2589.

Lumnitzera littorea (Jack.) Voigt.

Small trees. Occasional in mangrove swamps and tidal forests. Chatam Island: 15-12-2001, CSR 2590.

Lumnitzera racemosa Willd.

Small trees. Occasional in mangrove swamps and tidal forests. Swamp Island: 9-12-2001, CSR 2525.

Terminalia bialata Steud. (White Chuglam, Safed Chuglam)

A tall trees, up to 30-40 m high. Common in moist deciduous forests. Buchanan Island: 12-12-2001, CSR 2558.

Terminalia catappa L. (Badam)

Trees to 15 m high. Scattered in littoral forests. Lamia bay: 22-12-2001, CSR 2641.

Terminalia manii King (Black Chuglam, Kala Chuglam)

Trees to 40 m high, with a straight clean bole. Common in moist deciduous forests. Lamia bay: 22-11-2001, CSR 2219.

Endemic.

Terminalia procera Roxb.

Trees to 40 m high. Branches in horizontal whorls. Common in moist deciduous forests. Lakshmipur: 6-4-1977, Balak 5494. Saddle peak (100 m): 24-11-2001, CSR 2249.

Endemic.

Quisqualis indica L. (Rangoon creeper) is planted in gardens.

MYRTACEAE

Acmena acuminatissima (Bl.) Merr. (*Eugenia acuminatissima* (Bl.) Kurz)

Trees. Occasional in inland forests. Swarajgram: 25-12-2001, CSR 2723.

Decaspermum parviflorum (Lam.) Scott.

Trees. Rare in hill top forests. Saddle peak (720 m): 15-8-1991, PLN 20771. Saddle peak (700 m): 28-11-2001, CSR 2324.

Rhodamnia dumetorum (DC.) Merr. et Perry

Trees. Rare in hill top forests. Saddle peak (720 m): 14-12-1990, Ellis 14192.

Syzygium claviflorum (Roxb.) Wall.ex Cowan et Cowan (*Eugenia claviflora* Roxb.)

Trees. Karmatonng-10: 5-12-1992, BK Sinha 16250; Saddle peak: 1-12-1976, Balak and Nair 4770.

Syzygium cumini (L) Skeels

Trees to 20 m high. Scattered along forest streams. Mohanpur: 11-12-2001, CSR 2547.

Syzygium kurzii (Duthie) Balakr. var. **andamanica** (King) Balakr.

Trees. Rare in inland forests. Buchanan Island: 12-12-2001, CSR 2560.

Endemic.

Syzygium manii (King) Balakr.

Trees. Rare in inland forets. Point Island: 16-11-2001, CSR 2140.

Endemic.

Syzygium samarangense (Bl.) Merr. et Perry

Trees. Lamia Bay: 12-5-1982, MKV 9014; Interview Island: 5-3-1990, PLN 15101.

Callistemon lanceolatus DC., **Eucalyptus** spp., **Psidium guajava** L. (Guava, Jama) are cultivated for their fruits.

BARRINGTONIACEAE

Barringtonia asiatica (L.) Kurz

Trees. Occasional in littoral forests. Lamia Bay: 29-3-1977, Balak 5323. Interview Island: 24-2-1991, PLN 15689.

Barringtonia racemosa (L.) Kurz

Trees. Occasional in littoral forests. Stewart Island: 19-12-2001, CSR 2631.

Planchonia valida (Bl.) Bl. (*P. andamanica* King) (Lal Bombay)

Trees. Occasional in inland forests. Interview Island: 2-3-1991, PLN 20735.

MELASTOMATACEAE

Melastoma malabathricum L.

Evergreen shrubs to 2 m tall. Radhanagar: 25-12-2001, CSR 2722.

Obseckia chinensis L. **var. minor**

Herbs. Occasional in high elevation of Saddle peak. Saddle peak (700 m): 28-11-2001, CSR 2323.

Sonerila andamanensis Stapf et King

Herbs. Rare in hill forests. Saddle peak: 2-12-1976, Balak and Nair 4791. 28-11-2001, CSR 2321.

Endemic.

MEMECYLACEAE

Memecylon andamanicum King

Shrubs. Rare in inland forests. Karmatong-10: 8-12-1992, BK Sinha 16300.

Endemic.

Memecylon collinum Cogn.

Shrubs. Saddle peak (650 m): 1-12-1976, Balak and Nair 4749; 20-9-1982, RKP 9213. 28-11-2001, CSR 2322.

Endemic.

Memecylon edule Roxb.

Shrubs or small trees to 4 m tall. Rare in inland forests. Saddle peak (300m): 27-11-2001, CSR 2318.

Memecylon intermedium Bl.

Shrubs. Rare in inland forests. Interview Island: 17-11-1990, PLN 15639.

Memecylon scutellatum (Lour.) Naud.

Small trees. Occurs in hill forests. Saddle peak (600 m): 1-12-1976, Balak and Nair 4763; 21-9-1982, RKP 9244. 27-11-2001, CSR 2316.

LYTHRACEAE

Ammania baccifera L.

Herbs. Grown in marshy places and dried up paddy fields. Radhanagar: 25-12-2001, CSR 2721.

Lagerstroemia hypoleuca (Pema, Jarul)

Large trees. Common in all moist deciduous forests. Chatham Island: 15-12-2001, CSR 2580.

Endemic.

Pemphis acidula J.R. et G. Forst.

Shrubs. Rare in east coast forests. Point Island: 16-11-2001, CSR 2139.

Rotala illecebroides Koehne

Herbs. Grown in marshy places and dried up paddy fields. Radhanagar: 25-12-2001, CSR 2722.

Lagerstroemia indica L., **Lagerstroemia reginae** Roxb. are cultivated in gardens. **Lawsonia inermis** L. (Henna) is cultivated for henna, used to stain the finger nails, palm etc.

PUNICACEAE

Punica granatum L.

A glabrous, deciduous large shrub. Planted in house-compounds for its fruits.

SONNERATIACEAE

Duabanga grandiflora (Roxb.ex DC.) Walp. (*D. sonneratioides* Buch.-Ham.)

Trees. Scattered in evergreen forests. Mohanpur to Austin no.2: 13-12-2001, CSR 2570.

Sonneratia caseolaris (L.) Engl. (*S. acida* L.f.)

Trees. Occasional in mangroves. Stewart Island: 19-12-2001, CSR 2633.

ONAGRACEAE

Ludwigia hyssopifolia (G. Don) Exell

Herbs of marshy places. Keralapuram: 5-12-1976, Balak and Nair 4975.

Ludwigia octovalvis (Jacov.) P.H. Raven

Perennial herbs. Occassional in moist places. Mohanpur: 13-12-2001, CSR 2569.

Ludwigia perennis L.

Herb to 60 cm. Common in moist places. Mohanpur: 13-12-2001, CSR 2568.

Ludwigia prostrata Roxb.

Herb to 60 cm. Common in moist places. Aerial Bay: 23-12-2001, CSR 2678.

BEGONIACEAE

Begonia andamanensis Parish

Herbs. Rare on moist rocky crevices. Saddle peak (500 m): 20-9-1982, RKP 4859. 27-11-2001, CSR 2315.

Endemic.

DATISCACEAE

Tetrameles nudiflora R.Br. ex Benn. (Tippok. Pippok)

Trees with buttresses. Occasional in forests. Mohanpur: 13-12-2001, CSR 2567.

TURNERACEAE

Turnera ulmifolia L.

Erect herbs. Occasional weed. Diglipur: 30-11-2001, CSR2351.

PASSIFLORACEAE

Passiflora foetida L.

Climbing herbs. Common weed, foetid when bruised. Mohanpur: 13-12-2001, CSR 2566.

Modecca cardiophylla Mast.

Climbers. Rare in forests. Lamia Bay: 29-3-1977, Balak 5331.

Passiflora edulis Sims is planted in house compounds and gardens Fruits edible (Passion fruit).

CARICACEAE

Carica papaya L. (Papaya) is cultivated in home gardens for its edible fruits.

CUCURBITACEAE

Trichosanthes bracteata (Lam.) Voigt (*T. palmata* Roxb.)

Climbers. Rare in forests. Aerial Bay: 21-11-1976, Balak 4859. Mohanpur: 13-12-2001, CSR 2565.

The following species are cultivated for vegetable.

Coccinia grandis (L.) J. Voigt

Momordica charantia L.

Momordica dioica Roxb. ex Willd.

Mukia maderaspatana (L.) M. Roemer,

Benincasa hispida (Thunb.) Cogn.

Citrullus lanatus (Thunb.) Matsum. et Nakai

Cucumis melo L. (Karbuja)

Cucumis sativus L.

Cucurbita maxima Duchesne ex Lam.

Lagenaria siceraria (Molina) Standley

Luffa acutangula (L.) Roxb.

Trichosanthes anguina L.

APIACEAE

Centella asiatica (L.) Urban

Creeping herbs, rooting at nodes. Common in moist localities in forests. Hathilevel: 12-11-2001, CSR 2081.

Eryngium foetidum L.

Aromatic herb of 30 cm high. Weed in moist, shady localities. Diglipur: 30-11-2001, CSR 2354.

Coriandrum sativum L. cultivated for economic value.

ARALIACEAE

Schefflera venulosa (W.& A.) Harms.

Trees. Rare in evergreen forests. Austin No.4: 8-12-2001, CSR 2496.

ALANGIACEAE

Alangium salvifolium (L.f.) Wang. ssp. **hexapetalum** (Lam.) Wang.

A deciduous, small tree with spinescent branches and grey bark. Rare in deciduous forests.

GAMOPETALAE

RUBIACEAE

Anthocephalus chinensis (Lam.) Rich. ex Walp. (*A. cadamba* (Roxb.) Miq.)

Trees. Scattered in moist deciduous forests. Mohanpur: 7-12-2001, 2492.

Argostemma tavoyanum Wall. ex Benn.

Herbs. Rare, near streams. Lamia Bay to Saddle peak: 2-12-1976, Balak and Nair 4799.

Canthium dicoccum Merr.

Shrubs. Rare in moist deciduous forests. Mohanpur: 7-12-2001, 2494.

Canthium glabrum Bl.

Shrubs. Occasional in moist deciduous forests. Lakshmipur: 24-11-1976, Nair 4900. Hathilevel: 12-11-2001, CSR 2082.

Canthium gracilipes Kurz

Armed shrubs. Occasional in moist deciduous forests. Sitanagar: 15-5-1982, MKV 9027.

Endemic.

Ceriscoides turgida (Roxb.) Tirveng.

Small, deciduous tree. Occasional in moist deciduous forests. Mohanpur: 7-12-2001, 2499.

Chassalia curviflora (Wall.) Thw.var. **ophioxyloides** (Wall.) Deb et Krishna

Trees to 4 m high. Interview Island: 17-11-1990, PLN 15643. Diglipur: 30-11-2001, CSR 2355.

Chomelia asiatica (L.) O. Kuntze. (*Tarenna asiatica* (L.) Sant. et Merch.)

A shrub. Rare, in forests. Shyamnagar: 11-11-2001, CSR 2078.

Coffea leberica Bull.ex Hiern. emend Cheval

Shrubs. Inland forests. Interview Island: 13-2-1959, Thoth. 9212.

Dentella repens (L.) J.R. et G. Forst.

Herbs. Grows in moist, shady places. Radhanagar: 8-11-2001, CSR 2041.

Diplospora andamanica Balak. et Nair

Shrubs.Balakrishnan and Nair reported it from Saddle peak.

Endemic.

Discospermum abnormae

Shrubs. Rare in forests. Lamia Bay: 12-5-1982, MKV 8994.

Gardenia coronaria Ham.

Trees. Occasional in open moist deciduous forests. Saddle peak (200 m): 3-5-1982. MKV 8928. Shyamnagar: 11-11-2001, CSR 2079.

Geophila repens (L.) Johnston (*G. reniformis* D.Don)

Herbs. Rare in moist, shady places. Swarajgram: 25-12-2001, CSR 2728.

Guettarda andamanica Goel et Mehrotra

Trees. Rare in Littoral forests. Reported from North Andamans.

Endemic.

Guettarda speciosa L.

Trees. Occasional in littoral forests. Snake Island: 13-12-2001, CSR 2562.

Hedyotis diffusa Willd.

Diffuse herb to 30 cm. Occasional in among grasses. Narcondam Island: 22-9-1977, Nair 6220.

Hedyotis puberula (G.Don.) Arn.

Erect herb to 20 cm tall. Rare in waste lands. Shyamnagar: 11-11-2001, CSR 2077.

New record.

Hydnophytum formicarum Jack.

Herbs. Rare in forests. Saddle peak: 31-3-1977, Balak 5381.

Ixora arborea Roxb. ex Sm.

Trees to 8 m high. Rare in forests along the streams. Austin no.2: 13-12-2001, CSR 2564.

Ixora barbata Roxb.

Shrubs. Rare in forests. Karmatong-10, 5-12-1992, BK Sinha 16252.

Endemic.

Ixora brunnescens Kurz

Trees to 4.5 m high. Rare in coastal forests. Interview Island: 24-2-1991, PLN 15688.

Ixora cuneifolia Roxb.

Shrubs. Rare in forests. Lamia bay 28-3-1977, Balak 5195.

Endemic.

Ixora finlaysoniana Wallich ex Don

Shrubs. Rare in forests. Diglipur: 30-11-2001, CSR 2356.

Ixora grandifolia Z.& M.

Shrubs. Occasional in inland forests. Saddle peak (250m): 27-11-2001, CSR 2314.

Ixora nigricans R.Br.ex Wight et Arn.

Trees to 4 m high. Rare in coastal forests. Interview Island: 25-4-1964, Ellis and Ramam. 18948.

Ixora rosella Kurz

Shrubs to 3 m high. Rare in forests. Radhanagar: 8-11-2001, CSR 2043.

Ixora pubirama Bremek.

Shrubs. Rare in forests. Kalipur: 1-4-1977, Balak 5438.

Ixora thwaitesii Hook.f.

Shrubs. Rare in forests. Saddle peak (720 m): 16-10-1987, Ellis 12745.

Knoxia sumatrensis (Retz) DC.

Undershrubs. Inland forests. Mohanpur: 7-12-2001, 2498.

Mitragyna rotundifolia (Roxb.) O.K. (Karmi)

Trees. Common in moist deciduous forests. Aerial bay: 20-11-1976, Nair 4829. Radhanagar: 8-11-2001, CSR 2044.

Morinda citrifolia L. var. **bracteata** (Roxb.) Hook.f.

Trees. Occasional in beach forests. Lamia Bay: 12-5-1982, Nair 4829; Interview Island: 7-3-1990, PLN 15171.

Nauclea gageana King

Trees. Rare, in evergreen forests. Recorded from Saddle peak.

Nauclea purpurascens Roxb.

Small trees. Occasional in inland forests. Diglipur: 30-11-2001, CSR 2354.

Ophiorrhiza trichocarpa Bl.

Shrubs. Occasional in inland forests. Keralapuram: 21-11-1976, Nair 4858.

Oxyceros longiflora (Lam.) Yama. (*Randia longiflora* Lam.)

Straggler. Occasional in inland forests. Diglipur: 30-11-2001, CSR 2357.

Paederia scandens (Lour.) Merr. (*P. foetida* auct.non L.)

Twining shrub, foetid when bruised. Occasional, in open moist deciduous forests. Interview Island: 16-2-1959, Thoth. 9223; Radhanagar: 8-11-2001, CSR 2045. Diglipur: 30-11-2001, CSR 2358.

Psychotria balakrishnanii Deb et Gang.

Shrubs. Rare in forests. Saddle peak: 10-5-1982, MKV 8948.

Endemic.

Psychotria kurzii Deb et Gang.

Shrubs. Rare in inland forests. Saddle peak (200 m): 28-3-1977, Balak 5407.

Endemic.

Psychotria platyneura Kurz.

Small trees. Saddle peak (150 m): 20-9-1982, RKP 9217; 14-5-1991, PLN 20751. Saddle peak (150m): 27-11-2001, CSR 2313.

Psychotria sarmentosa Bl.

Small tree. Saddle peak (400 m): 1-12-1976, Balak and Nair 4743.

Pubistylus andamanensis Thoth.

Trees, 3-4 m tall. Rare in inland forests, near to the east coast of Interview Island. 12-2-1959, K. Thothathri 9204 A (Holotype-CAL), and 9204 (Isotypes 4 sheets- CAL); 5-3-1990, PLN 15128; 22-2-1991, PLN 15665.

Endemic genus reported from evergreen forests of Interview Island.

Rothmania pulcherrima (Kurz) Tirv.

Small trees. Saddle peak: 9-5-1982, MKV 8926.

Saprosma ternatum Hook.f.

Shrubs. Diglipur-4: 13-11-1976, Nair 4814; Laxmipur: 22-11-1976, Nair 4866.

Spermacoce articularis L.f. (*Borreria articularis* (L.f.) Will.

Herbs. Frequent in moist places. Aerial Bay: 5-12-1976, Balak and Nair 4988. Diglipur: 30-11-2001, CSR 2360.

Spermacoce hispida L.

Hispid, diffuse herb to 60 cm. Common weed. Radhanagar: 8-11-2001, CSR 2046.

Spermacoce pusilla Wallich

Erect herb. Weed of cultivation. Diglipur: 30-11-2001, CSR 2359.

Tarenna weberaefolia (Kurz) Balakr. (*Webera kurzi* Hook.f.)

Trees. Ray hills: 27-4-1964, Ellis and Ramamurthy 18962; Saddle peak: 10-5-1982, MKV 8942.

Timonius jambosella Thw.

Shrubs. Kalpong-10: 23-12-1986, Chakrab.& Kindo 12357; Saddle peak: 1-12-1976, Balak and Nair 4738.

Wendlandia andamanica Cowan

Small trees. Rare in forests. Saddle peak: 27-3-1977, Balak 5173.

Endemic.

Ixora coccinea L. **Hamelia patens** Jacq., **Mussaenda hirsutissima** (Hook.f.) Hutch. ex Gamble, **Mussaenda philippinum**, **Pentas lanceolata** (Forsk.) K. Schum. **Pseudomussaenda flava** Verdc. are commonly planted in gardens.

ASTERACEAE

Adenostemma lavenia (L.) O.K.

Herbs. Occurs along streams of forests. Hathilevel: 26-11-2001, CSR 7230.

Ageratum conyzoides L.

Annual, erect, foetid herbs. Common weed in disturbed places. Sitanagar: 15-5-1982, MKV 9053. Radhanagar: 8-11-2001, CSR 2047.

Blainvillea acmella (L.) Philipson

Annual erect herbs. Weed of disturbed localities. Hathilevel: 26-11-2001, CSR 7231.

Blumea balsamifera (L.) DC.

Annual, erect herbs, to 1 m tall. Occurs in shady places of forests. Mohanpur: 26-12-2001, CSR 2736.

Blumea eriantha DC.

Annual erect, foetid herbs, to 1 m tall. Weed of moist places, near by open forests. Shyamnagar: 11-11-2001, CSR 2075.

Blumea lacera (Burm.f.) DC.

Erect, annual, aromatic herbs. Diglipur: 25-3-1973, Balak 71. Kafeedera: 18-11-2001, CSR 2151.

Blumea lanceolaria (Roxb.) Druce

Herbs. Sitangar: 15-5-1982, MKV 9047. Interview Island: 8-3-1990, PLN 15184.

Blumea mollis (D. Don.) Merr.

Erect, annual, aromatic herbs, with sticky and glandular hairs. Hathilevel: 26-11-2001, CSR 7235.

Blumea obliqua (L.) Druce

An erect or decumbent, woody, deep-rooted, villous, annual herb. Diglipur: 30-11-2001, CSR 2361.

Blumea virens Wall. ex DC.

Annual erect herbs to 1.5 m tall. Ganesh nagar: 17-11-2001, CSR 2148.

Caesulia axillaris Roxb.

Erect or decumbent herbs. Rare in dried up paddy fields. Mohanpur: 26-12-2001, CSR 2737.

New Record.

Chromolaena odorata (L.) King et Robinson

Erect herbs to 2 m tall. Aerial bay: 5-12-1976, Balak and Nair 4991. Radhanagar: 8-11-2001, CSR 2048.

New Record.

Crassocephalum crepidioides (Benth.) Moore (*Gynura crepidioides* Benth.)

Erect herbs. Stems fleshy. Common in forest clearings of Smith island. 18-11-2001, CSR 2168.

Eclipta prostrata L.

Erect or decumbent herbs, about 30 cm high. Lamia bay: 29-3-1977, Balak 5345. Ganeshnagar: 17-11-2001, CSR 2149.

Elephantopus scaber L.

A small, erect, rhizomatous, scapigerous herbs, to 40 cm tall. Occasional forest under growth. Shyamnagar: 11-11-2001, CSR 2074.

Emilia sonchifolia (L.) DC.

Annual erect herbs to 40 cm tall. Common weed. Ganeshnagar: 17-11-2001, CSR 2148.

Parthenium hysterophorus L.

Annual, erect herbs to 1.5 m tall. Aggressive weed. Diglipur: 30-11-2001, CSR 2362.

Spilanthes paniculata Wallich ex DC.

Herbs up to 50 cm high. Rare, in moist places. Diglipur: 30-11-2001, CSR 2363.

Spilanthes radicans Jacq.

Erect herbs. Rare weed of moist fields. Ganeshnagar: 17-11-2001, CSR 2150.

Synedrella nodiflora (L.) Gaertn.

Herbs. Common in waste places. Arial Bay: 5-12-1976, Balak and Nair 4998.

Tithonia rotundifolia (Mill.) Blake

Herbs, spreading on bushes in coastal forests. Mohanpur to Austin no.1: 26-12-2001, CSR 2738.

Tridax procumbens L.

Annual decumbent herbs to 30 cm high. Common weed. Diglipur: 30-11-2001, CSR 2364.

Vernonia andamanica Balak et Nair

Herbs. Rare and reported from Saddle peak. Saddle peak (500m): 27-11-2001, CSR 2312.

Endemic.

Vernonia cinerea (L.) Less.

Annual erect herbs to 75 cm tall. Interview Island: 8-3-1990, PLN 15185. Point Island: 15-11-2001. CSR 2128.

Wedelia biflora (L.) DC.

Spreading herbs. Occurs along roadsides. Mohanpur: 26-12-2001, CSR 2739.

Cosmos sulphureus Cav., **Chrysanthemum cinerariaefolium**, **C. indicum**, **Tagetes patula** L., **Dahlia rosea**, **Helianthus annuus** L., **Kleinia grandiflora** (Wall.ex. DC.) N. Rani, **Solidago canadensis** L., **Vernonia divergens** (Roxb.) Edgew. **V. elaeagnifolia** DC. and **Wedelia chinensis** (Osbeck.) Merr are cultivated as ornamentals.

GOODENIACEAE

Scaveola sericea Vahl (*S. taccada* Roxb.).

Trees. Scattered in littoral forests. Chatham Island: 115-12-2001. CSR 2588.

CAMPANULACEAE

Lobelia alsinoides Lam.

Prostrate herb to 15 cm, rooting at nodes. Occurs in perennially moist, calyey soils. Mohanpur: 26-12-2001, CSR 2740.

Lobelia heyneana R.& S.

Prostrate herb. Occurs in perennially moist, calyey soils. Diglipur: 30-11-2001, CSR 2363.

SPHENOCLEACEAE

Sphenoclea zeylanica Gaertner

Annual, erect, succulent herb to 75 cm. Rare, in marshy ground and paddy fields. Diglipur: 30-11-2001, CSR 2364.

PLUMBAGINACEAE

Plumbago auriculata Lam. is an handsome, undershrub with bluish-voilet flowers; planted in gardens.

MYRSINACEAE

Aegiceros corniculatum (L.) Blanco

Trees. Common in mangrove forests. Entance Island: 11-12-2001, CSR 2548.

Ardisia attenuata Wall.

Shrubs, to 1.5 m high. Inland forests. Interview Island: 7-3-1990, PLN 15160.

Ardisia involucrata Kurz.

Shrubs. Rare in forests. Interview Island: 26-2-1991, PLN 20704.

Ardisia solanacea Roxb.

Shrubs. Rare in forests along streams. Mohanpur: 26-12-2001, CSR 2741.

Ardisia oxyphylla Wall. ex DC.

Shrubs. Rare in forests. Saddle peak: 9-5-1982, MKV 8927; 21-9-1982, RKP 9238; Interview Island: 17-11-1990, PLN 15642.

Maesa andamanica Kurz.

Small trees. Austin-9: 3-12-1992, BK Sinha 16218. Interview Island: 25-4-1964, Ellis and Ramam. 18956.

Endemic.

Maesa ramentacea Wall.

Small trees. Interview Island: 12-2-1959, Thoth. 9203. Saddle peak (150m): 27-11-2001, CSR 2311.

PRIMULACEAE

Anagallis arvensis L.

Ascending herb to 30 cm; Rare. Found in marshy soils. Mohanpur: 26-12-2001, CSR 2742.

SAPOTACEAE

Diploknema butyracea (Roxb.) Lam. **var. andamanensis** Royen (*Madhuca butyracea*)

Trees. Occasional in inland forests. Laxmipur: 6-4-1977, Balak 5489. Shyamnagar: 11-11-2001, CSR 2073. Point Island: 15-11-2001. CSR 2129.

Endemic.

Manilkara littoralis Kurz (Kari-Mohwa)

Trees. Common in littoral forests. Lamia bay: 29-3-1977, Balak 5329.

Planchonella longipetiolatum (King et Prain) Lam. (*Sideroxylon longipetiolatum* King et Prain) (Lambapatti)

Trees. Occasional in inland forests. Point Island: 16-11-2001, CSR 2137.

Planchonella obovata (R.Br.) Pierre (*Sideroxylon ferrugineum* Hk. et Arn.)

Trees. Scattered in littoral forests. Saddle peak: 27-3-1977, Balak 5158.

Mimusops elengi L. (Pogada), **Manilkara zapota** (L.) P. Royen (Sapota) are cultivated for fruits.

EBENACEAE

Diospyros kurzii Hiern. (Lohda, Marbel)

Trees. Common in moist deciduous forests. Interview Island: 23-2-1991, PLN 15679. Saddle peak (250m): 27-11-2001, CSR 2310.

Diospyros marmorata Parker (*D. oocarpa* auct.non Thw.) (Kala-lakdi)

Trees. Common in moist deciduous forests. Ganeshnagar: 17-11-2001, CSR 2145.

Endemic.

Diospyros montana Roxb.

Trees. Scattered in moist deciduous forests. Shyamnagar: 11-11-2001, CSR 2072.

Diospyros pilosiuscula (*D. pilosula* Wall. ex Hiern.) (Chotapatti Kendu)

Trees. Common in moist deciduous forests. Laxmipur: 23-11-1976, Nair 4879.

Diospyros pyrrhocarpa Miq. **var. andamanica** Kurz (Gab, Bada Kendu)

Trees. Common in semievergreen and moist deciduous forests. Paget Island: 6-11-2001, CSR 2008.

Endemic.

Diospyros ridleyi Bakh.

Trees. Rare in moist deciduous forests. Sitanagar: 15-5-1982, MKV 9041; Laxmipur: 24-11-1976, Nair 4899. Interview Island: 17-11-1990, PLN 15157.

Diospyros undulata Wall. ex G.Don

Trees. Occasional in moist deciduous forests. Saddle peak: 27-3-1977, Balak 5145; 9-5-1982, MKV 8936. Interview Island: 5-3-1990, Ellis and Ramam. 15121.

SYMPLOCACEAE

Symplocos racemosa Roxb.

Climbers. Scattered in evergreen forests. Mohanpur: 26-12-2001, CSR 2743.

OLEACEAE

Chionanthus terniflora Wall.

Small trees. Scattered in moist deciduous forests. Ramnagar: 29-3-1977, Balak 5336.

Chionanthus parkinsonii (Hutch.) Benn. et Raiz.

Small trees. Rare in inland forests. Shyamnagar: 11-11-2001, CSR 2071.

Endemic.

Jasminum angustifolium (L.) Willd.

Climbers. Occasional in moist deciduous forests. Saddle peak: 7-4-1989, Ellis and Sanjappa 14119.

Jasminum arborescens Roxb.

Climbers. Rare in forests. Saddle peak: 31-3-1977, Balak 5392.

Jasminum cordifolium Wall. ex Don ssp. **andamanicum** Srivastava et Kapoor

Climbers. Rare in forests. Saddle peak (400m): 25-11-2001, CSR 2270.

Endemic.

Jasminum flexile Vahl var. **ovata**

Climbers. Rare in forests. Laxmipur: 22-11-1976, Nair 4862.

Jasminum subglandulosum Kurz

Climbers. Rare in forests. Hathilevel: 12-11-2001, CSR 2083.

Jasminum unifoliolatum Balak. et Nair

Shrubs. Rare and threatened. Kalpong-10: Chakrab. and Kindo12350. Saddle peak (400m): 25-11-2001, CSR 2262.

Endemic.

Ligustrum perottetii DC.

Small trees. Rare in forests. Saddle peak: 14-12-1990, Ellis 14199. Saddle peak (400m): 25-11-2001, CSR 2269.

Jasminum grandiflorum L., **Jasminum multiflorum** (Burm.f.) Andrews, **Jasminum officinale** L. and **Jasminum sambac** (L.) Ait. are planted in gardens and house compounds.

NYCTANTHACEAE

Nyctanthes arbor-tristis L. (Parijat) is a large shrub or small tree, planted in gardens.

APOCYNACEAE

Aganosma marginata (Roxb.) G. Don

Woody climbers. Scattered in inland forests. Interview Isalnd: 16-11-1991, PLN 15634.

Alstonia kurzii Hook.f. (Haldi ped)

Trees. Occasional in inland forests. Point Island: 16-11-2001, CSR 2135.

Endemic.

Alyxia reinwardtii Bl. **var. lucida** Markgr.

Trees. Rare in inland forests. Saddle peak: 27-3-1977, Balak 5184; 31-3-1977, Balak 5385.

Anodendron paniculatum (Roxb.) A.DC.

Liana to 10 m. Occasional in moist deciduous forests. Lamia Bay: 12-5-1982, MKV 8998. Hathilevel: 12-11-2001, CSR 2084.

Carissa spinarum L.

A bushy, diffuse, evergreen shrub. Rare in deciduous forests. Ganeshnagar: 17-11-2001, CSR 2144.

Catharanthus roseus (L.) Don.

Prostrate or erect perennial herb. Grown in house compounds; runs wild. Hathilevel: 12-11-2001, CSR 2085.

Hunteria zeylanica (Retz) Gardens

Trees. Rare in inland forests. Saddle peak (200 m): 9-5-1982, MKV 8931. 27-11-2001, CSR 2309.

Ichnocarpus volubilis (Lour.) Merr.

Found as a climber in forests. Mohanpur: 26-12-2001, CSR 2745.

Ochrosia oppositifolia (Lam.) Schum. (*Neisosperma oppositifolium* (Lam.) Fosb. et Sach.

Trees. Rare in inland forests. Karmatong-10: 7-12-1992, Sinha 16274. Interview Island: 24-2-1991, PLN 15687.

Parameria levigata (Juss.) Mold.

Climbers. Rare in forests. Lamia bay: 1-4-1977, Balak 5434.

Parsonsia inodora (Lour.) Almeida et Almeida

Climbers. Inland forests. Interview Island: 5-3-1990, PLN 15107; 2-3-1991, PLN 20731.

Rauvolfia serpentina (L.) Benth. ex. Kurz

Shrub. Rare in moist deciduous forests. Hathilevel: 12-11-2001, CSR 2087.

Rauvolifia sumatrana Jacq.

Small trees. Inland forests. Interview Island: 7-3-1990, PLN 15156; 25-2-1991, PLN 15696.

Tabernaemontana crispa Roxb.

Trees. Occasional in moist deciduous forests. Lamia bay: 12-5-1982, MKV 9010. Ganeshnagar: 17-11-2001, CSR 2143.

Wrightia arborea (Dennst.) Mabb. (*W. tomentosa* R.& S.)

Small, deciduous tree. Common in semievergreen and deciduous forests. Paget Island: 6-11-2001, CSR 2009. Point Island: 15-11-2001. CSR 2130.

Allamanda cathartica L., **Alstonia. scholaris** (L.) R.Br., **Cascabela thevetia** (L.) Lippold. **Ervatamia divaricata** (L.) Burkill, **Nerium oleander** L., **Plumeria rubra** L., **Plumeria alba** L., **Rauvolfia tetraphylla** L. and **Vallaris solanacea** (Roth) Kuntze are planted in gardens.

ASCLEPIADACEAE

Asclepias curassavica L.

Erect herbs to 1 m tall. Occurs along marsh lands. Hathilevel: 12-11-2001, CSR 2088.

Calotropis gigantea (L.) R.Br.

Large bushy shrubs to 3 m tall, covered with white soft wool. Occurs in roadsides, waste lands and cultivated fields. Vidyasagarpalli: 23-12-2001, CSR 2675.

Cryptolepis buchanani Roem. et Schult.

Large, climbing shrub with plenty of milky latex. Occasional in forests. Hathilevel: 12-11-2001, CSR 2090.

Dischida nummularia R.Br.

Large, rooted creepr. Common in littoral forests. Saddle peak: 28-3-1977, Balak 5415. Point Island: 16-11-2001, CSR 2131.

Dischida major (Vahl.) Merr.

Climbers. Saddle peak: 27-3-1977, Balak 5156; Saddle peak (500m): 25-11-2001, CSR 2277. Interview Island: 8-3-1991, PLN 15181.

Gymnema latifolium Wall.

Large, climbing shrub. Sibpur: 28-11-1976, Balak and Nair 4728. Ganeshnagar: 17-11-2001, CSR 2142.

Hoya diversifolia Bl.

Climbers. Rare in forests of Narcondum Island. It was included under the authority of Parkinson.

Hoya parasitica Wall.

Climbers. Common in littoral forests. Lamia bay: 12-5-1982, MKV 9012.

Pergularia daemia (Forssk.) Chiov.

Straggler with foetid smell. Rare in thickets. Entrance Island: 11-12-2001, CSR 2547.

Sarcolobus carinatus Wall.

Climbers. Common in littoral forests. Point Island: 16-11-2001, CSR 2131.

Tylophora indica (Burm. f.) Merr. (*T. asthmatica* (L.f.) Wt.)

Climber. Common amongst bushes, in hedges, on open patches in the forests. Mohanpur: 26-12-2001, CSR 2749.

Wattakaka volubilis (L.f.) Stapf

Extensive, perennial woody, stout, large, hoary pubescent twiner, with watery sap. Common in hedges and open forests. Diglipur: 26-11-1976, Nair 4920; Lamia bay: 29-3-1977, Balak 5330; Ray hills: 27-4-1964, Ellis and Ramam. 18964. Hathilevel: 12-11-2001, CSR 2089.

LOGANIACEAE

Strychnos minor Dennst. **(***S. acuminata* Wall.ex DC.**)**

Climbers. Common in inland forests. Mayabander: 8-12-1992, Sinha 16289. Point Island: 16-11-2001, CSR 2133. Saddle peak (200m): 25-11-2001, CSR 2267.

Strychnos narcondamensis Hill

Climbers. Very rare in inland forests. Narcondum Island: March 1891, coll. Prain. Known from type collection only.

Strychnos wallichiana Steud. ex DC.

Climbers. Frequent in inland forests. Sitanagar: 15-5-1982, MKV 9034. Ganeshnagar: 17-11-2001, CSR 2141. Saddle peak (250m): 27-11-2001, CSR 2308.

GENTIANACEAE

Canscora diffusa (Vahl) R.Br.ex Wall.

Herbs. Occasional in moist places. Jarlait Bay: 8-2-1969, Thoth 9196.

HYDROPHYLLACEAE

Hydrolea zeylanica (L.) Vahl

Spreading herb, rooting at lower nodes. Common in low lying moist places. Diglipur: 5-12-1976, Balak and Nair 4958.

BORAGINACEAE

Argusia argentea (L.f.) Heine **(***Tournefortia argentea* L.f.**)**

Small trees. Rare in coastal forests. Point Island: 16-11-2001, CSR 2134.

Coldenia procumbens L.

Prostrate herbs, densely hirsute. Most common weed in dried ponds, wet lands and along the bunds of fields. Mohanpur: 26-12-2001, CSR 2748.

Heliotropium indicum L.

Herb with woody base. Common near drainage canals. Keralapuram: 5-12-1976, Balak and Nair 4983.

Heliotropium keralense Siv. et Manilal

Herbs. Occasional near marshes. Kalipur: 30-3-1977, Balak 5572.

Rotula aquatica Lour.

Climbers. Grown near by srteams of forests, in rocky areas. Mohanpur: 26-12-2001, CSR 2749.

Tournefortia ovata Wall. ex Don

Shrubs. Rare in inland forets. Lamia bay to Kalipur: 30-3-1977, Balak 5355; Laxmipur: 6-4-1977, Balak 5484.

CORDIACEAE

Carmona retusa (Vahl) Masam.

Shrub to 3 m. Occasional in footh hills and stunted hill top evergreen jungles. Kafeedera: 20-11-2001, CSR 2171.

Cordia dichotoma Forst. (Irika)

A deciduous tree upto 10 m tall. Rare in inland forests. Mohanpur: 26-12-2001, CSR 2740.

Cordia subcordata Lam. (Balla)

Trees. Scattered in littoral forests. Kafeedera: 20-11-2001, CSR 2172.

CONVOLVULACEAE

Argyreia capitata (Vahl) Choisy

Climbers. Occasional in open forests. Laxmipur: 24-11-1976, Nair 4895.

Argyreia hookeri Cl.

Climbers. Occasional in open forests. Kalipur: 3-12-1976, Balak and Nair 4935.

Argyreia mollis (Burm.f.) Choisy **(***A. championii* Benth.**)**

Climbers. Occasional in open forests. Saddle peak: 1-12-1976, Balak and Nair 4751.

Argyreia nervosa (Burm.f.) Boj. (*A. speciosa* (L.f.) Sweet)

Climbers. Rare in open forests. Ganesh nagar: 19-11-2001, CSR 2149.

Erycibe griffithii Cl.

Climbers. Occasional in open forests. Interview Island: 1-3-1991, PLN 20721.

Erycibe peguensis (Cl.) Prain

Climbers. Occasional in open forests. Way to Saddle peak: 20-9-1982, RKP 9206. Interview Island: 26-2-1991, PLN 20705.

Evolvulus alsinoides (L.) L.

Spreading or ascending much branched hairy herbs. Common weed in waste places, cultivated fields and forests. Radhanagar: 8-11-2001, CSR 2049.

Evolvulus nummularius (L.) L.

A prostrate herb, rooting at nodes. Common in moist places. Diglipur: 20-11-2001, CSR 2180.

Hewittia scandens (Milne) Mabb. (*H. sublobata* (L.f.) O.K.)

Woody climbers. Occasional in open forests. Durgapur: 28-11-1976, Balak and Nair 4704.

Ipomoea aquatica Forsskal

Large herb, rooting at nodes. Common in stagnant ponds, lakes, irrigation canals etc. Durgapur: 20-11-1976, Nair 4847.

Ipomoea carnea Jacq. ssp. **fistulosa** (Choisy) D. Austin

Large erect shrubs to 3 m tall. Common in hedges, road sides and river banks. Diglipur: 21-11-2001, CSR 2178.

Ipomoea campanulata L.

Climbers. Occasional in open forests. Kafeedera: 20-11-2001, CSR 2174.

Ipomoea mauritiana Jacq.

Climbers. Occasional in marshy places of forests. Diglipur: 21-11-2001, CSR 2179.

Ipomoea obscura (L.) Ker.-Gawl.

A slender twining herbs. Common in moist places. Lamia bay: 22-11-2001, CSR 2209.

Ipomoea pes-caprae (L.) R. Br.

A prostrate herb, rooting at nodes. Common in beach forests. Narcondam Island: 22-9-1977, Nair 6206.

Ipomoea quamoclit L.

Vine. Rare in hedges. Lamia bay: 22-11-2001, CSR 2211.

Ipomoea sepiaria Roxb.

Vine. Rare in forests. Mohanpur: 26-12-2001, CSR 2742.

Jacquemontia paniculata (Burm.f.) Hallier. f.

Vine. Occasional in forests. Mohanpur: 26-12-2001, CSR 2743.

Merremia hederacea (Burm.f.) Hallier f.

Profusely branched twiner. Common in thickets of forests. Interview Island: 15-2-1959, Thoth. 9218. Radhanagar: 8-11-2001, CSR 2050.

Merremia turpethum (L.) Shah et Bhat (*Operculina turpethum* (L.) S. Manso)

A stout climber; branchlets winged. Occasional twiner in forests. RK Gram: 20-11-2001, CSR 2178.

Merremia umbellata (L.) Hall. **ssp. orientalis** (Hall.f.) Oost.

Profusely branched twiner. Frequent in thickets of forests. Lamia Bay to Ramnagar: 29-3-1977, Balak 5351. Interview Island: 7-3-1990, PLN 15153.

Merremia vitifolia (Burm.f.) Wall.f.

Profusely branched twiner. Common in thickets of forests. RK Gram: 20-11-2001, CSR 2176.

Operculina riedeliana (Oliv.) Oost.

Vine. Occasional in coastal forests. Arial Bay: 23-11-1976, Nair 4892.

Ipomoea batatas (L.) Lam. cultivated for its edible tubers and **Ipomoea cairica** (L.) Sweet is cultivated as an ornamental plant.

SOLANACEAE

Datura metel L.

Erect, annual-perennial, undershrubs. Weed of disturbed places. Diglipur: 21-11-2001, CSR 2195.

Nicotiana plumbaginifolia Viv.

Erect viscid, annual herbs to 75 cm tall. Rare; a weed, near human inhabitations. Diglipur: 20-11-2001, CSR 2177.

Physalis angulata L.

Erect, widely branched, annual herbs to 75 cm tall. Common weed. Diglipur: 20-11-2001, CSR 2174.

Solanum anguivi Lam.

Erect undershrubs. Weed of disturbed places. Diglipur: 6-4-1977, Balak 5504.

Solanum americanum Miller

An annual, erect herb. Common weed.The description of **Solanum nigrum** in regional floras of India is actually of **S. americanum. Solanum nigrum** may not be found in Andaman. Aerial Bay: 21-11-2001, CSR 2678.

Solanum ferox L.

Herbs. Weed of disturbed places. Kalpong-10: 22-12-1986, Chakrab. and Kindo 12336.

Solanum surattense Burm. f.

A perennial, very prickly, diffute herb. Aggressive colonizer. Diglipur: 1-12-2001, CSR 2388.

Capsicum frutescens L. (Mirchi), **Lycopersicon esculentum** Mill. (Tomato), are commonly cultivated (vegetable crops.) **Cestrum nocturnum** L. (Night queen), **Petunia x hybrida** Vilmor are cultivated for their flowers, **Nicotiana tabacum** L. for tobacco.

SCROPHULARIACEAE

Bacopa floribunda (R.Br.) Wettst.

Herbs. Weed of paddy fields. Mohanpur: 26-12-2001, CSR 2760.

Bacopa monnieri (L.) Pennell

Prostrate, subsucculent herb, rooting at nodes. Occasional in perennially moist localities and dried up paddy fields. Mohanpur: 26-12-2001, CSR 2761.

Limnophila chinensis (Osb.) Merr. var. **scaberrima** Kurz

Herbs. Occasional in perennially moist localities. Diglipur: 30-11-2001, CSR 2348.

Lindernia antipoda (L.) Als. (*L. anagallis* auct.non Pennell)

Herbs. Occasional in open moist localities. Smith Island: 14-12-2001, CSR 2579.

Lindernia ciliata (Colsm.) Pennell

Herbs. Rare in open moist, sandy localities. Kafeedera: 20-11-2001, CSR 2167.

New record.

Lindernia crustacea (L.) F. Muell.

Erect herbs, often rooting at basal nodes. Occasional weed in moist places. Mohanpur: 26-12-2001, CSR 2763.

Lindernia viscosa (Horn.) Bold.

Erect herbs, often rooting at basal nodes. Occasional in moist places. Radhanagar: 30-11-2001, CSR 2340.

Mecardonia procumbens (Miller) Small

Prostrate herb. Occasional in riverbeds sandy moist places. Kalipur: 22-12-2001, CSR 2642.

Scoparia dulcis L.

Erect herb. Common weed in waste lands and drainages. Kalipur: 22-12-2001, CSR 2643.

Striga asiatica (L.) O. Kuntze

A small, scabrous herb to 30 cm tall. Rarely found in fields. Radhanagar: 30-11-2001, CSR 2339.

Torenia violacea (Aza. ex Blanco) Pennell (*T. peduncularis* Benth.)

Erect herb. Occasional in moist shady localities. Mohanpur: 26-12-2001, CSR 2766.

Angelonia biflora Benth. is planted in gardens.

LENTIBULARIACEAE

Utricularia exoleta R.Br.

Slender aquatic herb. Branches filiform with tiny bladders. Rare in still waters. Diglipur: 30-11-2001, CSR 2349.

GESNERIACEAE

Aeschynanthus griffithii R.Br.

Climbers. Rare, in the high elevation forest of Saddle peak. Saddle peak (720 m). 22-12-2001, CSR 2668.

BIGNONIACEAE

Dolichandrone spathacea (L.f.) Sch. (D. rheedii (Spreng.) Seem.) (Dongi, Hathipanjar)

Trees. Common in inland forests. Paget Island: 6-11-2001, CSR 2010.

Fernandoa adenophylla (Don) Steen. (*Heterophragma adenophyllum* (DC.) Dop.) (Chappal)

Trees. Common in inland forests. Mohanpur: 26-12-2001, CSR 2767.

Oroxylum indicum (L.) Kurz

Trees. Common in moist deciduous forests. Mohanpur: 26-12-2001, CSR 2768.

Pajanelia longifolia (Willd.) Sch. (*P. rheedii* Wight) (Jhingam)

Trees. Occasional in inland forests. Mohanpur: 26-12-2001, CSR 2774.

Jacaranda mimosifolia D. Don, **Kigelia africana** (Lam.) Benth., **Millingtonia hortensis** L.f., **Pyrostegia venusta** (Ker Gawler) Micrs, **Spathodea campanulata** P. Beauv. **Tabebuia rosea** (Bortol.) DC, **Tecoma stans** (L.) Kunth, and **Tecomaria capensis** (Thunb.) Lindley, are planted in gardens.

PEDALIACEAE

Sesamum orientale L. is cultivated for edible oil.

ACANTHACEAE

Acanthus ebracteatus Wall.

Shrubs. Rare in mangroves. Hathilevel: 12-11-2001, CSR 2090.

Acanthus ilicifolius L.

Shrubs. Rare in mangroves. Shyamnagar: 11-11-2001, CSR 2077.

Andrographis laxiflora (Bl.) Linda.

Erect herbs to 50 cm tall. Rare in forests. Saddle peak: 22-12-2001, CSR 2659.

Andrographis paniculata (Burm.f.) Wall. ex Nees

Erect herb to 50 cm tall. Occasional in forests of Mayabander (Narial Tikri): CSR 6-12-2001, CSR 2477.

New Record.

Asystasia gangetica (L.) T. And.

Climbers. Rare in open forests. Kalipur: 22-12-2001, CSR 2650.

Barleria prionitis L.

Armed undershrubs to 1 m. Occasional in waste lands and hedges. Vidyasagarpalli: 1-12-2001, CSR 2390.

Daedalacanthus suffruticosus And.

Herbs. Inland forests. Interview Island: 6-3-1990, PLN 15134.

Hygrophila erecta (Burm.f.) Hochr.

An erect herb upto 50 cm tall. Occasional weed moist places. Kalipur: 22-12-2001, CSR 2649.

Hygrophila salicifolia (Vahl) Nees

An erect herb upto 40 cm tall. Occasional weed moist places. Smith Island: 14-12-2001, CSR 2575.

Hypoestes andamanensis Thoth.

Herbs. Rare in forests. Reported from Saddle peak.

Endemic.

Pseuderanthemum album (Nees) Merr.

Herbs. Occasional in littoral forests. Lamia Bay: 12-5-1982, MKV 9009, 22-12-2001, CSR 2647; Durgapur: 20-11-1976, Nair 4830.

Rostellularia andamanica Vasud.

Herbs. Rare in forests of Saddle peak. Endemic to North Andaman.

Rungia pectinata (L.) Nees

Herbs. Inland forests. Interview Island: 7-3-1990, PLN 15172.

Staurogyne lanceolata (Bl.) O.K.

Herbs. Occasional in forests along streams. Austin No: 2: 26-12-2001, CSR 2775.

Strobilanthes glandulosus Kurz

Herbs. Occasional in forests. Interview Island: 7-3-1990, PLN 15109.

Endemic.

Crossandra infundibuliformis (L.) Nees, **Ecbolium viride** (Forsskal) Alston, **Gendarussa vulgaris** Nees, **Thunbergia alata** Boj. ex. Simsin, **Graptophyllum pictum** (L.) Griffith, **Eranthemum nervosum** (Vahl) R.Br. ex Roem. et Schult. are cultivated in fences of gardens.

THUNBERGIACEAE

Thunbergia laurifolia Lindl. (Harabel)

Climbers. Occasional in forests. Saddle peak (100 m): 28-3-1977, Balak 5421; 16-10-1987, Ellis 12729. Interview Island: 5-3-1990, PLN 15109.

AVICENNIACEAE

Avicennia alba Blume (Dudhi Baen)

Trees. Rare, in Mangrove forests. Austin No: 2: 26-12-2001, CSR 2778.

Avicennia marina (Forssk.) Vierh

Trees. Rare, in Mangrove forests. Kafeedera: 24-11-2001, CSR 2256.

VERBENACEAE

Callicarpa arborea Roxb.

Trees. Rare in forests. Austin No: 2: 26-12-2001, CSR 2776.

Clerodendrum inerme (L.) Gaertn.

Climbers. Occasional in littoral forests. Lamia Bay: 22-11-2001, CSR 2219.

Clerodendrum infortunatum L. (*C. viscosum* Vent.)

Shrubs. Occasional in open moist deciduous forests. Kalipur: 30-3-1977, Balak 5358.

Gmelina elliptica Sm. (Emne)

Trees. Scattered in moist deciduous forests. Saddle peak: 28-11-2001, CSR 2321.

Lantana camara L.

Straggling, armed shrubs. Common along the hedges and waste places. Aggressive colonizer. Kalipur: 22-12-2001, CSR 2648.

Phyla nodiflora (L.) E. Greene

Prostrate herb. Common in moist ground. Mohanpur: 26-12-2001, CSR 2778.

Premna obtusifolia R.Br. (*P. corymbosa* (Burm.f.) Rottl. et Willd.)

Small trees. Occasional in the edges of littoral forests. North Island: 17-12-2001, CSR 2625.

Sphenodesme involucrata (Presl.) Rob. (*S. unguiculata* (Sch.) Munir.)

Climbers. Occasional in moist deciduous forests. Interview Island: 22-2-1991, PLN 15691.

Stachytarpheta urticaefolia (Salisb.) Sims

A tall herb to 1 m tall. Naturalised in open forests. Kalipur: 22-12-2001, CSR 2647.

Vitex glabrata R.Br.

Small trees. Rare in coastal forests. Kafeedera: 19-11-2001, CSR 2163.

Vitex leucoxylon L.f.

Small trees. Rare in inland forests. Kafeedera: 19-11-2001, CSR 2170.

Vitex negundo L.

Shrubs to 4 m tall. Common in fences. Lamia bay to Aerial bay: 21-11-2001, CSR 2190.

Vitex pinnata (Lour.) Will.

Small trees. Rare in inland forests. Mohanpur: 26-12-2001, CSR 2777.

Vitex trifolia L.

Shrubs to 4 m tall. Rare in fences. Kafeedera: 20-11-2001, CSR 2173.

Vitex urceolata Cl.

Small trees. Rare in inland forests. Ganeshnagar: 17-11-2001, CSR 2148.

Citharexylum subserratum Sw., **Clerodendrum indicum** (L.) O. Kuntze, **Clerodendrum philippinum** Schuaer, **Duranta repens** L. **Holmskioldia sanguinea** Retz. **Tectona grandis** L.f., and **Verbena laciniata** (L.) Briq. are planted in gardens.

LAMIACEAE (*nom. alter.* LABIATAE)

Anisomeles indica (L.) Kuntze

Perennial, pubescent, aromatic, bushy under shrubs to 1.5 m. Kalipur: 22-12-2001, CSR 2651.

Anisomeles malabarica (L.) R. Br. ex Sims

Perennial, white wooly, aromatic, under shrubs to 1.5 m. Occasional in distrubed localities of forests. Kafeedera: 20-11-2001, CSR 2174.

Gomphostemma javanicum (Bl.) Benth.

Undershrubs. Occasional in open forests. Kalpong-10: 22-12-1986, Chakrab. and Kindo 12324.; Karmatong-9: 7-12-1992, Sinha 16289. Saddle peak: 22-12-2001, CSR 2660.

Hyptis brevipes Poit.

Herbs. Introduced weed, naturalized along road sides and forests paths. Mohanpur: 26-12-2001, CSR 2778.

Hyptis capitata Jacq.

Herbs. Introduced weed, naturalized along road sides and forests paths. Lamia bay to Kalipur: 22-12-2001, CSR 2646.

Hyptis suaveolens (L.) Poit.

Aromatic, erect, undershrubs to 3 m tall. Abundant weed; aggressive colonizer by distrubed places. Mohanpur: 26-12-2001, CSR 2779.

Leucas biflora (Vahl) R.Br.

Trailing herbs to 40 cm. Rare in cultivated fields and human inhabitations. Shyamnagar: 27-12-2001, CSR 2780.

Leucas cephalotes (Roth) Spreng.

Erect, stout herbs to 40 cm tall. Rare weed. Kafeedera: 19-11-2001, CSR 2168.

Ocimum canum Sims

Erect herbs to 50 cm. Common weed. Kalipur: 22-11-2001, CSR 2220.

Ocimum gratissimum L.

Aromatic, profusely branched subshrub to 1 m. Rare, weed. Kafeedera: 19-11-2001, CSR 2169.

Ocimum basilicum L., **Ocimum tenuiflorum** L., **Origanum majorana** L. (Davanam), **Salvia coccinea** Buch-Ham. are planted in gardens.

MONOCHLAMYDEAE

NYCTAGINACEAE

Boerhavia diffusa L.

Diffuse, ascending, much branched herbs. Common weed. Shyamnagar: 27-12-2001, CSR 2785.

Boerhavia erecta L.

Annual, erect herbs up to 60 cm tall. Rare weed. Mohanpur: 26-12-2001, CSR 2780.

Commicarpus chinensis (L.) Heimel

Diffuse, subscandent, glabrescent herbs to 23 cm tall. Rare in hedges. Kafeedera: 20-11-2001, CSR 2180.

Mirabilis jalapa L.

An erect, much branched herb or under shrub. Commonly cultivated in gardens; runs wild in villages. Shyamnagar: 27-12-2001, CSR 2789.

Pisonia aculeata L.

Armed climbers. Rare in forests. Kafeedera: 20-11-2001, CSR 2179.

Pisonia umbellifera (Forst.) Seem. (**P. excelsa** Bl.) (Baniya)

Trees. Occasional in evergreen forests. Mohanpur: 7-12-2001, CSR 2498.

Bougainvillea spectabilis Willd. is cultivated as a decorative climber.

AMARANTHACEAE

Achyranthes aspera L.

Undershrubs. Rare along fences. Shyamnagar: 27-12-2001, CSR 2784.

Aerva lanata (L.) Juss.

A suberect or diffuse herb. Rare weed. Diglipur: 30-11-2001, CSR 2366.

Allmania nodiflora (L.) R.Br. ex Wight

Erect herbs. Rare in waste lands, along road sides. Shyamnagar: 27-12-2001, CSR 2786.

Alternanthera philoxeroides (Mart.) Griseb.

Aquatic or marshy prostrate herbs, with many erect branches. Frequent weed in driedup paddy fields, stagnant waters and pools. Diglipur: 30-11-2001, CSR 2365.

New Record.

Alternanthera pungens Kunth

Spiny, prostrate herb to 50 cm. Common weed. Diglipur: 30-11-2001, CSR 2363.

Alternanthera sessilis (L.) R.Br. ex. DC.

Very variable prostrate to erect herb, to 50 cm tall, rooting from the basal nodes. Common weed of damp places along tanks ditches and dried rice fields. Keralapuram: 5-12-1976, Balak and Nair 4981.

Amaranthus graceizans L.

Diffuse, prostrate herb. Common weed of cultivated fields. Keralapuram: 5-12-1976, Balak and Nair 4966.

Amaranthus spinosus L.

Erect herbs to 1 m tall. Common weed in fields and waste lands. Diglipur: 30-11-2001, CSR 2572.

Amaranthus viridis L.

Erect herb to 1 m tall. Common weed in fields and waste lands. Radhanagar: 9-11-2001, CSR 2051.

Celosia argentea L. var. **argentea**

Erect herb to 1 m. Common weed in dry fields. Diglipur: 1-12-2001, CSR 2371.

Digera muricata (L.) Mart.

Erect herbs to 1 m tall. Common weed in moist fields. Diglipur: 1-12-2001, CSR 2372.

Gomphrena serrata L.

Erect or ascending herbs to 50 cm. Rare weed. Shyamnagar: 27-12-2001, CSR 2781.

Nothosaerva brachiata (L.) Wight

Erect or diffuse, well branched herb to 70 cm tall. Rare weed in moist places. Shyamnagar: 27-12-2001, CSR 2782.

Alternanthera bettzickiana (Regel) Nicolson, is commonly planted as a hedge plant. **Amaranthus hybridus** L. **Celosia argentea** L. var. **cristata** (L.) O. Kuntze and **Gomphrena globosa** L. are planted in gardens.

POLYGONACEAE

Polygonum chinense L.

Shrubs. Rare in open forests. RK Gram: 27-12-2001, CSR 2785.

Polygonum glabrum Willd.

Erect shrubs to 1 m tall. Common in water logged places and margins of river banks. Diglipur: 1-12-2001, CSR 2376.

Muehlenbeckia platyclada (F.V. Muell. ex Hook.) Meissn. is planted as an ornamental plant. **Rumex vesicarius** L. is cultivated as a leafy vegetable.

CHENOPODIACEAE

Antigonon leptopus Hook. et Arn. planted in gardens.

Chenopodium album L.

Erect herbs to 50 cm. Occasional weed in dried up paddy fields. Diglipur: 1-12-2001, CSR 2374.

Spinacia oleracea L. cultivated as leafy vegetable.

BASELLACEAE

Basella alba L.

Succulent vine; Commonly cultivated as leafy vegetable.

ARISTOLOCHIACEAE

Aristolochia tagala Cham.

Climbers. Inland forests. Interview Island: 23-2-1991, PLN 15686.

Aristolochia grandiflora Sw. is planted in gardens.

PIPERACEAE

Peperomia pellucida (L.) H.B.K.

Fleshy herbs. Common in shady localities. Diglipur: 1-12-2001, CSR 2375.

Piper betle L. var. **densum** DC.

Climbers. Inland forests. Interview Island: 2-3-1991, PLN 20734.

Piper betle L, **Piper longum** L., **Piper nigrum** L. are cultivated for economic purposes.

MYRISTICACEAE

Horsfieldia glabra (Bl.) Warb. **(***Myristica glabra* Bl.**)**

Trees, ca. 15 m tall. Inland forests. Interview Island: 2-3-1991, PLN 20732.

Horsfieldia irya (Gaertn.) Warb. **(***Myristica irya* Gaertn.**)**

Trees, ca. 10 m tall. Inland forests. Diglipur: 19-11-1976, Nair 4804. Interview Island: 1-3-1991, PLN 20722.

Knema andamanica (Warb.) de Wilde

Trees. Rare, in evergreen forests. Smith Island: 22-11-1976, Nair 4869; Saddle peak: 1-12-1976, Balak and Nair 4732. Interview Island: 6-3-1990, PLN 15131; 22-2-1991, PLN 15655. Diglipur: 1-12-2001, CSR 2376.

Myristica andamanica Hook.f. (Kala Jaiphal)

Trees. Occasional in evergreen forests. Durgapur: 20-11-1976, Nair 4843.

Myristica glaucescens Hook. f. (Lal Jaiphal)

Trees. Occasional in evergreen forests. Diglipur: 1-12-2001, CSR 2379.

LAURACEAE

Cassytha filiformis L.

Leafless twiner. Common parasitic twiner on thickets. Point Island: 16-11-2001, CSR 2136.

Cryptocarya andamanica Hook.f.

Trees. Rare in evergreen forests. Diglipur: 1-12-2001, CSR 2376.

Endemic.

Cryptocarya amygdalina Nees

Small trees. Occasional in evergreen forests. Saddle peak: 9-5-1982, MKV 8932; Interview Island: 23-2-1991, PLN 15685.

Dehaasia kurzii King

Small trees. Occasional in evergreen forests. Interview Island: 7-3-1990, PLN 15159; 22-2-1991, PLN 15631.

Litsea glutinosa (Lour.) C. Robinson

Evergreen tree, to 12 m. Rare in forests. Shyamnagar: 27-12-2001, CSR 2782.

Litsea kurzii King

Small trees. Occasional in semievergreen forests. Interview Island: 5-3-1990, PLN 15108; 7-3-1990, PLN 15151.

Endemic.

Neolitsea zeylanica (Nees et T.Nees) Merr.

Small trees. Rare in evergreen forests. Near Kalpong dam: 3-12-2001, CSR 2407.

New record.

HERNANDIACEAE

Gyrocarpus americanus Jacq. (*G. jacquini* Roxb.)

Trees. Frequent in beach forests. North Island: CSR 2782.

Hernandia peltata Meissn.

Trees. Frequent in beach forests. Interview Island: 24-2-1991, PLN 15692.

Illigera appendiculata Bl.

Climbers. Occasional in forests. Keralapuram: 5-4-1977, Balak 5477. Diglipur: 1-12-2001, CSR 2379.

ELAEAGNACEAE

Elaeagnus conferta Roxb.

Climbers. Rare in forests. Mohanpur: 26-12-2001, CSR 2767.

LORANTHACEAE

Dendrophthoe falcata (L.f.) Ettingsh.

Large, bushy, glabrous branched parasites. Saddle peak: 2-12-1976, Balak and Nair 4787. Interview Island: 17-11-1990, PLN 15646.

EUPHORBIACEAE

Acalypha indica L.

Erect herbs to 75 cm; weed of dumped ground. Diglipur: 1-12-2001, CSR 2381.

Acalypha lanceolata Willd.

Annual, erect herbs, up to 50 cm tall. Rare in disturbed places. Diglipur: 1-12-2001, CSR 2380.

Actephila excelsa (Dalz.) Muell.-Arg.

Shrubs. Common in inland forests. Saddle peak (150 m): 20-9-1982, RKP 9208. Interview Island: 8-3-1990, PLN 15187.

Antidesma bunius (L.) Spreng.

Small trees. Rare in forests. Swamp Island: 9-12-2001, CSR 2522.

Antidesma ghaesembilla Gaertner

Small tree to 10 m. Rare in forests. Shyamnagar: 27-12-2001, CSR 2787.

Aporusa dioica (Roxb.) Muell.-Arg. (*A. villosula* Kurz)

Small trees. Occasional in inland forests. Saddle peak: 9-5-1982, MKV 8933; Interview Island: 23-2-1991, PLN 15682.

Baccaurea ramiflora Lour. (*B. sapida* (Roxb.) Muell.-Arg.)

Small trees. Frequent in forests. Interview Island: 16-2-1959, Thoth 9235; Durgapur: 5-4-1977, Balak 5463.

Baliospermum montanum (Willd.) Muell.-Arg.

Stout undershrubs to 2 m. Occasional in forest paths. Mohanpur: 7-12-2001, CSR 2498.

Bischofia javanica Bl. (Ye-Padauk, Pani Padauk)

Trees. Rare in forests. Lamia bay: 29-3-1977, Balak 5332; Ramnagar: 3-12-1976, Balak and Nair 4928. Diglipur: 1-12-2001, CSR 2381.

Blachia andamanica (Kurz) Hook.f.

Trees. Frequent in moist deciduous forests. Interview Island: 15-2-1959, Thoth 9220. Smith Island: 25-11-1976, Nair 4916.

Endemic.

Breynia retusa (Dennst.) Alston

Bushy shrub to 3 m. Rare in damp forests. Diglipur: 1-12-2001, CSR 2382.

Breynia vitis-idaea (Burm.f.) Fischer (*B. rhamnoides* (Willd.) Muell.-Arg.)

Shrubs or small trees. Frequent in open forests. Mayabander: 8-12-1992, Sinha 16293.

Bridelia griffithii Hook.f.

Shrubs. Occasional in littoral forests. Durgapur: 26-11-1976, Nair 4922. Interview Island: 5-3-1990, PLN 15102.

Bridelia ovata Decne

Shrubs. Inland forests. Interview Island: 24-2-1991, PLN 15693.

Bridelia stipularis (L.) Bl.

Climbers. Inland forests. Diglipur: 1-12-2001, CSR 2383.

Bridelia tomentosa Bl.

Small trees. Occasional in open forests. Interview Island: 22-2-1991, PLN 15653.

Chrozophora rottleri (Geis.) Spreng.

Erect, fulvous-tomentose herbs to 75 cm tall. Weed of driedup fields. Diglipur: 1-12-2001, CSR 2384.

Claoxylon indicum (Reinw.ex Bl.) Hook.

Trees, to 15 m tall. Inland forests. Interview Island: 13-2-1959, Thoth. 9213.

Endemic.

Claoxylon rostratum Airy Shaw

Small tree. Saddle peak: 31-3-1977, Balak 5374.

Cleidion javanicum Bl.

Trees, to 6 m tall. Inland forests. Interview Island: 22-2-1990, PLN 15651.

Cleidion nitidum Thw.

Trees. Frequent, in evergreen forests. Durgapur: 20-11-1976, Nair 4839. Interview Island: 28-2-1991, PLN 20715. Diglipur: 1-12-2001, CSR 2385.

Cleistanthus myrianthus (Hassk.) Kurz

Small trees, to 8m tall; Rare in forests. Laxmipur: 6-4-1977, Balak 5498; Sitanagar: 15-5-1982, MKV 9017. Interview Island: 7-3-1990, PLN 15158.

Croton bonplandianum Bail.

Annual, erect much branched herbs to 75 cm tall. Common weed. Diglipur: 1-12-2001, CSR 2386.

Croton sublyratus Kurz (*Croton argyratus* Bl.)

Trees. Occasional in forests. Sitanagar: 15-5-1982, MKV Rao 9035.

Drypetes andamanica (Kurz) Pax et Hoffm.

Medium sized evergreen tree. Interview Island: 13-2-1959, Thoth. 9216. Entrance Island: 8-11-1990, PLN 15207.

Endemic.

Drypetes assamica (Hook.f.) Pax et Hoffm. (*Cyclostemon assamicus* Hk.f.)

Medium sized evergreen tree. Ramnagar: 29-3-1977, Balak 5353. Interview Island: 17-1-1991, PLN 15638; 22-2-1991, 15670.

Drypetes longifolia (Bl.) Pax. et Hoffm.

Trees, 4-10 m tall. Inland forests. Interview Island: 16-11-1990, PLN 15632; 22-2-1991, PLN 15670.

Endospermum malaccense Benth. ex Muell.-Arg. (Bakota)

Medium sized evergreen tree. Common in forests. Shyamnagar: 27-12-2001, CSR 2784.

Euphorbia epiphylloides Kurz

Small tree. Occasional in hilltop stunted evergreen forests. Saddle peak (400 m): 1-12-1976, Balak and Nair 475528-3-1977, Balak 5412; 10-3-1982, MKV Rao 8961; Kalpong: 23-12-1986, Chakrab. and Kindo 12364.

Euphorbia heterophylla L.

Herb to 75 cm. Rare weed. Shyamnagar: 27-12-2001, CSR 2788.

Euphorbia hirta L.

Erect or decumbent herbs to 50 cm tall. Common weed. Swamp Island: 10-12-2001, CSR 2535.

Euphorbia prostrata Ait.

Herbs. Rare in dried up paddy fields. Shyamnagar: 27-12-2001, CSR 2786.

Euphorbia thymifolia L.

Herbs. Weed of disturbed, shady places. Diglipur: 1-12-2001, CSR 2387.

Euphorbia trigona Hawort (*E. barnhartii* Croiz.)

Trees, 3-4 m tall. Scattered in moist deciduous forests. Interview Island: 8-3-1990, Thoth 15191.

Excoecaria agallocha L.

Shrubs. Scattered in coastal forests. Shyamnagar: 27-12-2001, CSR 2788.

Glochidion airy-shawii Balakr. et Chakrab.

Trees. Rare and reported from Saddle peak.

Endemic.

Glochidion bilobulatum Chakrab. et Gang.

Trees. Rare and reported from Saddle peak.

Endemic.

Glochidion calocarpum Kurz

Trees. Rare, in evergreen forests. Kafeedera: 20-11-2001, CSR 2166.

Endemic.

Glochidion hirsutum (Roxb.) Voigt

Small tree to 10m. Occasional in forests. Swarajgram: 14-11-2001, CSR 2100.

Homonoia riparia Lour.

Gregarious, bushy shrub to 3 m tall, dioecious; Rare in rocky crevices of streams. Kafeedera: 20-11-2001, CSR 2176.

Jatropha curcas L.

Shrub or small tree to 5 m. Often planted by fences; runs wild. Diglipur: 1-12-2001, CSR 2388.

Jatropha gossypifolia L. var. **elegans**

Shrub to 3 m. Common in waste lands and fields. Diglipur: 1-12-2001, CSR 2389.

Macaranga indica Wight

Trees. Rare, in evergreen forests. Kalighat: 2-12-2001, CSR 2394.

Macaranga peltata (Roxb.) Muell.-Arg.

Trees. Rare, in evergreen forests. Kafeedera: 20-11-2001, CSR 2165.

Macaranga tanarius (L.) Muell.-Arg.

Trees. Rare, in evergreen forests. Lamia Bay: 29-3-1977, Balak 5338. Lamia Bay to Kalipur: 22-12-2001, CSR 2653.

Mallotus oblongifolius (Miq.) Muell.-Arg.

Trees. Occasional in inland forests. Saddle peak (100m): 22-12-2001, CSR 2659.

Endemic.

Mallotus peltatus (Geis.) Muell.-Arg.

Trees. Saddle peak (250 m): 28-3-1977, Balak 5418; Keralapuram: 21-11-1976, Nair 4853. Interview Island: 15-2-1959, Thoth. 9219.

Mallotus philippensis (Lam.) Muell.-Arg.

Evergreen trees to 15 m. Rare in forests. Swarajgram: 14-11-2001, CSR 2101.

Mallotus resinosus (Blanco) Merr.

Trees. Sitanagar: 15-5-1982, MKV 9037; Ray hills: 8-12-1992, PLN 15625. Interview Island: 12-2-1959, Thoth. 9207.

Phyllanthus andamanicus Balakr. et Nair

Shrub. Rare, in Saddle peak. Rare and threatened.

Endemic.

Phyllanthus amarus Schum. et Thonn.

Ascending or erect herbs to 70 cm. Common weed. Mayabandar: 1-8-1974, Bhargava 1979. Kalipur: 22-12-2001, CSR 2654.

Phyllanthus columnaris Muell.-Arg.

Erect shrub to 4 m. Rare in forests. Recorded from Saddle peak.

Phyllanthus emblica L.

Deciduous trees to 15 m. Planted in house compounds for fruits. Shyamnagar: 27-12-2001, CSR 2789.

Phyllanthus maderaspatensis L.

Erect herb to 70 cm. Rare, weed. Kalighat: 2-12-2001, CSR 2393.

Phyllanthus reticulatus Poiret

Scandent or erect shrub to 4 m. Common in waste lands, fences. Kishorinagar: 5-12-2001, CSR 2464.

Phyllanthus sanjappai Chakrab. et Gang.

Shrub. Rare, reported from Saddle peak. Rare and threatened.

Endemic.

Phyllanthus urinaria L.

Erect herb to 50 cm tall. Common in forests as undergrowth. Kalipur: 22-12-2001, CSR 2655.

Phyllanthus virgatus Forster f.

Erect herb to 75 cm. Rare, weed along roadsides. Kishorinagar: 5-12-2001, CSR 2462.

Sauropus quadrangularis (Willd.) Muell.-Arg.

Shrub to 1m tall, monoecious. Rare in forests. Kishorinagar: 5-12-2001, CSR 2461.

Securinega virosa (Willd.) Bail.

Shrub to 5 m tall. Rare in forests. Narcondum Island (near Police Camp): 22-9-1977, Nair 6213.

Sphyranthera airy-shawii Chakrab. et Vasud.

Trees. Rare, in Saddle peak. Endemic genus to North Andamans.

Sphyranthera lutescens (Kurz) Pax et Hoffm.

Trees, ca. 4 m tall. Inland forests. Interview Island: 23-2-1991, PLN 15675. Endemic genus to North Andamans.

Suregada multiflora (Juss.) Baill.

Shrubs. Inland forests. 5-3-1990, PLN 15114 and 26-2-1991, PLN 20701.

Synadenium grantii Hook. f.

Succulent shrub to 5 m tall. Planted as an hedge plant; runs wild. Kishorinagar: 5-12-2001, CSR 2460.

Tragia involucrata L.

Twining hispid herb, with stinging hairs. Occasional along forest out skirts. Swarajgram: 14-11-2001, CSR 2102.

Trigonostemon aurantiacus (Kurz ex Tiejsm et Binnend.) Boerl. **var. rubriflorus** Balakr. et Chakrab.

Small tree. Occasional in evergreen forests. Kalighat: 2-12-2001, CSR 2398.

Trigonostemon viridissimus (Kurz) Airy Shaw

Small tree. Occasional in evergreen forests. Saddle peak: 28-3-1977, Balak 5418; Sitanagar: 15-5-1982, MKV Rao 8919; Interview Island: 6-3-1990, PLN 15147; Chatham Island: CSR.

Acalypha hispida Burm.f., **Acalypha wilkesiana** Muell.-Arg. **Euphorbia millii** Des, **Euphorbia pulcherrima** Willd. ex Klotzsch, **Jatropha multifida** L., **Jatropha panduraefolia** Andr., **Manihot esculenta** Crantz, **Pedilanthus tithymaloides** (L.) Poit., **Phyllanthus acidus** (L.) Skeels are planted in gardens. **Ricinus communis** L. (Castor) is cultivated for castor oil.

URTICACEAE

Boehmeria nivea (L.) Gaud.

Small trees. Inland forests. Kishorinagar: 5-12-2001, CSR 2459.

Elatostemma cuneatum Wight

Dwarf herbs. Rare in shady places of forests on moist rocks. Narcondum Island: 22-9-1977, Nair 6201. Paget Island: 6-11-2001, CSR 2011.

Elatostemma integrifolium (Don) Wedd.

Herbs. Saddle peak foot hill: 11-5-1982, MKV 8976.

Pilea microphylla (L.) Liebm.

Diffuse herb to 15 cm. Common potweed in gardens.

Pipturus argenteus (Forst.f.) Wedd. (*P. velutinus* Wedd.)

Small trees. Inland forests. Interview Island: 18-2-1959, Thoth. 9215.

Pouzolzia zeylanica (L.) Bennett

Erect herbs to 30 cm. Rare, in crevices of moist rocks. Kalipur: 22-12-2001, CSR 2656.

MORACEAE

Antiaris toxicaria (Pers.) Lesch

Trees. Occasional in evergreen forests. Kalighat: 2-12-2001, CSR 2391.

Artocarpus chaplasha Roxb.

Trees. Occasional in evergreen forests. Diglipur: 1-12-2001, CSR 2390.

Artocarpus gomeziana Wall. ex Trec.

Trees. Occasional in evergreen forests. Kalighat: 2-12-2001, CSR 2392.

Artocarpus lakoocha Roxb.

Trees. Occasional in evergreen forests. Kishorinagar: 5-12-2001, CSR 2458.

Ficus benjamina L.

Trees. Rare in coastal forests. Interview Island: 8-3-1990, PLN 15177.

Ficus exasperata Vahl

Small trees. Scarce in inland forests. Kafeedera: 18-11-2001, CSR 2156.

Ficus hederacea Roxb. (*F. scandens* Roxb.)

Scandent trees. Occasional along rocky places near streams of forests. Radhanagar: 9-11-2001, CSR 2053.

Ficus hispida L.f.

Small trees. Occasional along roadsides near villages. Durgapur: 5-4-1977, Balak 5462. Interview Island: 7-3-1990, PLN 15177. Kalipur: 22-12-2001, CSR 2657.

Ficus laevis Bl.

Trees. Rare in inland forests. Kishorinagar: 5-12-2001, CSR 2457.

Ficus microcarpa L.f.

Densely foliaceous tree to 15 m. Rare, in forests; often cultivated. Kalipur: 22-12-2001, CSR 2658.

Ficus mollis Vahl

Trees with rusty foliage. Occasional in forests. Kafeedera: 18-11-2001, CSR 2157.

New Record.

Ficus nervosa Heyne ex Roth

Densely foliaceous tree to 15 m. Rare, in forests. Saddle peak: 27-3-1977, Balak 5170. Interview Island: 7-3-1990, PLN 15170.

Ficus obscura Blume **var. borneensis** (Miq.) Corner

Trees. Rare, in forests. Saddle peak foot hill: 11-5-1982, MKV 8975.

Ficus racemosa L.

Tree to 10 m. Planted in gardens and near wells; wild in outskirts of villages. Lamia bay: 29-3-1977, Balak 5334. Swarajgram: 14-11-2001, CSR 2103.

Ficus tinctoria Forster f. ssp. **gibbosa** (Bl.) Corner

Trees, 3 m tall. Inland forests. Interview Island: 26-2-1991, PLN 15700.

Ficus tinctoria Forster f. ssp. **parasitica** (Willd.) Corner

Tree or straggler to 8 m. Rare in forests. Interview Island: 7-3-1990, PLN 15169.

Ficus variegata Bl.

Trees. Rare in forests. Point Island: 16-11-2001, CSR 2134.

Ficus virens Aiton

Tree to 15 m. Occasional in forests. Kalighat: 2-12-2001, CSR 2393.

Streblus asper Lour.

Stiff shrub or a tree to 8 m. Common in moist deciduous forests and near outskirts of villages. Interview Island: 13-2-1959, Thoth. 9211; 22-2-1991, PLN 15672. Lamia Bay: 30-3-1977, Balak 5356; Saddle peak foot hill: 11-5-1982, MKV 8978.

Streblus taxoides (Heyne) Kurz

Small trees. Inland forests. Kishorinagar: 5-12-2001, CSR 2456.

Artocarpus heterophyllus Lam. (Jack fruit) is planted for its fruits. **Ficus benghalensis** L., **Ficus pumila** L., **Ficus religiosa** L. are planted in avenues, gardens and temples; runs wild in outskirts of villages, **Morus alba** L. (Mulberry) is cultivated to feed silkworms.

ULMACEAE

Celtis philippensis Blanco **var. wightii** (Planch.) Scop. (*C. wightii* Planch.)

(Tej Patti)

Trees. Common in inland forests. Paget Island: 6-11-2001, CSR 2012.

Celtis timorensis Span. (*C. cinnamomea* Lindl.ex Planch.)

Trees. Frequent in inland forests. Kafeedera: 18-11-2001, CSR 2160.

Trema orientalis (L.) Blume

Evergreen tree to 15 m. Common in moist deciduous forests. Kafeedera: 18-11-2001, CSR 2158.

Trema tomentosa (Roxb.) Hara (*T. amboinensis* (Willd.) Bl.)

Evergreen tree to 15 m. Occasional in moist deciduous forests. Kalighat: 2-12-2001, CSR 2395.

CASUARINACEAE

Casuarina equisetifolia Forster et Forster f.

Tall trees to 15m. Commonly planted in gardens and avenues.

CERATOPHYLLACEAE

Ceratophyllum demersum L.

LILIOPSIDA

HYDROCHARITACEAE

Blyxa aubertii L.C. Rich.

Aquatic herbs. Grown in ditches. Shibpur: 28-11-1976, Balak and Nair 4725.

Blyxa octandra (Roxb.) Planchon ex Thwaites

Acaulescent, submerged, tufted herbs. Occasional in stagnant water bodies. Diglipur: 1-12-2001, CSR 2352.

.

Hydrilla verticillata (L.f.) Royle

Caulescent, submerged herb. Common in ponds, ditches, lakes etc., Diglipur: 1-12-2001, CSR 2353.

ORCHIDACEAE

Aerides emericii Reichb.f.

Epiphyte. Common in forests. Kafeedera: 18-11-2001, CSR 2165.

Ascocentrum ampullaceum (Roxb.) Sch.

Epiphyte. Stunted hill forests. Saddle peak (400-500 m): 31-3-1977, Balak 5400.

Bulbophyllum lepidum (Bl.) Sm.

Epiphyte. Common in forests. Kafeedera: 18-11-2001, CSR 2161.

Bulbophyllum tenuifolium (Bl.) Lindl.

Epiphytes. Rare, on moist tree trunks clothed with mosses especially on *Memecylon* spp. Kalpong (Nabagram): Base of rock hill dam (250 m): 23-1-2000, PVS and VSK 17634.

Coelogyne thailandica Seidenf.

Epiphyte. Occassional in Stunted hill forests. Saddle peak (400-500 m): 5-12-1976, Balak and Nair 4297.

Cleisostoma andamanicum Balakr.

Epiphyte. Rare, on moist tree trunks. Kalpong: 24-12-1986, Chakrab. & Kindo 12375.

Endemic.

Cymbidium aloifolium (Lindl.) Schltr.

Epiphyte. Occasional in forests. Kalighat: 2-12-2001, CSR 2396.

Dendrobium anceps Sw.

Epiphyte. Stunted hill forests. Saddle peak (400-500 m): 31-3-1977, Balak 5380.

Dendrobium cruementum Sw.

Epiphyte. Inland forests. Interview Island: 5-3-1990, PLN 15111.

Dendrobium formosum Roxb.

Epiphyte. Stunted hill forests. Saddle peak (400-500 m): 31-12-1976, Balak and Nair 4750.

Dendrobium gunnarii PSN Rao

Reported from Saddle peak. Endemic.

Dendrobium macrostachyum Lindl.

Epiphyte. Rare in inland forests. Saddle peak (400 m): 23-11-2001, CSR 2222.

Dendrobium secundum (Bl.) Lindl. **(***Phraetia secunda* (Bl.) Lindl.)

Epiphyte. Stunted hill forests. Saddle peak (400-500 m): 27-3-1977, Balak 5163; Laxmipur: 6-4-1977, Balak 5163. Interview Island: 22-2-1991, PLN 15658.

Eria bractescens Lindl. **var. affinis** (Griff.) Hk.f.

Epiphyte. Laxmipur: 6-4-1977, Balak 5493. Kalipur to Lamia Bay: 22-12-2001, CSR 2658.

Flickengeria fimbriata (Bl.) Haw.

Epiphyte. Lamia bay to Kalipur: 30-3-1977, Balak 5372. Kalighat: 2-12-2001, CSR 2397.

Gastrochilus calceolaris Don

Epiphyte. Saddle peak: 27-3-1977, Balak 5181; 11-5-1982, MKV 8981.

Geodorum densiflorum (Lam.) Schltr.

Terrestrial herbs. Occasional in dry deciduous forests in shade. Radhanagar: 9-11-2001, CSR 2054.

Habenaria andamanica Hook.f.

A terrestrial orchid to 40 cm long. Rare in forests. Kalighat: 2-12-2001, CSR 2398. Endemic.

Kingidium deliciosum (Reicb.) Sweet.

Epiphyte. Rare in forests. Entrance Island: 8-11-1990, PLN and Ray 15199.

Luisia teretifolia Gaud.

Epiphytes. Rare on the trees. Chatham Island: 15-12-2001, CSR 2587.

Luisia zollingeri Reichb.f.

Epiphytes. Rare on the trees. South Island: CSR 2601.

Malaxis andamanica (K.& P.) Balakr. et Vasud.

A terrestrial orchid to 40 cm long. Rare in forests. Saddle peak: 22-12-2001, CSR 2659.

Malaxis latifolia Sm.ex Rees

A terrestrial orchid to 40 cm long. Occasional in forests. Kalighat: 2-12-2001, CSR 2399.

Nervilia aragoana Gaud.

A terrestrial orchid to 20 cm long. Rare, in evergreen forests. Kalighat: 2-12-2001, CSR 2397.

Oberonia iridifolia (Roxb.) Lindl.

Epiphytes. Occasional in forests. Kalighat: 2-12-2001, CSR 2392.

Papilionanthe teres (Roxb.) Schl. (*Vanda teres* Lindl.)

Epiphytes. Inland forests. Saddle peak (500 m): 27-3-1977, Balak 5185. Interview Island: 7-3-1990, PLN 15156. Kalighat: 2-12-2001, CSR 2400.

Peristylus parishii Reichb.f.

Terrestrial orchids. Rare in inland forests. Saddle peak (300 m): 26-11-2001, CSR 2298.

Pholidota imbricata (Roxb.) Lindl. (*P. pallida* Lindl.)

Epiphyte. Occasional in inland forests. Kalpong: 24-12-1986, Chakrab and Kindo 12374. Swarajgram: 14-11-2001, CSR 2106.

Pomatocalpa wendlandorum (Reichb.f.) J.J. Sm.

Epiphyte. Inland forests. Laxmipur: 6-4-1977, Balak 5501. Interview Island: 5-3-1990, PLN 15166.

Porpax meirax Par. et Reichb.f. (*Eria muscicola* Ridl.)

Lithophyte. Stunted hill forests. Saddle peak (400-500 m): 21-9-1982, RKP 9227.

Pteroceras muriculatus (Reichb.f.) Hunt

Epiphyte. Inland forests. Sitanagar: 15-5-1982, MKV 9021. Interview Island: 23-2-1991, PLN 15678.

Rhynchostylis retusa Bl.

Epiphyte. Inland forests. Karmatong-10: 6-12-1992, Sinha 16264. Interview Island: 7-3-1990, PLN 15154. Paget Island: 6-11-2001, CSR 2014.

Schoenorchis minutiflora (Ridl.) Sm.

Epiphyte. Saddle peak: 1-12-1976, Balak and Nair 4774.

Spathoglottis plicata Bl. is cultivated as ornamental.

ZINGIBERACEAE

Costus speciosus (J. Koenig) Smith

Plants to 2 m tall. Abundant, in damp localities of forests. Aerial Bay: 5-12-1976, Balak and Nair 4985. Swarajgram: 14-11-2001, CSR 2107.

Curcuma petiolata (Koen.) Sm.

An erect herb to 70 cm. Common in shady localities of forests. Sitanagar: 15-5-1982, MKV 9032. Radhanagar: 9-11-2001, CSR 2055.

Globba marantina L.

Herbs. Rare in forests. Swarajgram: 14-11-2001, CSR 2108.

Globba pauciflora King ex Baker

Herbs. Rare in forests. Kalpong: 3-12-2001, CSR 2401.

Kaempferia siphonantha King ex Baker

Herbs. Rare, in moist deciduous forests. Mayabandar: 31-7-1974, Bhargava 1942.

Endemic.

Zingiber acuminatum Valeton

A tuberous herb to 2 m. Rare in damp places of forests. Saddle peak (700 m): 2-12-1976, Balak and Nair 4790.

Zingiber spectabile Griff.

A tuberous herb to 2 m. Rare in damp places of forests. Kalpong: 3-12-2001, CSR 2402.

CANNACEAE

Canna indica L.

Perennial erect herb to 1.5m. Commonly grown as ornamental, also as an escape in moist places in villages. Swarajgram: 14-11-2001, CSR 2109.

MUSACEAE

Musa acuminata Colla

Stout herbs. Rare in inland forests. Swarajgram: 14-11-2001, CSR 2102.

Musa paradisiaca L. is cultivated for its delicious fruits. **Ravenala madagascariensis** Sonn. is cultivated in parks and gardens as an ornament.

MARANTACEAE

Donax canneformis K. Schum. (*Clinogyne grandis*) (Paitra)

Herbs. Common in moist deciduous forests. Saddle peak foot hill: 11-5-1982, MKV 8986.

Phrynium pubinerve Bl.

Herbs. Rare, in moist deciduous forests. Swarajgram: 14-11-2001, CSR 2103.

AMARYLLIDACEAE

Crinum asiaticum L.

A stout herbs. Rare in coastal forests; also planted in gardens. Ross Island: 26-12-2001, CSR 2641.

Crinum latifolium L. and **Zephyranthes grandiflora** Lindl. are cultivated in gardens.

HYPOXIDACEAE

Curculigo orchioides Gaertner

A plant with stout root stock. Rhizome vertical, elongate, 5-20 cm. Frequent on the floor of forests. Swarajgram: 14-11-2001, CSR 2110.

Molineria latifolia (Dryand. ex Ait.) Herb.ex Kurz (*Curculigo latifolia* Dryand. ex Ait.)

A plant with stout root stock. Frequent on the floor of forests. Kalpong: 3-12-2001, CSR 2405.

TACCACEAE

Tacca leontopetaloides (L.) O. Kuntze

Erect herbs, about 1 m high; root stocks tuberous. Very rare in forests. Kalpong: 3-12-2001, CSR 2406.

DIOSCOREACEAE

Dioscorea bulbifera L.

Vine. Tubers solitary or 2-3, large, globose. Stems twining to left; Occasional in forests. Swarajgram: 14-11-2001, CSR 2111.

Dioscorea glabra Roxb.

Vine. Arial Bay: 19-11-1976, Nair 4802; Austin-9: 3-12-1992, Sinha 16225. Kalpong: 3-12-2001, CSR 2407.

Dioscorea nummularia Lam.

Climbers. Inland forests. Interview Island: 5-3-1990, PLN 15120 and 22-2-1991, PLN 15656.

Dioscorea oppositifolia L.

Vine. Rhizomes branched with many long cylindric roots. Occasional in forests. Diglipur-4: 19-11-1976, Nair 4805.

Dioscorea pentaphylla L.

Vine; branchlets twining to left, sometimes prickled. Rare, in forests. Saddle peak (300 m): 26-11-2001, CSR 2300.

Dioscorea vexans Prain et Burk.

Vine. Branchlets twining to left, tomentose, rarely sparsely prickled. Rare, in forests. Kalpong-10: 22-12-1986, Chakrab. and Kindo 12337.

Endemic.

LILIACEAE

Asparagus racemosus Willd.

Armed vine to 8 cm; branchlets glabrous. Occasional in forests and hedges. Austin no.-4: 2-12-2001, CSR 2741.

Gloriosa superba L.

Climber; branchlets glabrous. Rare in open forests. Kalpong: 3-12-2001, CSR 2408.

Allium cepa L. (Onion), **Allium sativum** L. (Garlic) are cultivated for bulbs. **Asparagus densiflorus** and **A. plumosus** are cultivated in gardens.

SMILACACEAE

Smilax lanceaefolia Roxb.

Climbers. Inland forests. Interview Island: 25-2-1991, PLN 15699. Swarajgram: 14-11-2001, CSR 2112.

Smilax odoratissima Bl. (*S. aspericaulis* Wall.)

Climbers. Inland forests. Saddle peak (300 m): 26-11-2001, CSR 2299.

Smilax ovalifolia Roxb. (*S. macrophylla* Roxb.)

Vine to 10 m; prickles recurved. Rare in forests. Kalpong: 3-12-2001, CSR 2409.

AGAVACEAE

Dracaena angustifolia Roxb.

Shrubs. Inland forests. Interview Island: 5-3-1990, PLN 15125.

Dracaena elliptica Thunb.

Herbs. Saddle peak (700 m): 31-3-1977, Balak 5376. Saddle peak (700m): 25-11-2001, CSR 2280.

Dracaena pachyphylla Kurz

Herbs. Frequent in inland forests. Laxmipur: 22-11-1976, Nair 4868. Kalpong: 3-12-2001, CSR 2410.

Dracaena pendula Kurz

Herbs. Frequent in inland forests in rocky loam soils. Durgapur: 5-4-1977, Balak 5464.

Agave americana L. and **Sansevieria roxburghiana** Schult. et Schult.

Are planted in gardens.

COMMELINACEAE

Aneilema ovatum (Hassk.) Wall. ex Cl. (*Diplospermum ovatum*)

Herbs. Keralapuram: 21-11-1976, Nair 4854. Saddle peak (200m): 26-11-2001, CSR 2281.

Commelina attenuata J. Koenig ex Vahl.

Spreading herbs. Occasional in marshy places. Shyamnagar: 27-12-2001, CSR 2811.

Commelina benghalensis L.

Spreading herbs. Root stock with cleistogamous flowers. Common weed in moist fields.

Commelina diffusa Burm.f.

Spreading herbs. Shyamnagar: 27-12-2001, CSR 2810.

Murdannia nudiflora (L.) Brenan.

Spreading herbs. Shyamnagar: 27-12-2001, CSR 2812.

Murdannia spirata (L.) Brueckner

Decumbent or diffuse herbs. Occasional by moist exposed rocks. Durgapur: 5-12-1976, Balak and Nair 4969. Swarajgram: 14-11-2001, CSR 2113.

Pollia thyrsiflora (Bl.) Steud.

Herbs. Occasional in moist deciduous forests. Ramnagar: 3-12-1976, Balak and Nair 4932. Kalpong: 3-12-2001, CSR 2411.

ARECACEAE (PALMAE)

Areca triandra Roxb. (Jangli Supari)

Small trees. Frequent in inland forests. Kalapahar hills (50 m): 1-4-1977, Balak 5475. Interview Island: 7-3-1990, PLN 15167.

Arenga pinnata Merr.

Small trees. Rare in inland forests. Radhanagar: 12-2-1999, PSN Rao 15855.

Calamus andamanicus Kurz

Climbers. Occasional in evergreen forests. Kalpong: 3-12-2001, CSR 2412.

Endemic.

Calamus longisetus Griff.

Canes. Occasional in inland forests. Mayabander: 4-5-1974, Balak 1350. Kalpong: 3-12-2001, CSR 2430.

Calamus palustris Griff. (Malai bet)

Canes. Occasional in inland forests. Saddle peak (200m): 25-11-2001, CSR 2282.

Calamus pseudo-rivalis Becc.

Canes. Occasional in inland forests. Shyamnagar: 27-12-2001, CSR 2809. Endemic.

Calamus rotang L.

Armed, scandent shrub/cane. Occasional clumps in forests. Kalpong: 3-12-2001, CSR 2413.

Calamus viminalis Wendl. **var. fasciculatus** Becc.

Canes. Occasional in inland forests. Shyamnagar: 27-12-2001, CSR 2803.

Caryota mitis Lour.

Tree to 15 m; stem annulate, shining. Scattered in forests. Diglipur: 1-12-2001, CSR 2354.

Daemonorops kurzianus Hook.f.

Armed, scandent shrub. Rare in forests. Shyamnagar: 27-12-2001, CSR 2809.

Daemonorops manii Becc.

Armed, scandent shrub. Occasional clumps in forests. Saddle peak (200m): 25-11-2001, CSR 2283.

Licuala peltata Roxb. (Silai Patti)

Arborescent, shrubs to 8 m. Occasional clumps in forests. Lamia Bay: 15-5-1982, MKV 9015.

Nypa fruticans Wurmb. **(**Dani Palm, Khajur)

Shrub. Occasional clumps in mangroves. Mayabandar: 3-5-1974, Balak 1302.

Phoenix paludosa Roxb.

Shrub. Occasional clumps in mangroves. Mayabandar: 3-5-1974, Balak 1303.

Areca catechu L. (Areca nut), **Caryota urens** L., **Cocos nucifera** L. (Coconuts), **Elaeis guinensis** are cultivated. **Roystonia regia** (H.B.& K) F. Cook. (Royal Palm) is planted in gardens.

PANDANACEAE

Pandanus andamanensis Kurz (Kewda)

Much branched shrub or small tree. Rare, in stunted hilltop evergreen forests. Kalpong: 3-12-2001, CSR 2420.

Pandanus odoratissimus L.f. (Chatai Patti, Kewda)

Much branched shrub or small tree. Rare, in evergreen forests. Kalpong: 3-12-2001, CSR 2415.

Pandanus tectorius Soland. (Chatai Patti, Kewda)

Trees. Common in littoral forests. Lamia bay: 12-5-1982, MKV 9002. Paget Island: 6-11-2001, CSR 2015.

ARACEAE (AROIDEAE)

Amorphophallus campanulatus (Roxb.) Bl. ex Decne

Cormous herbs. Occasional in forests. Kalpong: 3-12-2001, CSR 2421.

Amorphophallus carnosus Engl.

Cormous herbs. Occasional in forests. Shyamnagar: 27-12-2001, CSR 2808.

Colocasia virosa Kunth

Cormous herbs. Common in inland forests. Saddle peak (150m): 25-11-2001, CSR 2286.

Pothos roxburghii de Vriese (*P. scandens* L.)

Climbers. Common on trunks of the trees. Paget Island: 6-11-2001, CSR 2016. Saddle peak (150m): 27-11-2001, CSR 2301.

Rhaphidophora laciniata (Burm.f.) Merr. **(***R. pertusa* Schott**)**

Climbers. Scattered on trunks of the trees. Saddle peak (200m): 25-11-2001, CSR 2287.

Scindapsus officinalis Schott (Pipal)

Climbers. Scattered in forests. Smith Island: 25-11-1976, Nair 4905; Kalpong: 3-12-2001, CSR 2422; South Island: 18-12-2001, CSR.

Typhonium divaricatum (L.) Decne

Climbers. Shyamnagar: 27-12-2001, CSR 2807.

Acorus calamus L., **Caladium bicolor** (Aiton) Vent., **Colocasia esculenta** (L.) Schott, **Dieffenbachia picta** (Lodd.) Schott, **Rhaphidophora pertusa** (Roxb.) Schott are cultivated in gardens.

PONTEDERIACEAE

Eichhornia crassipes (C. Martius) Solms-Loub.

Free floating or rooted, stolons, slender, roots elongate, fibrous. Occasional in still or slow floating waters; aggressive colonizer. Aerial Bay: 21-11-2001, CSR 2184.

Monochoria vaginalis (Burm.f.) C. Presl.

Rooted, aquatic herb, 20-70 cm; root stock thick, erect. Occasional in still waters and marshy places. Shyamnagar: 27-12-2001, CSR 2806.

LEMNACEAE

Lemna perpusilla J. Torrey

Minute floating herbs. Fronds in groups of 2 or 3 sometimes solitary. Common in stagnant waters. RK Gram: 21-11-2001, CSR 2145.

NAJADACEAE

Najas indica (Willd.) Cham.

Herbs. Rare, in stagnant waters. RK Gram: 21-11-2001, CSR 2144.

FLAGELLARIACEAE

Flagellaria indica L.

Climbers. Frequent in coastal forests. Mohanpur: 7-12-2001, CSR 2492.

ERIOCAULACEAE

Eriocaulon truncatum Buch.-Ham. ex Mart.

A tufted, annual herb. Common in marshy places by thin layer of soil and forests streams. Swarajgram: 14-11-2001, CSR 2114.

CYPERACEAE

Abildgardia ovata (Burm.f.) Kral

Stem densely tufted, 5-25 cm, slender. Common weed in plains. Mohanpur: 7-12-2001, CSR 2494.

New record.

Bulbostylis barbata (Rottb.) Clarke

Stem 6-30 cm. Common weed. Mohanpur: 7-12-2001, CSR 2493.

Cyperus bifax Clarke

Stems to 35 cm tall, rigid. Rare, in moist sandy places. Mohanpur: 7-12-2001, CSR 2490.

Cyperus compactus Retz

Stems to 30 cm. Lamia bay to Kalipur: 1-4-1977, Balak 5432; Austin no-9: 3-12-1992, Sinha and Tigga 16203.

Cyperus compressus L.

Stems to 35 cm tall, rigid. Rare, in moist places prefarably nearby paddy fileds. Radhanagar: 9-11-2001, CSR 2064.

Cyperus difformis L.

Stem 8-30 cm, erect, to 2.5 mm wide, slender, triquetrous. Occasional weed in moist fields. Mohanpur: 7-12-2001, CSR 2494.

Cyperus diffusus Vahl

Stem 8-30 cm, erect. Austin no-9: 3-12-1992, Sinha and Tigga 16207.

Cyperus haspan L.

Stem solitary, often tufted, to 50 cm, erect, to 3 mm wide, slender, triquetrous, obscurely winged. Occasional in moist ditches and along streams.

Cyperus iria L.

Stem sparsely tufted or solitary, to 75 cm, erect, triquetrous. Common in wet fields. Keralapuram: 5-12-1976, Balak and Nair 4984.

Cyperus javanicus Houtt.

Stems to 1 m. Diglipur: 19-11-1976, Nair 4818; Narcondum Island: 22-9-1977, Nair 6211.

Cyperus kyllinga Endl.

Stems to 15 cm tall. Karmatong-9 (Mayabandar): 7-12-1992, Sinha 16280. Radhanagar: 9-11-2001, CSR 2060.

Cyperus odoratus L.

Stems to 30 cm, odorous. Occasional in moist grassy places. Kalpong: 3-12-2001, CSR 2428.

Cyperus pangorei Rottb.

Stem sparsely tufted or solitary, to 75 cm, erect, triquetrous. Common in wet fields. Swarajgram: 14-11-2001, CSR 2115.

Cyperus pilosus Vahl

Stem sparsely tufted, 1m, erect, to 7 mm wide rigid, triquetrous. Rare, in ditches, edges of tanks and ricefields. Lamia Bay to Ramnagar: 29-3-1977, Balak 5343. Keralapuram: 5-12-1976, Balak and Nair 4972.

Cyperus platystylis R.Br.

Stems to 35 cm tall. Ganeshnagar: 17-11-2001, CSR 2156.

Cyperus pumilus L.

RK Gram: 21-11-2001, CSR 2149.

Cyperus rotundus L.

Stem sparsely tufted, to 20 cm, erect, to 1 mm wide, rigid, triquetrous. Common weed in moist places. Diglipur: 5-12-1976, Balak and Nair 4968.

Cyperus sanguinolentus Vahl

Stems to 25 cm tall. Karmatong-10 (Mayabandar): 6-12-1992, Sinha 16257.

Cyperus tenuiculmis Boeckeler

Stem solitary, to 40 cm high, ribbed, triquetrous. Occasional along margins of paddy fields. Mohanpur: 7-12-2001, CSR 2495.

Cyperus tenuispica Steudel

Stem tufted, 5-40 cm. Rare weed of moist localities. RK Gram: 21-11-2001, CSR 2147.

Eleocharis acutangula (Roxb.) Schultes

Stem to 60 cm, erect, stout, tufted, finely ribbed. Common weed of dried up rice fields and marsh localities. Mohanpur: 7-12-2001, CSR 2496.

Eleocharis atropurpurea (Retz.) C. Presl.

Stem dwarf, to 10 cm, to 0.5 mm across, slender, angular, smooth. Common weed of dried up rice fields and marsh localities. Mohanpur: 7-12-2001, CSR 2497.

Eleocharis dulcis (Burm.f.) Hans.

Stems to 75 cm tall, angular, smooth. Rare, nearby stagnant water bodies. Shyamnagar: 27-12-2001, CSR 2812.

Eleocharis geniculata (L.) Roemer et Schultes

Stem to 30 cm. to 1 mm across, erect, angular, smooth. Common weed of dried up rice fields and marsh localities. Swarajgram: 14-11-2001, CSR 2116.

Fimbristylis acuminata Vahl

Sedge, to 50 cm tall. Rare in dried up rice fields. Diglipur: 5-12-1976, Balak and Nair 4960. Mayabandar: 6-12-2001, CSR 2471.

Fimbristylis aestivalis (Retz) Vahl

Stem to 30 cm. Mohanpur: 7-12-2001, CSR 2498.

Fimbristylis argentea (Rottb.) Vahl

Stem tufted, to 7 cm, suberect, 0.5 mm wide, trigonous, striate. Rare along streams.

Fimbristylis bisumbellata (Forsskal) Bubani

Stem tufted, 7-30 cm, erect, 1 mm wide, slender, trigonus, scabrid. Common weed of moist places.

Fimbristylis cymosa R. Br.

Stem densely clustered, 10-20 cm, rigid, 1.5 mm wide, trigonous or subterete. Occasional in sandy moist places.

Fimbristylis dichotoma (L.) Vahl

Stem caespitose, to 70 cm, rigid, 2.5 mm wide. Rare, in moist places.

Fimbristylis ferruginea (L.) Vahl

Stem to 30 cm. Sibpur: 28-11-1976, Balak and Nair 4717.

Fimbristylis miliacea (L.) Vahl (*F. quinquangularis* Kunth)

Stem caespitose, 10-30 cm, rigidulous, 1 mm wide. Occasional in marshy places. Laxmipur: 6-4-1977, Balak 5482; Diglipur: 5-12-1976, Balak and Nair 4949; Lamia bay to Ramnagar: 29-3-1977, Balak 5340.

Fimbristylis ovata (Burm.f.) Kern.

Stem to 30 cm. Mohanpur: 7-12-2001, CSR 2499.

Fimbristylis schoenoides (Retz.) Vahl

Stem tufted, to 35 cm, slender, 0.8 mm wide, base incrassate. Occasional in rice fields. Diglipur: 5-12-1976, Balak and Nair 4951.

Fimbristylis tenera Schult. (*F. monticola* Hochst. ex Steud.)

Caespitose herbs to 25 cm tall. Rare, weed. Mayabandar: 6-12-2001, CSR 2471.

Fimbristylis tetragona R. Br

Stem densely caespitose, to 40 cm, slender, 1 mm wide, rigidulous, 4-angular. Rare in moist places.

Fuirena ciliaris (L.) Roxb.

Stem to 40 cm, rigid, angular, scaberulous. Occasional weed of rice fields. Swarajgram: 14-11-2001, CSR 2117.

Fuirena wallichiana Kunth

Stem to 60 cm, erect, 3 mm wide, rigid, sharply trigonous, hollow, smooth. Rare, weed of rice fields.

Hypolytrum nemorum (Vahl) Spreng.

Stem to 30 cm. Laxmipur: 22-11-1976, Nair 4865; Aerial Bay: 5-4-1977, Balak 5471.

Kyllinga brevifolia Rottb.

Stem spaced, 12-20 cm, erect, 1 mm wide, slender, triquetrous. Common in marshy places. Kalpong: 3-12-2001, CSR 2402.

Kyllinga triceps Rottb.

Stem densely tufted, 4-12 cm, erect, 0.5 mm wide, slender, triquetrous. Common weed of sandy localities. Mohanpur: 7-12-2001, CSR 2500.

Mariscus clarkei (Cooke) T. Koyama

Perennial, rhizomatous, stoloniferous herb. Rare, on wet soil. Diglipur: 21-11-2001, CSR 2200.

Mariscus paniceus (Rottb.) Vahl

Perennial, rhizomatous, stoloniferous herb. Rare, on wet soil. Radhanagar: 9-11-2001, CSR 2067.

Pycreus polystachyos (Rottb.) Beauv.

Stem tufted, to 40 cm tall, triquetrous, smooth. Rare weed of marshy places. Kalpong: 3-12-2001, CSR 2403.

Pycreus pumilus (L.) Nees

Annual, densely trufted herb, to 15 cm tall. Occasional weed of dried up rice fields and marsh localities. Mayabandar: 6-12-2001, CSR 2472.

Schoenoplectus articulatus (L.) Palla

Perennial, robust herb to 80 cm tall. Occasional in water logged areas. Diglipur: 20-11-1976, Nair 4846. 30-11-2001, CSR 2348.

Schoenoplectus littoralis (Schrad.) Palla (*Scirpus littoralis* Schrad.)

Perennial, rhizomatous herb. Occasional in marshy places. Kalpong: 26-12-1986, Chakrab. and Kindo 12606.

Schoenoplectus supinus Palla (*Scirpus supinus* L.)

Annual, tufted herb to 30 cm tall. Rare, in moist places. Diglipur: 21-11-2001, CSR 2201.

Scirpus juncoides Roxb.

Stem to 30 cm. Rare, in moist places. Mayabandar: 6-12-1992, Sinha 16267.

Scleria biflora Roxb. (*Scleria lithosperma* (L.) Sw.)

Annual, closely tufted herb, to 50 cm tall. Occasional in sandy places of forests. Saddle peak: 1-12-1976, 1-12-1976, Balak and Nair 4756.

Scleria scrobiculata Nees et Mey. ex Nees

Stem to 30 cm. Occasional in sandy places of forests. Laxmipur: 6-4-1977, Balak 5478; Austin no.4: 2-12-2001 CSR 2080.

Scleria terrestris (L.) Fassett.

Perennial herb with short, creeping woody rhizome, to 1.5 m tall. Occasional in forests. Kalpong: 3-12-2001, CSR 2429.

POACEAE (*nom. alter.* GRAMINEAE)

Alloteropsis cimicina (L.) Stapf

Annual, delicate grass with tufted culms, to 40 cm. Occasional weed in fields. Swarajgram: 14-11-2001, CSR 2118.

Andropogon pumilus Roxb.

Annual, with tufted culms to 50 cm. Occasional weed in fields. Mayabandar: 6-12-2001, CSR 2473.

Apluda mutica L.

A perennial, tufted grass. Rare, near moist rocky places. Diglipur: 21-11-2001, CSR 2189.

Aristida adscensionis L.

Erect herbs to 60 cm tall. Common weed in plains. Kalpong: 3-12-2001, CSR 2411.

Aristida funiculata Trin. et Rupr.

Slender erect, tufted herbs, about 50 cm high. Occasional in cultivated fields.

Aristida hystrix L.f.

Culms tufted to 50 cm. Rare weed. Kalpong: 3-12-2001, CSR 2403.

Aristida setacea Retz.

Culms tufted, stout, robust, to 1 m. Rare, in waste lands. Shyamnagar: 27-12-2001, CSR 2805.

Arthraxon prionoides (Stend.) Dandy

A perennial, tufted grass. Rare, near moist rocky places. Shyamnagar: 27-12-2001, CSR 2804.

Brachiaria distachya (L.) Stapf

A much branched herb. Frequent in waste places. Shyamnagar: 27-12-2001, CSR 2803.

Brachiaria eruciformis (Smith) Griseb.

An erect annual grass to 75 cm high. Common weed. Radhanagar: 9-11-2001, CSR 2056.

Brachiaria ramosa (L.) Stapf

An erect, annual grass to 70 cm high. Weed of cultivation. Diglipur: 21-11-2001, CSR 2199.

Cenchrus ciliaris L.

Annual, ascending grass. Common weed. Shyamnagar: 27-12-2001, CSR 2802.

Centotheca lappacea (L.) Desv.

Annual, ascending grass. Common weed. Shibpur: 28-11-1976, Chakrab and Kindo 4712; Austin no-9: 3-12-1992, B.K. Sinha 16201.

Chloris barbata Sw.

Perennial herbs to 1 m tall. Common weed. Mayabandar: 6-12-2001, CSR 2474.

Chrysopogon aciculatus (Retz.) Trin.

Culms to 50 cm, tufted. Occasional weed. Keralapuram: 5-12-1976, Balak and Nair 4980. Swarajgram: 14-11-2001, CSR 2119.

Chrysopogen fulvus (Sprengel) Chiov.

Culms to 60cm, densely tufted at base. Occasional weed. Ramnagar: 4-12-2001, CSR 2431.

Coix lacryma-jobi L.

A tall, perennial leafy, monoecious grass. Rare in moist localities near forests. Diglipur: 21-11-2001, CSR 2198.

Cymbopogon nardus (L.) Rendl.

Culms to 1.5 m. Rare, in forests. Austin No.:2: 25-12-2001, CSR 2600.

Cynodon dactylon (L.) Pers.

A perennial, stoloniferous grass. Common weed. Radhanagar: 9-11-2001, CSR 2057.

Cyrtococcum accrescens (Trin.) Stapf

An annual grass. Diglipur-4: 19-11-1976, Nair 4809.

Dactyloctenium aegyptium (L.) Beauv.

An annual grass. Culms erect or creeping. Very common weed. Swarajgram: 14-11-2001, CSR 2120.

Dendrocalamus strictus (Roxb.) Nees

Culms solid or with a small cavity, up to 15 m tall. Occasional in forests; often planted in gardens. Shyamnagar: 27-12-2001, CSR 2795.

Dichanthium annulatum (Forsskal) Stapf

Culms to 70 cm, erect or ascending. Common in rice fields. Shyamnagar: 27-12-2001, CSR 2794.

Dichanthium pertusum (L.) Clayton (*Bothriochloa pertusa* (L.) A. Camus)

Perennial, tufted, erect grass to 70 cm long. Occasional weed. Diglipur: 21-11-2001, CSR 2197.

Digitaria bicornis (Lam.) Roemer et Schultes

An annual tufted grass. Common in fields. Ramnagar: 4-12-2001, CSR 2432.

Digitaria ciliaris (Retz.) Koeler.

An annual tufted grass. Occasional weed. Shyamnagar: 27-12-2001, CSR 2792.

Digitaria longiflora (Retz.) Pers.

An annual tufted grass. Occasional in waste lands. Mayabandar: 6-12-2001, CSR 2474.

Digitaria tomentosa (Willd.) Henr.

A perennial tufted grass. Rare weed of fields. Diglipur: 21-11-2001, CSR 2196.

Dimeria ornithopoda Trin.

Saddle peak: 2-12-1976, Balak and Nair 4788. Ramnagar: 4-12-2001, CSR 2433.

Dinebra retroflexa (Vahl) Panz.

An annual. Common weed, nearby paddy fields. Shyamnagar: 27-12-2001, CSR 2791.

New record.

Dinochloa scandens (Bl.ex Nees) O.K. (*D. andamanica* Kurz)

Scandent bamboo. Occasional in forests. Saddle peak: 27-3-1977, Balak 5186; Kalpong: 23-12-1986, Chakrab and Kindo 12352. Saddle peak (100 m): 24-11-2001, CSR 2244.

Echinochola colona (L.) Link

An annual, tufted grass to 75 cm. Common in cultivated wet fields. Diglipur: 5-12-1976, Balak and Nair 4954. Radhanagar: 9-11-2001, CSR 2058.

Echinochloa crusgalli (L.) Beauv.

An annual, tufted grass to 75 cm. Occasional in rice fields. Ramnagar: 4-12-2001, CSR 2434.

Echinochola stagnina (Retz) Beauv.

Keralapuram: 5-12-1976, Balak and Nair 4974.

Eleusine indica (L.) Gaertn.

An annual, tufted grass to 50 cm tall. Common near bunds of rice fields. Diglipur: 5-12-1976, Balak and Nair 4956.

Eragrostiella bifaria (Vahl) Bor

A perennial, tufted grass. Occasional in rice fields. Diglipur: 21-11-2001, CSR 2195.

Eragrostis atrovirens (Dest.) Trin. ex Steud.

A perennial, tufted grass to 1m tall. Rare, in waste places. Shyamnagar: 27-12-2001, CSR 2791.

Eragrostis tenella (L.) P. Beauv. ex Roemer et Schultes

Culms 20-60 cm. Abundant in cultivated fields. Narcondum Island: 22-9-1977, Nair 6205. Diglipur: 21-11-2001, CSR 2194.

Eragrostis unioloides (Retz.) Nees ex Steud.

Culms 15-70 cm. Occasional in waste lands and cultivated fields. Keralapuram: 5-12-1976, Balak and Nair 4979. Radhanagar: 9-11-2001, CSR 2059.

Eriochloa procera (Retz) Hubb.

Culms to 50 cm. Occasional in paddy fields. Ramnagar: 4-12-2001, CSR 2434.

Heteropogon contortus (L.) P. Beauv. ex Roemer et Schultes

A tufted, gregarious grass, above 1 mm height. Frequent in hedges and fields. Narcondum Island: 22-9-1977, Nair 6212.

Imperata cylindrica (L.) Raensch.

An erect, perennial grass with solid leafy stems and creeping root stocks. Common in forest clearings and cultivated fields. Mayabandar: 6-12-2001, CSR 2475.

Ischaemum indicum (Houtt.) Merr.

Perennials. Culms to 50 cm, decumbent. Occasional in fields. Keralapuram: 5-12-1976, Balak and Nair 4979. Kalpong: 25-12-1986, Chakrab and Kindo 12385.

Ischaemum rugosum Salisb.

Perennials. Culms to 60 cm, decumbent. Common in wet fields. Sibpur: 28-11-1976, Balak and Nair 4926; Diglipur: 5-12-1976, Balak and Nair 4953. Ramnagar: 4-12-2001, CSR 2436.

Iseilema anthephoroides Hackel

Annual. Occasional in waste lands. Diglipur: 21-11-2001, CSR 2193.

Iseilema laxum Hackel

Annuals; culms to 30 cm, decumbent. Occasional in waste lands. Shyamnagar: 27-12-2001, CSR 2790.

Leptochloa chinensis (L.) Nees

An annual, tufted, marshy grass to 70 cm. Occasional on bunds of rice fields and near the water courses. Shyamnagar: 27-12-2001, CSR 2803.

Microchloa indica (L.f.) Beauv.

An annual, tufted grass to 20 cm. Occasional in fields. Shyamnagar: 27-12-2001, CSR 2802.

Mnesithea granularis (L.) Koning et Sosef (*Hackelochloa granularis* (L.) Kuntze)

Culms to 50 cm. Frequent along water canal sides in cultivated fields. Diglipur: 21-11-2001, CSR 2192.

Ophiuros exaltatus (L.) Kuntze

Culms to 1.5 m. Occasional, along water courses. Diglipur: 21-11-2001, CSR 2192.

Oplismenus burmannii (Retz.) Beauv.

An annual, slender grass. Culms up to 50 cm tall. Common grass in shade, moist places in forests. Laxmipur: 28-11-1976, Nair 4887.

Oplismenus compositus (L.) Beauv.

An annual, slender grass. Culms up to 50 cm tall. Common grass in shade, moist places in forests. Laxmipur: 24-11-1976, Nair 4893.

Oryza rufipogon Griff.

A perennial, stoloniferous, aquatic grass. Occasional near the tanks, pools and ditches.

Panicum psilopodium Trin.

Culms 30-60 cm tall. Occasional in marshy areas. Ramnagar: 4-12-2001, CSR 2438.

Panicum repens L.

Culms to 1 m. Occasional, near margins of tanks. Diglipur: 21-11-2001, CSR 2191.

Paspalidium flavidum (Retz.) A. Camus

Culms tufted, 10-70 cm, decumbent. Occasional in fields. Diglipur: 21-11-2001, CSR 2189.

Paspalum commersonii Lam.

Culms to 1 m. Occasional in fields. Kalipur: 30-3-1977, Balak 5367.

Paspalum conjugatum Berg.

Diglipur: 5-12-1976, Balak and Nair 4957.

Paspalum paspalodes (Michx.) Scribner

Culms up to 50 cm. Occasional in marshy areas. Diglipur: 21-11-2001, CSR 2188.

Paspalum scrobiculatum L.

Culms 10-60 cm. Commonly found in rice fields. Ramnagar: 4-12-2001, CSR 2439.

Pennisetum purpureum Schum.

Culms decumbent, to 1 m. Occasional in waste lands. Diglipur: 22-11-2001, CSR 2199.

Phragmites karka (Retz.) Trin. ex Steud.

An erect, tall grass up to 3 m high. Occasional along streams in forests. Diglipur: 21-11-2001, CSR 2190.

Pseudosorghum fasciculare (Roxb.) Camus

An annual, tufted grass. Culms upto 75 cm tall, decumbent below. Rare, in forests. Shyamnagar: 27-12-2001, CSR 2801.

Rottboellia exaltata L.f.

Culms to 30 cm, erect. Rare, near water canals and ditches. Shyamnagar: 27-12-2001, CSR 2800.

Sacciolepis indica (L.) Camus

Culms to 50 cm. Common weed. Diglipur: 5-12-1976, Balak and Nair 4948. Diglipur: 20-11-2001, CSR 2175.

Saccharum spontaneum L.

Culms to 3 m; nodes waxy. Rare, in moist localities along stream sides. Aerial Bay: 23-12-2001, CSR 2673.

Setaria glauca (L.) Beauv.

Culms to 50 cm. Common weed. Diglipur: 20-11-2001, CSR 2174.

Setaria pallide-fusca (Schum.) Stapf et Hubb.

Culms to 50 cm. Common weed. Diglipur: 21-11-2001, CSR 2189.

Setaria pumila (Poiret) Roemer et Schultes

Culms to 50 cm. Common weed. Diglipur: 21-11-2001, CSR 2188.

Setaria verticillata (L.) P. Beauv.

An annual, tufted grass to 50 cm tall. Occasional, weed of cultivated fields. Mayabandar: 6-12-2001, CSR 2476.

Spinifex littoreus (Burm.f.) Merr.

Coastal grass. Ross Island: 14-12-2001, CSR 2641.

Sporobolus coromandelianus (Retz.) Kunth,

Culms 10-15 cm. Occasional weed. Diglipur: 21-11-2001, CSR 2185.

Sporobolus diander (Retz.) P.Bauv.

Culms 20-60 cm. Occasional weed. 21-11-2001, CSR 2187.

Themeda triandra Forsk.

Culms to to 1m. Grown in open forests. Diglipur: 5-12-1976, Balak and Nair 4955. 21-11-2001, CSR 2186.

Thuarea involuta (Forst.) R.Br.ex R.& S.

Coastal prostrate grass, rooting at nodes. Ross Island: 14-12-2001, CSR 2640.

Thysanolaena maxima (Roxb.) O.K.

Culms to to 1m. Grown in open forests. Diglipur: 5-12-1976, Balak and Nair 4940; 1-12-2001, CSR 2379.

Urochloa panicoides P. Beauv.

Culm to 30 cm. Occasional weed. Diglipur: 21-11-2001, CSR 2184.

Vetiveria zizanioides (L.) Nash

Culms tufted, to 2m. Rare, along water courses. Diglipur: 21-11-2001, CSR 2183.

Bambusa arundinacea (Retz.) Willd. is raised in plantations by Forest Department. **Oryza sativa** L. var. **sativa** (Paddy, Rice), **Pennisetum americanum** (L.) Leeke (Bajra), **Sorghum vulgare** Pers. (Jowar) and **Zea mays** L. (Maize) are cultivated for food grains. **Saccharum officinarum** L. (Sugar Cane), is cultivated for cane.

GYMNOSPERMS

CYCADACEAE

Cycas rumphii Miq.

Palms. Occasional in beach forests. Ross Island: 26-12-2001, CSR 2639.

GNETACEAE

Gnetum scandens Roxb.

Lianas. Scattered in evergreen forests. Saddle peak (250 m): 27-3-1977, Balak 5175; 1-12-1976, Balak and Nair 4745; 9-5-1982, MKV Rao 8935; 27-11-2001, CSR 2302. Interview Island: 13-2-1959, Thoth. 9214 and 5-3-1990, PLN 15119.

PODOCARPACEAE

Podocarpus neriifolius D. Don (Titmin)

Conifer. Rare, in hill top evergreen forests. Kalpong: 3-12-2001, CSR 2402.

PTERIDOPHYTA
(Ferns and fern allies)

Acrostichum aureum L. **(Acrostichaceae)**

Mangrove fern. Occasional. Diglipur-4: 5-12-1976, Balak and Nair 4941, Mohanpur: 26-12-2001, 2650.

Adiantum lunulatum Burm.f. **(Adiantaceae)**

Narcondum Island: 22-9-1977, Nair 6208. Mayabandar: 6-12-2001, CSR 2477.

Adiantum philippense L. **(Adiantaceae)**

Austin No.-9: 3-12-1992, Sinha 16205.

Angiopteris evecta (Forst.) Hoff. **(Angiopteridaceae)**

Karmatong-10: 4-12-1992, Sinha 16243.

Antrophyum reticulatum (Forst.) Kaulf. **(Vittariaceae)**

Ramnagar: 3-12-1976, Balak and Nair 4930; Saddle peak (200m): 28-3-1977, Balak 5405.

Asplenium nidus L. **(Aspleniaceae)**

Saddle peak range (200m): 28-3-1977, Balak 5403. Saddle peak (150m): 27-11-2001, CSR 2303.

Bolbitis appendiculata (Willd.) Iwatsuki **ssp. vivipara** (Hook.) Henn. **(Lomariopsidaceae)**

Saddle peak range (200m): 28-3-1977, Balak 5404.

Ceratopteris thalictroides Brongn. **(Ceratopteridaceae)**

Ferns of marshy localities. Kalpong: 3-12-2001, CSR 2404.

Cheilanthes tenuifolia (Burm.f.) Sw. **(Adiantaceae)**

Saddle peak (250 m): 21-9-1982, RKP 9229.

Cyclosorus parasiticus (L.) Farwell (*Christella parasitica* (L.) Lev.) **(Thelypteridaceae)**

Kishorinagar: 5-12-2001, CSR 2452.

Diplazium esculentum (Retz.) Sw. (**Athyriaceae)**

Kishorinagar: 5-12-2001, CSR 2451.

Drymoglossum piloselloides Presl (**Polypodiaceae)**

Laxmipur: 6-4-1977, Balak 5491.

Drynaria quercifolia (L.) J. Sm. **(Drynariaceae)**

Kalpong: 3-12-2001, CSR 2405.

Humata repens (L.f.) Diels **(Davalliaceae)**

Saddle peak (500 m): 1-12-1976, Balak and Nair 4777.

Onychium siliculosum (Desv.) C.Chr. **(Pteridaceae)**

Narcondum Island: 22-9-1977, Nair 6217.

Phymatosorus nigrescens (Bl.) Pinch. **(Polypodiaceae)**

Saddle peak (200 m): 28-3-1977, Balak 5402.

Pteris biaurita L. **(Pteridaceae)**

Karmatong-10: 6-12-1992, Balak 5199.

Pteris martensioides Willd. **(Pteridaceae)**

Saddle peak (100 m): 28-3-1977, Balak 5199. 27-11-2001, CSR 2304.

Pteris pellucida Presl. **(Pteridaceae)**

Kishorinagar: 5-12-2001, CSR 2453.

Pteris quadriaurita Retz. **(Pteridaceae)**

Kalpong: 3-12-2001, CSR 2406.

Pteris vittata L. **(Pteridaceae)**

Interview Island: 13-2-1959, Thoth. 9202 and 25-4-1964, Ellis and Ramam. 18955.

Lycopodium phlegmaria L. **(Lycopodiaceae)**

Saddle peak (500 m): 31-3-1977, Balak 5391.

Lygodium circinatum (Burm.f.) Sw. **(Lygodiaceae)**

Saddle peak (175m): 20-9-1982, RKP 9218; 10-5-1982, MKV 8965. Saddle peak (100 m): 24-11-2001, CSR 2243.

Microsorium pteropus (Bl.) Ching **var. minor** (Bedd.) Balakr. **(Polypodiaceae)**

Rare on rocks at lower elevations. Narcondum Island: 1-10-1977, Nair 6225.

Microsorium punctatum (L.) Copel **(Polypodiaceae)**

Kalpong: 3-12-2001, CSR 2408.

Marsilea quadrifolia L. **(Marsileaceae)**

Kishorinagar: 5-12-2001, CSR 2455.

Schizaea digitata (L.) Sw. **(Schizaeaceae)**

Saddle peak (400 m): 1-12-1976, Balak and Nair 4747. Saddle peak (300 m): 24-11-2001, CSR 2249.

Selaginella ciliaris (Retz.) Sprg. **(Selaginellaceae)**

Saddle peak (500 m): 1-12-1976, Balak and Nair 4775. Saddle peak (500m): 27-11-2001, CSR 2320.

Stenochlaena palustris (Burm.) Bedd. (Deka fern): **(Blechnaceae)**

Leaves used as a vegetable. Kalpong: 3-12-2001, CSR 2410.

Tectaria polymorpha (Wall.ex Hook.f.) Copel **(Aspidiaceae)**

Radhanagar: 25-12-2001, CSR 2712.

Tectaria polymorpha (Wall.ex Hook.f.) Copel **var. cuneifolia** Nonap **(Aspidiaceae)**

Ramnagar: 3-12-1976, Balak and Nair 4927.

Acknowledgements

I am thankful to Dr. V. Jayaraman, Director, NRSC, Hyderabad and Dr. P.S. Roy, Director, IIRS, Dehradun for suggestions and encouragement. I am grateful to Mr. P.R.C. Prasad, JRF, NRSC for associating in the field work. I owe my thanks to Forest Department staff of Andaman and Nicobar Islands for support and cooperation in various ways.

References

Anonymous. 1983. Hundred years of forestry in Andamans. The Centenary celebration of Forest Department, Andaman and Nicobar Islands.

Balakrishnan, N.P. 1989. Andaman Islands – Vegetation and floristics. In: Saldanha, C.J. (ed.), Andaman, Nicobar and Lakshadweep: An Environmental Impact Assessment. Oxford and IBH Publishing House, New Delhi.

Dagar, J.C. 1989. Endemic plant species of the Bay Islands. J. Andaman Sci. Assoc. 5: 161-168.

Dagar, J.C. and Singh, N.T. 1999. Plant Resources of the Andaman and Nicobar Islands. Vol. I-II. Bishen Singh Mahendra Pal Singh, Dehra Dun.

Ellis, J.L. 1989. Plant diversity in the Andamans with emphasis on endangered and endemic species. In: Saldanha, C.J. (ed.), Andaman, Nicobar and Lakshadweep: An Environmental Impact Assessment. Oxford and IBH Publishing House, New Delhi.

Hajra, P.K., Rao, P.S.N. and Mudgal, V. 1999. Flora of Andaman and Nicobar Islands. Vol. 1 (Ranunculaceae to Combretaceae). BSI, Calcutta.

Jain, S.K. 1990. Botanical regions and flora of India. Everyman's Science. 24(6): 213-223.

Lakshminarasimhan, P. and Rao, P.S.N. 1996. A supplementary list of angiosperms recorded (1983-1993) from Andaman and Nicobar Islands. J. Econ. Tax. Bot. 20: 175-185.

Mathew, S.P. 1998. A supplementary report on the flora and vegetation of the Bay Islands, India. J. Econ. Tax. Bot. 22: 249-272.

Parkinson, C.E. 1923. A Forest Flora of the Andaman Islands, Govt. of India, Shimla.

Rao, T.A. and Sastry, A.R.K. 1984. An ecological approach towards classification of coastal vegetation. Indian Forester 98: 594-607.

Rao, P.S.N. 1996. Phytogeography of Andaman and Nicobar Islands. Malayan Nature Journal 50: 57-79.

Reddy, C.S., Prasad, P., Murthy, M. and Dutt, C.B.S. 2004. Census of endemic flowering plants of Andaman and Nicobar Islands, India. J. Econ. Tax. Bot. 28(3): 712-728.

Reddy, C.S. and Dutt, C.B.S. 2005. Some interesting additions to the Flora of Andaman and Nicobar Islands from North Andaman. J. Bombay Nat. Hist. Soc. 102 (1): 133-135.

Reddy, C.S. and Raju, V.S. 2005. Invasion of Alligator weed *Alternanthera philoxeroides* (Mart.) Griseb. in Andaman Islands. J. Bombay Nat. Hist. Soc. 102 (1): 133.

Saldanha, C.J. (ed.) 1989. Andaman, Nicobar and Lakshadweep: An Environmental Impact Assessment. Oxford and IBH Publishing House, New Delhi.

Singh, V.P., Mall, L.P., Garge, A. and Pathak, S.M. 1986. Some ecological aspects of mangrove forests of Andaman Islands. J. Bombay Nat. Hist. Soc. 83(3): 525-537.

Subba Rao, N.V. 1989. Fauna of Andaman and Nicobar Islands: diversity, endemism, endangered species and conservation strategies. In: Saldanha, C.J. (ed.), Andaman, Nicobar and Lakshadweep: An Environmental Impact Assessment. Oxford and IBH Publishing House, New Delhi.

Upadhyay, V.P. 2000. The tropical rain forest ecosystem of Andamans. Sabujima 2000-2001: 69-80.

Vasudevarao, M.K. 1986. A preliminary report on the angiosperms of Andaman-Nicobar Islands. J. Econ. Tax. Bot. **8:** 107-184.

Chapter 6

Diversity of Bioresources of Nilgiris in the Southern Western Ghats, India

R. Sasi* and A. Rajendran

Department of Botany, School of Life Sciences,
Bharathiar University, Coimbatore – 641 046

Introduction

India, a land of physical, cultural, social and linguistic diversity, is also endowed by nature with an ecosystem and tremendous biodiversity, genetic as well as species. It is a veritable emporium of medicinal and aromatic plants and one among the 25 hot spots of the richest and highly endangered eco-regions of the world (Meyers *et al.,* 2000). It also possesses rich floristic wealth of over 15,000 species of higher plants and about one-third of this flora is endemic. Of these over 3,000 are of edible plants (Anon., 1994).

The study area comprises the major portion of the Western Ghats and Sri Lanka biodiversity hotspot (Myers *et al.,* 2000). Western Ghats is an area of exceptional biological diversity and conservation interest, and is "one of the major Tropical Evergreen Forest regions in India" (Rodgers and Panwar, 1988). Use of large number of wild species by the tribals to meet their diverse requirements is largely due to the prevalence of diversity of vegetation in the area.

---

* Corresponding Author: E-mail: sasibotany@gmail.com

Wild edible plants not only supplement the food quantity but also make significant contribution to the population's nutrition throughout the year (Grivetti and Ogle, 2000; Herzog *et al.,* 2012; Ogle, 2001; Ogle *et al.,* 2001; Ogle *et al.,* 2003). Although the principal role of these plants is to supplement the food cultivated in home gardens and other forms of agriculture. Many of the species grown or wild – harvested are reported to have both therapeutic and dietary functions (Ogle *et al.,* 2003).

Nutrients derived from plants are important to human health and complement other food sources. World over, tribal population stores a vast knowledge on utilization of local plants as food material and other specific uses (Sundriyal *et al.,* 1998). People living in most of the rural and remote areas do not produce enough food grains to meet yearly food requirements and at times food supplies are not easily available. Therefore, a large share of rural population is meeting their nutritional requirement through non conventional means, by consuming various wild plants and animal resources.

Study Area

The Nilgiri district in Tamil Nadu is a central focal point and is termed as the manipulation zone in the overall bioregion. Nature has bestowed Nilgiris with magnificent forests with varied and colourful plants and animal life. The Nilgiri Biosphere Reserve is home to a large number flora species comprises of 3300 species of flowering plants. Among these, 1232 are endemic. The genus *Baeolepis* is exclusively endemic to the Nilgiri Biosphere Reserve.

People

There are six major primitive tribes in Nilgiri districts are Todas, Kotas, Kurumbas, Paniyas, Irulas and Kattunayakas. *Todas* or *Tudas,* to whom the most sacred objects on this earth are a holy dairy-man are pastoral people and to a certain extent, nomadic, bred buffaloes that supplied the milk, *Kotas* generally known as Kotagiri or Kokkal were potters, ironsmiths and carpenters, *Kurumbas* were the 'people of the jungle' and lived in caves inside thick forests were the magico-religious sorcerers, *Irulas* have been derived from the word "Irul" (Darkness), the dwellers of the jungle hence the name "People of the darkness". Irulas were herbal doctors and the *Paniyas* literally means "workers" were slaves and bonded labourers, *Kattunayakas* literally means 'kadu' (forests) and nayakan (leader/chief) were hunting and gathering, especially honey.

Biogeography

Western Ghats is one of the eight hottest hot spots of the world are known for its rich biodiversity. The Western Ghats are a chain of highlands running along the western edge of the Indian subcontinent, from Bombay south to the southern tip of the peninsula, through the states of Maharashtra, Karnataka, Kerala and Tamil Nadu, covering an estimated area of 159,000 sq. km. The Nilgiri Biosphere Reserve is located in Southwest India, north of the Palghat gap between 10° 45′–12° 5′ N latitude and 76° 10′–77° 10′ E longitude. Nilgiris are at an elevation of 900 to 2636 meters above

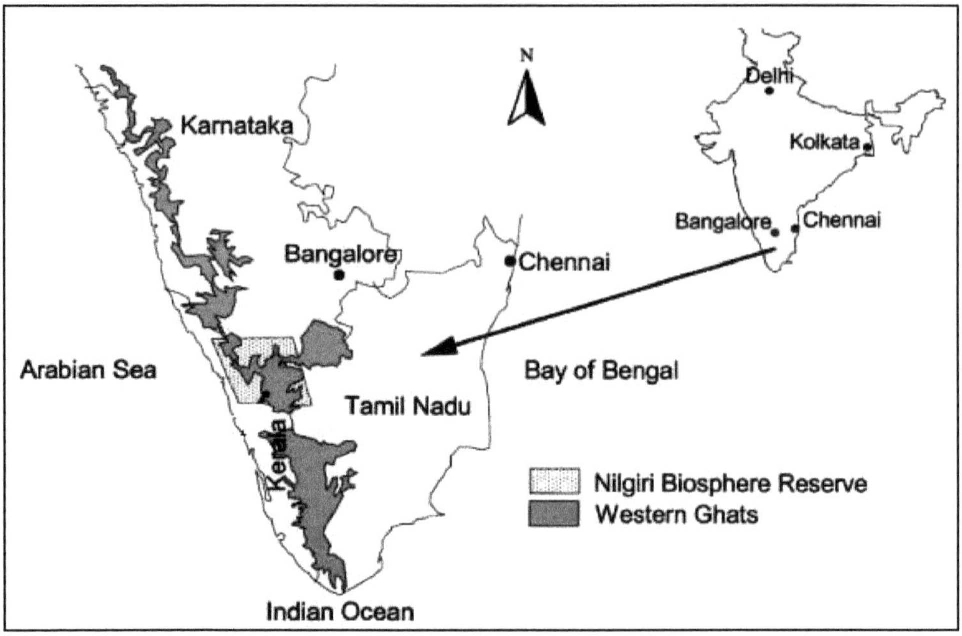

Figure 6.1: India–Location Map of Nilgiri Biosphere Reserve

MSL. The region generally receives 2000–7000 mm of rainfall and is rich in natural resources.

The Nilgiris includes all important forest types just peculiar to the belt such as Evergreen, Semi Evergreen, Moist Deciduous, Shola, Dry Deciduous, Dry Scrub and Grasslands. The soils of the Nilgiris are of lateritic and gneissic origin. The depth of the soil varies from zero to 75 cm only an average and that of subsoil from 3m to 5.2m. Soil is coarse, loamy mixed, isothermic type. Soil pH ranges from 5 to 6.5 and in some localities higher or lower values of pH occur. The annual rainfall ranges from approximately 600- 2000 mm. The temperature varies, the hottest part is from March to May, the coldest part is from December to February.

Enumeration

Acalypha alnifolia Klein *ex* Willd. Sp. Pl. 4: 525. 1805; Hook. f. Fl. Brit. India 5: 415. 1887; Gamble, Fl. Pres. Madras 2: 930. 1957 (repr. ed.); Chandrabose in Henry *et al.,* Fl. Tamil Nadu 2: 220. 1987.

Shrub. Branchlets pubescent. Leaves elliptic- ovate, membranous, margin serrarte, apex acute. Stipules triangular. Flowers, axillary, slender bisexual, spike inflorescence. Male portion slender, elongate, cylindric, usually in lower axils. Anther small, vermiculiform, sessile. Female portion short, bracts copular, dentate, enclosing the ovary. Styles small with filiform hairs. Capsule 3-locular.

Phenology: Feb- Nov.

Parts used: Leaves

Mode of use: Tender leaves are cooked and eaten as vegetable.

Therapeutic value: Crushed leaves are used for dysentery (Senthilkumar *et al.,* 2006).

Acalypha indica L. Sp. Pl. 1003. 1753; Wight, Ic. t. 877. 1844- 1845; Hook. f. Fl. Brit. India 5: 416. 1887; Gamble, Fl. Pres, Madras 2: 930. 1957 (repr. ed.); Chandrabose in Henry *et al.,* Fl. Tamil Nadu 2: 220. 1987.

Herbs. Branchlets apressed pubescent. Leaves apically clustered into flat rosette. Stipules triangular. Leaf blade rhombic-ovate or subovate, membranous, pilosulose along veins, base cuneate, margin serrate, apex acute. Flowers axillary, unbranched, pilosulose, bisexual, margin crenulate. Male portion short, slender. Bracts ovate-triangular or broadly triangular, allomorphic female flower at apex. Male flowers 5-7 per bract; sepals 4, ovate; stamens 8. Female flowers subsessile; sepals 3, triangular, ciliate. Ovary pilose; Styles 3, 2.5-3 mm, 5-laciniate. Capsule 3-locular.

Phenology: Mar-Oct.

Parts used: Twigs.

Mode of use: Tender leaves and shoots are cooked and eaten as vegetable.

Therapeutic value: Whole plant is crushed and the juice is given to cure bronchitis in children (Senthilkumar *et al.,* 2006).

Alangium salvifolium (L. f.) Wang. in Engler, Pflanzenr. 41:9. 1910, subsp. **salvifolim**; Gamble, Fl. Pres. Madras 1: 404. 1957 (repr. ed.); Rajan in Nair and Henry, Fl. Tamil Nadu 1: 183. 1983. *Grewia salvifolia* L.f. Suppl. Pl. 409. 1781. *Alangium lamarckii* Thw. Enum. Pl. Zeyl. 133. 1859; Clarke in Hook. f. Fl. Brit. India 2: 741. 1879. *A. decapetalum* Lam. Encycl. 1: 174. 1783; Wight, Ic. t. 194. 1839.

Trees. Leaves alternate, oblong- oblanceolate. Flowers bisexual, polygamous, or unisexual, usually arranged in 3- to many-flowered cymes, solitary or fascicled. Sepals 5, white, or lilac, free, valvate, abaxially hairy, adaxially usually glabrous, rarely hairy. Petals 5, mostly yellow or white, shorter than sepals. Androgynophore short, usually glabrous. Stamens many, free. Ovary 2-4-loculed; style simple; stigma swollen, peltate. Fruit drupe usually with 2 or 4 drupelets.

Phenology: Feb-Apr.

Parts used: Fruits.

Mode of use: Ripened raw fruits are edible.

Therapeutic value: Fruits are sweet, cooling and purgative and are useful in treating burning sensation and haemorrhages.

Aloe vera (L.) Burm. f. Fl. Ind. 83. 1768; Gamble, Fl. Pres. Madras 3: 1062. 1957 (repr. ed.); Bhargavan in Henry *et al.,* F. Tamil Nadu 3: 37. 1989. *Aloe perfoliata* L. var. *vera* L. Sp. Pl. 320. 1753.

Herbs, succulent. Stems short, suckering freely to form dense clumps. Leaves sub-basal, slightly distichous in seedlings and new shoots, erect, pale green, sometimes with pale spots in very young plants, linear-lanceolate, margin sparsely spiny-dentate, apex 2- or 3-dentate-pointed. Flowers racemes, numerous flowered. Bracts whitish, broadly lanceolate, apex acute. Perianth pale yellow mottled with red, slightly ventricose, outer lobes free, slightly recurved at apex. Stamens exserted. Style conspicuously exserted.

Phenology: Sep-Dec.

Parts used: Pulp of the leaves.

Mode of use: Raw pulp is eaten.

Therapeutic value: Fresh juice is used as cathartic and cooling. It is also used in treating fever and eye diseases (Senthilkumar *et al.,* 2006).

Alternanthera sessilis (L.) R. Br. *ex* DC. Cat. Hort. Monsp. 4: 77. 1813; Wight, Ic. t. 727. 1840; Hook. f. Fl. Brit. India 4: 731. 1885; Melville in Kew Bull. 13: 172. 1958; Kumari in Henry *et al.,* Fl. Tamil Nadu, 2: 191. 1987. *Gomphrena sessilis* L. Sp. Pl. 225. 1753. *Alternanthera triandra* Lam. Encyl. 1: 95. 1783; Gamble, Fl. Pres. Madras 2: 825. 1957 (repr. ed.).

Herbs. Leaf blade linear-lanceolate, oblong-obovate, or ovate-oblong, glabrous or pilose, base attenuate, margin entire or slightly serrate, apex acute or obtuse. Flowers in heads 1-4, axillary, sessile, at first globose, later cylindric. Flowers dense; rachis densely white hairy; Tepals white, ovate, 2-3 mm, glabrous, with a vein, apex acuminate or acute. Stamens 3, filaments, connate into a cup at base; anthers oblong, pseudostaminodes subulate, shorter than stamens, margin entire, apex acuminate, Style very short, stigma shortly parted. Seeds ovoid.

Phenology: May-Sep.

Parts used: Leaves and young twigs.

Mode of use: Tender leaves and shoots cooked and eaten.

Therapeutic value: It helps for improving eyesight (Balakrishnan *et al.,* 2009).

Amaranthus spinosus L. Sp. Pl. 991. 1753; Wight, Ic. t. 513. 1841; Hook. f. Fl. Brit. India 4: 718. 1885; Gamble, Fl. Pres. Madras 2: 819. 1957 (repr. ed.); Kumari in Henry *et al.,* Fl. Tamil Nadu, 2: 191. 1987.

Herb. Leaf blade ovate-rhombic or ovate-lanceolate, glabrous or slightly pubescent along veins when young, base cuneate, margin entire, apex obtuse. Flowers in terminal spike usually with all male flowers at or toward apex. Tepals green, transparent at margin and with green or purple median band, apex acute, with a mucro. Male flowers oblong. Female flowers oblong-spatulate; Filaments nearly as long as or slightly shorter than perianth. Stigmas 3(or 2). Seeds brownish black, subglobose.

Phenology: Jul.-Nov.

Parts used: Leaves.

Mode of use: Tender leaves cooked and eaten.

Therapeutic value: It is used as a laxative, abscesses, boils and wounds (Khan *et al.,* 2009).

Amaranthus tricolor L. Sp. Pl. 989. 1753. *Amaranthus mangostanus* L. Cent. Pl. 1: 32. 1755; Hook. f. Fl. Brit. India 4: 720. 1885; Gamble, Fl. Pres. Madras 2: 820. 1957 (repr. ed.); Kumari in Henry *et al.,* Fl. Tamil Nadu, 2: 191. 1987. *Amaranthus gangeticus* L. Syst. Nat. ed. 10. 1268. 1759 incl. var. *tristis*; Hook. f. Fl. Brit. India 4: 719. 1885; Gamble, Fl. Pres. Madras 2: 819. 1957 (repr. ed.). *Amaranthus oleraceus* Burm. f. Fl. Ind. 198. 1768, non L. 1753; Wight, Ic. t. 715. 1843.

Herb. Leaf blade green, red, purple, or yellow, ovate, ovate-rhombic, or lanceolate, glabrous, base cuneate, margin entire or undulate, apex obtuse or notched, with a mucro. Flowers in dense clusters at leaf axils or in spike at apex. Male and Female flowers in same inflorescences. Bracts and bracteoles ovate-lanceolate, transparent, apex long pointed. Stamens 3. Stigmas 3. Seeds brownish black, subglobose or obovoid.

Phenology: May -Sep.

Parts used: Leaves and twigs.

Mode of use: Leaves are edible and used in stir fries, soups and salads.

Therapeutic value: It increases bile and aid digestion (http://www.herbnature.tk/2011/02/spinachamaranthus- tricolor-l.html)

Amaranthus viridis L. Sp. Pl. ed. 2. 2: 1405. 1763; Gamble, Fl. Pres. Madras 2: 820. 1957 (repr. ed.); Kumari in Henry *et al.,* Fl. Tamil Nadu, 2: 191. 1987. *Euxolus caudatus* (Jacq.) Moq. in DC. Prodr. 13 (2): 274. 1849; Wight, Ic. t. 1773. 1852. *Chenopodium caudatum* Jacq. Coll. Bot. 2: 235. 1788, non L. 1753.

Herb. Leaf blade ovate, ovate-oblong, or ovate-elliptic, base broadly cuneate or subtruncate, margin entire or slightly undulate, apex notched or rounded, with a pointed mucro. Flowers in spikes, erect, slender, terminal ones longer than lateral ones. Bracts and bracteoles lanceolate, shorter than 1 mm, apex pointed. Tepals oblong or broadly oblanceolate, apex acute. Stamens shorter than perianth. Stigmas 3 or 2. Seeds black or brownish black, subglobose.

Phenology: Jun - Oct.

Parts used: Leaves and Young twigs.

Mode of use: Tender leaves and shoots cooked and eaten.

Therapeutic value: It is used for Constipation, inflammation, eczema, bronchitis, anemia, leprosy (Ashok Kumar *et al.,* 2009).

Argyreia elliptica (Roth) Choisy, Convolv. Orient 35. 1834; Chandrabose in Henry *et al.,* Fl. Tamil Nadu, 2: 102. 1987. *Ipomoea elliptica* Roth, Nov. Pl. Sp. 113. 1821. *Lettsomia elliptica* (Roth) Wight, Ic. t. 1356. 1848, in nota; Clarke in Hook. f. Fl. Brit. India 4: 192. 1883; Gamble, Fl. Pres. Madras 2: 640. 1957 (repr. ed.).

Climber. Branchlets sparsely pubescent. Leaves elliptic-ovate to obovate, glabrescent above, pubescent below. Flowers in cymes. Bracts deciduous. Calyx lobed 5, subequal, hardly enlarged. Corolla 5 lobed, pinkish, funnel shaped. Stamens 4, unequal, inserted in the corolla tube. Ovary 2- locular, ovules 2; Style long and stigma biglobose. Fruit berry, conical, fleshy, orange when ripe.

Phenology: Dec.- Jan.

Parts used: Fruit.

Mode of use: Fruits are taken as raw.

Artocarpus heterophyllus Lam. Encycl. 3: 210. 1789, "*heterophylla*"; Jarrett in J. Arnold Arbor. 40: 334. 1959. *Artocarpus integrifolia* auct. non L. f. 1781; Wight, Ic. t. 678. 1840; King in Hook. f. Fl. Brit. India 5: 541. 1888; Gamble, Fl. Pres. Madras 3: 957. 1957 (repr. ed.); Chithra in Henry *et al.,* Fl. Tamil Nadu, 2: 251. 1987.

Trees. Leaves spirally arranged; leaf blade elliptic to obovate, 7-15(or more), lobed on new growth of young trees, leathery, abaxially pale green and with scattered globose to ellipsoid resin cells, adaxially dark green, glabrous, and shiny, base cuneate, margin of mature leaves entire, apex blunt to acuminate, leaves on mature trees entire. Male inflorescences axillary on apical branchlet, sometimes axillary on axillary brachlets, cylindric to conic-ellipsoid, many-flowered but some sterile. Female inflorescences with a globose fleshy rachis. Male flowers: calyx tubular, apically 2-lobed, pubescent; filament straight in bud; anther ellipsoid. Female flowers: calyx tubular, apically lobed; ovary 1-celled; Fruiting syncarp pale yellow when young, yellowish brown when mature, ellipsoid, globose, or irregularly shaped.

Phenology: Feb-Mar.

Parts used: Fruits and seeds.

Mode of use: Ripened fruits are taken raw. Fried and boiled seeds are taken.

Therapeutic value: The young fruits are acrid, astringent, and carminative. The ripe fruits are sweet, cooling, laxative, aphrodisiac and also used as a brain tonic. The seeds are, diuretic, and constipating (Hemborn, 1996).

Artocarpus hirsutus Lam. Encycl. 3: 210. 1789, "*hirsuta*"; Wight, Ic. t. 1957. 1853; King in Hook. f. Fl. Brit. India 5: 541. 1888; Gamble, Fl. Pres. Madras 3: 957. 1957 (repr. ed.); Chithra in Henry *et al.,* Fl. Tamil Nadu, 2: 251. 1987.

Tree. Leaves simple, alternate, coriaceous, entire, ovate, acute and shortly acuminate at apex, rounded at base. Stipules yellow, hairy. Flowers yellowish, monoecious, crowded on globose, solitary, axillary, receptacles. Bracteoles thin, leafy, ovate- lanceolate. Male flower: Perianth in 2- 4 lobed; Stamen 1, filaments thin and anther small. Female flower: Perianth tubular, style exserted and stigma small, rounded; Fruit a large fleshy receptacle, covered with spines, seed one, membranous.

Phenology: Feb-Mar.

Parts used: Fruits and seeds.

Mode of use: Ripened fruits are taken raw. Fried and boiled seeds are taken.

Therapeutic value: Unripe fruit cause vitiation of tridosha and rakta, and cause indigestion and impotency. Ripe fruits are cooling and aphrodisiac. (http://ayurvedicmedicinalplants.com).

Asparagus racemosus Willd. Sp. Pl. 2: 152. 1799; Wight, Ic. t. 2056. 1853; Hook. f. Fl. Brit. India 6: 316. 1892; Gamble, Fl. Pres. Madras 3: 1060. 1957 (repr. ed.); Bhargavan in Henry *et al.,* Fl. Tamil Nadu, 3: 38. 1989.

Subshrubs. Cladodes in fascicles of 3-6(-8), linear, flat, midvein distinct; Leaf spur spinescent, spine straight or subrecurved. Inflorescences developing after cladodes, axillary, each a many-flowered raceme or panicle. Perianth campanulate. Stamens equal, anthers yellow, minute. Style terminal, apex 3-fid; Stigma 3, recurved. Fruit berry, globose.

Phenology: Oct-Dec.

Parts used: Tuber.

Mode of use: Tubers are cooked and eaten.

Therapeutic value: Root extract used in diarrhoea, dysentery and general debility (Senthilkumar *et al.,* 2006).

Asystasia gangetica (L.) T. And. in Thw. Enum. Pl. Zeyl. 235. 1860; Gamble, Fl. Pres. Madras 2: 744. 1957 (repr. ed.). *Justicia gangetica* L. Cent. Pl. 2: 3. 1756. *Asystasia coromandelina* Wight *ex* Nees in Wall. Pl. Asiat. Rar. 3: 89. 1832; Wight, Ic. t. 1506. 1850; Clarke in Hook. f. Fl. Brit. India 4: 493. 1884; Kumari in Henry *et al.,* Fl. Tamil Nadu 2: 141. 1987.

Herbs. Leaf blade ovate to elliptic, glabrous or sparsely pilose especially on veins, adaxially with numerous cystoliths, base truncate to rounded, margin entire or slightly crenulate, apex acuminate. Flowers in racemes axillary or terminal; bracts triangular, pilose. Bracteoles linear-lanceolate, pilose. Calyx lobes linear-lanceolate, outside glandular pilose, margin ciliate. Corolla yellow or white, outside gland-tipped pilose, inside glabrous; tube basally cylindric then gradually widened; lobes obovate to semicircular, middle lobe of lower lip with violet or maroon markings. Stamens included; filaments glabrous; anther thecae. Ovary ellipsoid, style velutinous; stigma slightly capitate, 2-lobed. Capsule pubescent. Seeds irregularly obovate in outline, tuberculate-rugose.

Phenology: Sep-Mar.

Parts used: Leaves.

Mode of use: Young leaves are made into curries and taken.

Therapeutic value: It is used for curing asthma (Akah *et al.,* 2003).

Azadirachta indica A. Juss. in Mem. Mus. Natl. Hist. Nat. 19: 221. t. 2. f. 5. 1830; Wight, Ic. t. 17. 1838; Gamble, Fl. Pres. Madras 1: 127. 1957 (repr. ed.); Nair and Rajan in Nair and Henry, Fl. Tamil Nadu 1: 66. 1983. *Melia azadirachta* L. Sp. Pl. 385. 1753; Hiern in Hook. f. Fl. Brit. India 1: 544. 1875.

Tree. Branches glabrous. Leaves imparipinnate, pulvinus at the base; leaflets alternate to opposite, broad, ovate, subsessile, acuminate. Flowers white, sweet-scented. Sepals obovate, puberulous, imbricate. Petals obovate to oblong, white, margin ciliate. Staminal tubular, puberulous, 10-striate, 10-toothed; teeth 2-lobed; anthers oblong, basifixed. Ovary sub-globose; style linear; stigma trifid. Drupe oblong, 1-seeded.

Phenology: Mar- Aug.

Parts used: Fruits.

Mode of use: Ripened fruits are edible.

Therapeutic value: It is used as purgative, anti-hemorrhoids and anthelmintic (Bhowmik *et al.,* 2010).

Bacopa monnieri (L.) Pennell in Proc. Acad. Nat. Sci. Philad. 98: 94. 1946; Philcox in Kew Bull. 33: 679. 1979; Henry and Chithra in Henry *et al.,* Fl. Tamil Nadu 2: 119. 1987. *Lysimachia monnieri* L. Cent. Pl. 2: 9. 1756. *Moniera cuneifolia* Michaux, Fl. Bor.- Amer. 2: 22. 1803, "*monniera*"; Gamble, Fl. Pres. Madras 2: 669. 1957 (repr. ed.). *Herpestis monnieria* (L.) H. B. K. Nov. Gen. Sp. 2: 366. 1818, nom. illeg.; Hook. f. Fl. Brit. India 4: 272. 1884. *Gratiola monnieri* (L.) L. Syst. Nat. ed. 10. 851. 1759, "*monnieria*"; Roxb. Pl. Cor. t. 178. 1805.

Herb. Leaves sessile, oblong-oblanceolate, margin entire or rarely dentate, apex rounded. Flowers axillary. Bracteoles 2, linear, below calyx. Sepals 5, lower and upper sepals ovate-lanceolate; lateral 2 sepals lanceolate to linear. Corolla blue, purple or white, obscurely 2- lipped; Stamens didynamous. Stigma capitate. Capsule ovoid, enveloped in persistent calyx. Seeds yellow-brown, ellipsoid.

Phenology: May- Dec.

Parts used: Leaves.

Mode of use: Young leaves are used as vegetables.

Therapeutic value: The leaves are used for curing abdominal pain and blood purifier (Jain and Srivastava, 2005).

Berberis tinctoria Lesch. in Mem. Mus. Natl. Hist. Nat. 9: 306. 1822; Fyson, Fl. South Indian Hill St. 2: t. 10. 1932; Gamble, Fl. Pres. Madras 1: 23. 1957 (repr. ed.); Chithra in Nair and Henry, Fl. Tamil Nadu 1: 9. 1983.

Shrub. Scandent branches bearing numerous slender leafy twigs; wood very tough, bright yellow in colour. Leaves when young purplish, obovate, entire or with a few spiny teeth, glabrous. Racemes of flowers drooping, sometimes branched. Petals 6, petaloid, notched, glandular in the inner surface. Stamens 6, anthers small. Stigma subsessile. Fruits berry, purplish red, turning to a dark-blue with glaucous bloom, with the dry style and large round stigma still attached.

Phenology: May- Sep.

Parts used: Fruits.

Mode of use: Raw fruits are taken raw and also used for the preparation of jams.

Boerhavia diffusa L. Sp. Pl. 3. 1753; Gamble, Fl. Pres. Madras 2: 814. 1957 (repr. ed.). Kumari in Henry *et al.,* Fl. Tamil Nadu 2: 188. 1987. *B. repens* L. Sp. Pl. 3. 1753; Hook. f. Fl. Brit. India 4: 709. 1885. *B. procumbens* Banks *ex* Roxb. Fl. Ind. 1: 148. 1820; Wight, Ic. t. 874. 1844- 1845.

Herb. Leaf blade ovate, both surfaces sparsely pubescent, abaxially gray- yellow, wrinkled when dry, base rounded or cuneate, margin undulate, with stout, multicellular hairs, apex obtuse or acute. Flowers in terminal, capitate- cymose panicle. Bracts small, lanceolate, pubescent. Perianth limb bright purple. Stamens 1- 3(-5), slightly exserted or included; Anthocarp clavate, 5- ribbed, sparse pubescence, apex rounded.

Phenology: Jul- Mar.

Parts used: Leaves.

Mode of use: Young leaves are cooked and eaten.

Therapeutic value: Leaf paste is applied on the cuts and wounds to stop bleeding (Senthilkumar *et al.,* 2006).

Bridelia retusa (L.) Spreng. Syst. Veg. 3: 48. 1826; Bedd. Fl. Sylv. t. 260. 1872; Hook. f. Fl. Brit. India 5: 268. 1887 p.p.; Gamble, Fl. Pres. Madras 2: 896. 1957 (repr. ed.). *Cluytia retusa* L. Sp. Pl. 1042. 1753, "*Cluytia*"; Chandrabose in Henry *et al.,* Fl. Tamil Nadu 2: 223. 1987.

Trees. Leaves simple, alternate, entire, pinnately veined, petiolate, stipulate. Flowers small, unisexual, monoecious, few to many flowers grouped into axile glomerules, pedicellate or sessile. Sepals 5, valvate. Petals 5, scale-shaped; Male flowers: stamens 5; filaments connate at base, forming staminal column, apical free and extended filaments surrounding rudimentary ovary. Female flowers: disk pulvinate or saucer-shaped lining flower with a conical or urceolate part, surrounding ovary, tearing as ovary expands; ovary 2-loculed; ovules 2 in each cell; styles 2, free or connate at base, bifid or entire at apex. Fruits drupe or rarely capsules.

Phenology: May- Jun.

Parts used: Fruit.

Mode of use: Ripened fruits are edible.

Borassus flabellifer L. Sp. Pl. 1187. 1753; Beccari and Hook. f. in Hook. f. Fl. Brit. India 6: 1892; Gamble, Fl. Pres. Madras 3: 1090. 1957 (repr. ed.); Bhargavan in Henry *et al.,* Fl. Tamil Nadu 3: 51. 1989. *B. flabelliformis* Murr. Syst. Veg. ed. 13. 827. 1774; Roxb. Pl. Cor. tt. 71 and 72. 1796.

Solitary palm. Stem with rough and black stem. Leaf blade segmented, linear-lanceolate, induplicate. Male inflorescence : Primary and secondary branches; sepals narrowly cuneate with truncate inflexed tips; petals shorter, obovate-spathulate; anthers subsessile, large. Female inflorescence: Sepals fleshy, reniform; petals smaller; ovary subtrigonous; stigmas sessile, recurved. Fruits broadly ovoid.

Phenology: Feb- Apr.

Parts used: Endosperm of fruit.

Mode of use: The endosperm of young fruit is juicy and edible as raw. It is also used for making jams and cordials.

Therapeutic value: The sugar-candy produced from the juice is used in coughs and pulmonary affection. The fruit is stomachic, aphrodisiac, antibilious, tonic, laxative, alexiteric; improves taste and allays thirst. Milky juice from immature fruit checks hiccup sickness. Pulp from the immature fruit is diuretic, demulcent and nutritive. (http://www.mpbd.info/plants/borassus-flabellifer.php).

Boussingaultia baselloides H. B. K. Nov. Gen. Sp. 7: 196. t. 645. 1825; Gamble, Fl. Pres. Madras 2: 830. 1957 (repr. ed.); Kumari in Henry *et al.,* Fl. Tamil Nadu 2: 195. 1987.

Climbing shrub. Leaf blade ovate to orbicular, base of larger leaf blades proximal to inflorescences cordate, base of small distal leaf blades tapering, apex obtuse to acute. Flowers in racemes or in branched panicles of racemes, flowers bisexual, usually functionally staminate. Sepals basally adnate to petals, cream-white, ovate to elliptic, apex obtuse. Petals basally connate, cream-white, ovate to elliptic, apex obtuse. Stamens fleshy; filaments basally connate and dilated; anthers early deciduous. Styles basally connate, stigmas clavate to capitate; style bases persistent. Fruit globose.

Phenology: Mar.- Jun.

Parts used: Leaves.

Mode of use: Leaves are used as vegetable.

Canna indica L. Sp. Pl. 1. 1753. *C. indica* L. var. *orientalis* (Roscoe) Baker in Hook. f. Fl. Brit. India 6: 260. 1892. *C. orientalis* Roscoe, Monandr. Pl. Scitam. t. 12. 1826; Gamble, Fl. Pres. Madras 3: 1045. 1957 (repr. ed.); Ramamurthy in Henry *et al.,* Fl. Tamil Nadu 3: 31. 1989.

Shrub. Leaf sheath green or purple; leaf blade adaxially green, abaxially and at margin green or purple, ovate-oblong to oblong. Flowers in raceme of cincinni. Bracts light purple, ovate. Flowers 1 or 2 per cincinnus. Sepals pale purplish green, lanceolate. Corolla tube apricot yellow, lobes erect, red to apricot yellow with purplish apex, lanceolate. Staminodes 2(or 3), erect, red with apricot yellow base, oblanceolate, 1 with apex emarginate; labellum red, sometimes with yellow spots, lanceolate, margin revolute, apex emarginate; fertile stamen with filament reddish apricot yellow, lanceolate, anther small. Ovary green, globose, style red to apricot yellow, narrow, base adnate to staminode column. Capsule broadly ovoid.

Phenology: Sep.- Oct.

Parts used: Rhizome.

Mode of use: Root decoction is used for the treatment of fever, dropsy, and dyspepsia. The crushed fresh root is given in fever (Choudhury *et al.,* 2010).

Cansjera rheedii Gmel. Syst. 1: 280. 1791; Wight, Ic. t. 1861. 1852; Mast. in Hook. f. Fl. Brit. India 1: 582. 1875; Gamble, Fl. Pres. Madras 1: 138. 1957 (repr. ed.); Srinivasan in Nair and Henry, Fl. Tamil Nadu 1: 70. 1983.

Straggler. Stem occasionally bears thorns, curved. Leaves ovate or oblong-lanceolate. Flowers in spike inflorescence. Calyx 4-lobed. Corolla 4-lobed, urceolate, yellowish, tomentose, connate. Stamens 4, shorter than petals. Ovule 1, pendulous, style short and stigma shallowly 4-lobed. Fruit drupe, ovoid, 1-seeded.

Phenology: Jul.- Dec.

Parts used: Leaves.

Mode of use: Leaves are cooked and eaten as vegetable.

Therapeutic value: The whole plant is used for the treatment of post-natal pain (Ravikumar and Vijayasankar, 2003). It is also used for the treatment of intermittent fever (Hosagoudar and Henry, 1996).

Capsicum annuum L. Sp. Pl. 189. 1753; Gamble, Fl. Pres. Madras 2: 661. 1957 (repr. ed.); Chithra in Henry *et al.,* Fl. Tamil Nadu 2: 112. 1987. *C. frutescens* sensu Clarke in Hook. f. Fl. Brit. India 4: 239. 1883, non L. 1753.

Herb. Stems glabrate, woody in age. Leaves solitary or paired, mostly ovate, apically short acuminate or acute, basally narrowed, entire, glabrate. Flowers solitary in leaf axils. Calyx cupular, undulate or denticulate. Corolla white, triangular, ovate. Stamens 5, exserted. Style as long as corolla; Stigma capitate. Berry red when ripe.

Phenology: May- Dec.

Parts used: Fruit.

Mode of use: Add in the preparation of curries.

Therapeutic value: Plant pacifies vitiated kapha, fever, gout, arthritis, sciatica, anorexia, flatulence, cough, malaria, cholera and edema (http://ayurvedicmedicinalplants.com).

Cardiospermum canescens Wall. Pl. Asiat. Rar. 1: 14. T. 14. 1829; Wight, Ic. t. 74. 1838; Hiern in Hook. f. Fl. Brit. India 1: 670. 1875; Gamble, Fl. Pres. Madras 1: 175. 1957 (repr. ed.); Chithra in Nair and Henry, Fl. Tamil Nadu 1: 83. 1983.

Climbers. Leaves biternate, triangular in outline, blades thinly papery, margin sparsely serrate or pinnately parted, abaxially sparsely villous, apex acuminate, lateral slightly smaller, ovate. Flowers in panicles of few flowered, slightly longer than leaves. Sepals 4, ciliate, outer 2 sub-orbicular, inner 2 narrowly obovate. Petals milky-white, obovate. Male Flower: Stamens as long as or slightly longer than petals; filaments sparsely long villous. Female Flower: Ovary obovoid or sometimes subglobose, tomentose. Seeds black, shiny.

Phenology: Throughout the year.

Parts used: Leaves.

Mode of use: Leaves are used as leafy vegetable.

Therapeutic value: Whole plant is used for rheumatism (Ganesan *et al.,* 2009).

Cardiospermum halicacabum L. Sp. Pl. 366. 1753; Wight, Ic. t. 508. 1841; Hiern in Hook. f. Fl. Brit. India 1: 670. 1875; Gamble, Fl. Pres. Madras 1: 175. 1957 (repr. ed.); Chithra in Nair and Henry, Fl. Tamil Nadu 1: 83. 1983.

Climbers. Leaves biternate, triangular in outline, leaflets subsessile; blades thinly papery, margin sparsely serrate or pinnately parted, apex acuminate, lateral slightly smaller, ovate. Flowers in panicles of few flowered, slightly longer than leaves. Sepals 4, ciliate, outer 2 ovate, inner 2 narrowly elliptic. Petals milky-white, obovate. Male Flower: Stamens as long as or slightly longer than petals; filaments sparsely long villous. Female flower: Ovary obovoid or sometimes subglobose, pubescent. Seeds black, shiny.

Phenology: Throughout the year.

Parts used: Leaves.

Mode of use: Leaves are used as leafy vegetable.

Therapeutic value: Whole plant is used for rheumatism (Ganesan *et al.,* 2009).

Carissa paucinervia A. DC. in DC. Prodr. 8: 333. 1844; Wight, Ic. t. 1290. 1848; Hook. f. Fl. Brit. India 3: 631. 1882; Gamble, Fl. Pres. Madras 2: 566. 1957 (repr. ed.); Srinivasan in Henry *et al.,* Fl. Tamil Nadu 2: 76. 1987.

Shrubs. Leaves opposite, coriaceous, spine-bearing pairs, elliptic-ovate, acute at both ends, mucronate, nerves 2- 3 pairs, oblique and indistinct. Flowers in terminal and axillary pedunculate cymes. Calyx 5- lobed, lanceolate, ciliate. Corolla pinkish, hypocrateriform, 5- lobed. Stamens 5, included in the corolla tube. Ovary 2- celled, style filiform, stigma fusiform. Fruit ellipsoid, purplish when ripe.

Phenology: Apr- Dec.

Parts used: Fruit.

Mode of use: Ripened raw fruits are edible.

Centella asiatica (L.) Urban in Mart. Fl. Bras. 11: 287. t. 78. f. 1. 1879; Gamble, Fl. Pres. Madras 1: 392. 1957 (repr. ed.); Chandrabose in Nair and Henry, Fl. Tamil Nadu 1: 179. 1983. *Hydrocotyle asiatica* L. Sp. Pl. 234. 1753; Wight, Ic. t. 565. 1842; Clarke in Hook. f. Fl. Brit. India 2: 669. 1879.

Herbs. Leaves petiolate, forming rosettes along the creeping stem, blade simple, entire or shallowly dentate, palmately veined. Flowers simple, umbels loose to subcapitate, few-flowered. Calyx teeth obsolete. Petals valvate, orbicular with a narrow inflexed apex. Stylopodium obsolete; styles short, equaling filaments. Fruit reniform or globose, entire.

Phenology: Throughout the year.

Parts used: Leaves.

Mode of use: Leaves are made into curries and eaten.

Therapeutic value: Used for the treatment of dysentry, cough, fever, tuberculosis, boils etc. (Jain and Sumita, 2005).

Chenopodium album L. Sp. Pl. 219. 1753; Hook. f. Fl. Brit. India 5: 3. 1886; Gamble, Fl. Pres. Madras 2: 827. 1957 (repr. ed.); Kumari in Henry *et al.,* Fl. Tamil Nadu 2: 195. 1987.

Herbs. Leaf blade rhombic-ovate to broadly lanceolate, adaxially usually glabrous, or sometimes reddish purple vesicular hairy on young leaves, base cuneate to broadly so, margin irregularly serrate, apex subobtuse or acute. Flowers bisexual. Perianth segments 5, broadly ovate to elliptic, abaxially longitudinally keeled, farinose, margin membranous, apex acute or slightly emarginated. Stamens 5; anthers exserted. Stigmas 2; Pericarp adnate to seed. Seed black, lineate.

Phenology: May-Oct.

Parts used: Leaves and young twigs.

Mode of use: Leaves and young twigs are used as leafy vegetables.

Therapeutic value: It is used for curing appetite, anthelmintic, laxative, diuretic (Yadav, 2007).

Cissus quadrangularis L. Mant. Pl. 39. 1767; Gamble, Fl. Pres. Madras 1: 167. 1957 (repr. ed.); Ramamurthy in Nair and Henry, Fl. Tamil Nadu 1: 81. 1983. *Vitis quadrangularis* (L.) Wall. *ex* Wight, Cat. 26. 1833 and in Wight and Arn. Prodr. 125. 1834; Wight, Ic. t. 51. 1838; Lawson in Hook. f. Fl. Brit. India 1: 645. 1875.

Climber. Stem thick, succulent, quadrangular, angles winged, constricted at nodes, glabrous or slightly downy, almost leafless when old. Leaves simple, ovate, entire or cordate, serrulate dentate, or crenate-serrate, 3-lobed, terminal lobe triangular or sub-spathulate, cuspidate, membranous, glabrous on both sides; Flowers compound umbelliform cymes; flower pink and white, hypanthium cuplike, truncate or obscurely lobed, green. Petals 4, distinct, ovate-oblong, acute, hooded at apex. Disc longer than the ovary. Ovary glabrous, style slender subulate, stigma small. Berry globose, red, succulent; Seed obovoid smooth.

Parts used: Young leaves and stems.

Mode of use: Young leaves and stems are made into chutneys and eaten.

Therapeutic value: It is used for the healing Bone fractures (Shaheen *et al.*, 2010).

Clausena dentata (Willd.) M. Roem. Syn. Hesper. 1: 44. 1846; Vajravelu in Nair and Henry, Fl. Tamil Nadu 1: 59. 1983. *Amyris dentata* Willd. Sp. Pl. 2: 337. 1799. *Clansena willdenovii* Wight et Arn. Prodr. 96. 1834; Wight, Ic. t. 14. 1838; Hook. f. Fl. Brit. India 1: 506. 1875; Gamble, Fl. Pres. Madras 1: 110. 1957 (repr. ed.).

Tree. Leaves imparipinnate; leaflets alternate, gland-dotted, ovate-lanceolate, oblique, crenate. Flowers in axillary cymose panicle inflorescence. Sepals, 4-lobed. Petals 5, free, membranous, gland dotted. Stamens 10, inserted in the disk; filaments dilated below the tip, alternately longer and shorter, anther large. Style thick; stigma lobed. Fruit globose berry.

Phenology: Sep-Mar.

Parts used: Fruit.

Mode of use: Ripened raw fruits are consumed.

Coccinia grandis (L.) Voigt, Hort. Sub. Calc. 59. 1845; Srinivasan in Nair and Henry, Fl. Tamil Nadu 1: 170. 1983. *Bryonia grandis* L. Mant. Pl. 126. 1767. *Coccinia indica* Wight et Arn. Prodr. 347. 1957 (repr. ed.). *Cephalandra indica* (Wight et Arn.) Naud. in Ann. Sci. Nat. Bot. ser. 5. 5: 16. 1866; Clarke in Hook. f. Fl. Brit. India 2: 621. 1879 excl. syn.

Climber. Leaves equally broad, cordate, usually entire to 5-angular with few glistering glands on both sides of the midrib towards the base. Flowers large, white, pedicellate, solitary, male and female on different plants. Fruit cylindrical or fusiform, bright scarlet when ripe, slightly beaked, many seeded. Seeds asymmetrically pyriform in outline, compressed.

Phenology: Dec.-June.

Parts used: Leaves and fruits.

Mode of use: Leaves and unripened fruits are cooked and consumed as a vegetable. Ripened fruits are edible as raw.

Therapeutic value: Powdered leaves or fresh leaves are taken in treating diabetes (Senthilkumar *et al.,* 2006).

Colocasia esculenta (L.) Schott in Schott et Endlicher, Melet. Bot. 18. 1832; Matthew, Ill. Fl. Tamilnadu Cartnatic tt. 766 and 979C. 1982; Bhargavan in Henry *et al.,* Fl. Tamil Nadu 3: 56. 1989. *Arum esculentum* L. Sp. Pl. 965. 1753. *Colocasia antiquorum* Schott in Schott et Endlicher, Melet. Bot. 18. 1832; Hook. f. Fl. Brit. India 6: 523. 1893; Gamble, Fl. Pres. Madras 3: 1102. 1957 (repr. ed.).

Herb. Leaf blade adaxially matte waxy-glaucous and water- shedding, oblong-ovate to suborbicular, base shallowly cordate, apex broadly amd shortly cuspidate. Spathe tuve green, limb open proximally, cream- coloured to golden yellow, lanceolate or elliptic, apex acuminate; Spadix. Female Part: conic, stigma subsessile, narrower than apex of ovary, sterile part narrowly cylindric; sterile flowers above elongate. Male Part: Cylindric, ápex narrowly conic. Berry green, sedes few.

Phenology: Feb.- Sep.

Parts used: Leaves and Petiole.

Mode of use: Leaves and Petiole are consumed as green leaves.

Therapeutic value: Cooked leaves are a source of dietary fibre for diabetics helping in lowering post-prandial blood glucose level. The pressed juice of the petioles is used as an astringent and styptic (Khare, 2007).

Commelina benghalensis L. Sp. Pl. 41. 1753; Wight, Ic. t. 2065. 1853; Hook. f. Fl. Brit. India 6: 370. 1892; Gamble, Fl. Pres. Madras 3: 1075. 1957 (repr. ed.); Bhargavan in Henry *et al.,* Fl. Tamil Nadu 3: 44. 1989.

Herb. Leaf sheaths, not auriculate, often with red hairs at summit, blade ovate to lanceolate-elliptic, apex rounded, obtuse or acute, pubescent. Flowers cyme often exserted and 1-flowered, sometimes vestigial, spathes often clustered, subsessile, funnelform, margins connate basally, pubescent. Flowers bisexual and staminate, subterranean cleistogamous. Petals of staminate flowers all blue, proximal smaller, lateral stamen filaments not winged. Staminodes 2-3; antherodes yellow, cruciform; stamen pollen yellow. Capsules 3-locular; Seeds 5, brown or blackish.

Phenology: Jun.- Sep.

Parts used: Leaves.

Mode of use: Leaves are consumed as leafy vegetable.

Therapeutic value: Whole plant is used to treat leprosy (Senthilkumar *et al.,* 2006).

Curcuma longa L. Sp. Pl. 2. 1753; Baker in Hook. f. Fl. Brit. India 6: 214. 1890; Gamble, Fl. Pres. Madras 3: 1036. 1957 (repr. ed.); Ramamurthy in Henry *et al.,* Fl. Tamil Nadu 3: 28. 1989. *C. domestica* Valeton in Bull. Jard. Bot. Buitenzorg ser. 2. 27: 31. 1981.

Herb. Rhizomes many branched, orange or bright yellow, aromatic. Leafblade oblong- elliptic, glabrous, base attenuate, apex shortly acuminate. Flowers in terminal spike. Fertile bracts pale green, ovate or oblong, apex obtuse; sterile bracts spreading, white or green, apex acute. Calyx white, puberulent, apex unequally 3- toothed. Corolla pale yellow, lobes deltoid, apex mucronate. Lateral staminodes shorter than labellum. Labellum yellowish with central, yellow band, obovate. Anther spurred at base. Ovary sparsely hairy.

Phenology: Aug.

Parts used: Rhizome.

Mode of use: Added as a flavouring agent.

Therapeutic value: It is taken as the blood purifier and is very useful in the common cold, leprosy, intermittent affections of the liver, dropsy, inflammation and wound healing. The rhizome of the turmeric plant is highly aromatic and antiseptic. It is even used for contraception, swelling, insect stings, wounds, whooping cough, inflammation, internal injuries, pimples, injuries, as a skin tonic. Sweetened milk boiled with the turmeric is the popular remedy for cold and cough. It is given in liver ailments and jaundice (http://www.ecoindia.com/flora/trees/turmeric-plant.html).

Digera muricata (L.) Mart. Beitr. Amar. 77. 1825; Kumari in Henry *et al.,* Fl. Tamil Nadu 2: 192. 1987. *Achyranthes muricata* L. Sp. Pl. ed. 2. 1: 295. 1762. *Digera arvensis* Forssk. Fl. Aeg.- Arab. 65. 1775; Hook. f. Fl. Brit. India 4: 717. 1885; Gamble, Fl. Pres. Madras 2: 818. 1957 (repr. ed.). *Desmochaeta muricata* (L.) DC. Cat. Hort. Monsp. 103. 1813; Wight, Ic. t. 732. 1843.

Herb. Leaves entire, ovate. Flowers small, in long-pedunculate, axillary, spike-like bracteate racemes, each bract subtending a very shortly pedunculate partial inflorescence consisting of a central fertile flower and two highly modified, sterile, unibracteolate lateral flowers. Perianth segments 4-5, the outer pair opposite and sheathing the remaining flower parts, the inner segments much more delicate and hyaline. Stamens 4-5, free, without intermediate staminode-like teeth, filaments filiform. Style filiform; stigmas 2, divergent. Fruit a hard, persistent perianth and falling together the sterile flowers and bracteoles.

Phenology: Nov.- Mar.

Parts used: Young twigs.

Mode of use: Young twigs are cooked and eaten.

Therapeutic value: During constipation, leaves used as a vegetable; for kidney stones, tender tips extract is administered daily once until cured (Reddy *et al.,* 2010).

Dioscorea alata L. Sp. Pl. 1033. 1753; Wight, Ic. t. 810. 1844; Hook. f. Fl. Brit. India 6: 296. 1892; Gamble, Fl. Pres. Madras 3: 1056. 1957 (repr. ed.); Ramamurthy in Henry *et al.,* Fl. Tamil Nadu 3: 35. 1989. *D. globosa* Roxb. Fl. Ind. 3: 797. 1832; Wight, Ic. t. 812. 1844; Hook. f. Fl. Brit. India 6: 296. 1892.

Climber. Leaves alternate or opposite, petiolate, simple or palmately compound. Flowers unisexual, arranged spirally in axillary, usually elongate spikes or racemes. Male flowers: stamens 6, 3 sometimes reduced to staminodes or absent. Female spikes, few flowered. Female flowers: staminodes 3, 6, or absent. Capsule 3-winged, dehiscent; Seeds with a membranous wing.

Phenology: Nov.- Feb.

Parts used: Tuber.

Mode of use: Tubers are cooked and consumed as a vegetable.

Therapeutic value: It is used for indigestion (Kala, 2005).

Dioscorea bulbifera L. var. **sativa** Prain et Burkill in J. Proc. Asiat. Soc. Bengal 10: 26. 1914; Gamble, Fl. Pres. Madras 3: 1055. 1957 (repr. ed.); Ramamurthy in Henry *et al.,* Fl. Tamil Nadu 3: 36. 1989.

Climber. Leaves alternate, simple, leaf blade broadly cordate, glabrous, margin entire or slightly undulate, apex caudate-acuminate. Male flowers: solitary, bract and bracteole ovate; perianth purple, lobes lanceolate; stamens 6, inserted at base of perianth, filaments nearly as long as anthers. Female spikes often 2 or more together, similar to male ones, 20-30 cm. Female flowers: staminodes 6, ca. 1/4 as long as perianth lobes. Capsule straw-coloured, densely purplish dotted, oblong-globose.

Phenology: Oct.- Feb.

Parts used: Tuber.

Mode of use: Tubers are cooked and consumed as a vegetable.

Therapeutic value: It is used by tribal ladies as contra receptive (Choudhary *et al.,* 2008).

Dioscorea pentaphylla L. Sp. Pl. 1032. 1753, var. **pentaphylla**; Wight, Ic. t. 814. 1844; Hook. f. Fl. Brit. India 6: 281. 1892; Gamble, Fl. Pres. Madras 3: 1056. 1957 (repr. ed.); Ramamurthy in Henry *et al.,* Fl. Tamil Nadu 3: 36. 1989. *D. pentaphylla* L. var. *linnaei* Prain and Burkill in Proc. Asiat. Soc. Bengal 10: 23. 1914; Gamble, Fl. Pres. Madras 3: 1056. 1957 (repr. ed.).

Climber. Leaves alternate, palmately 3-7-foliolate, densely reddish brown pubescent; leaflets ovate to lanceolate, adaxially glabrous, abaxially appressed pubescent, sometimes glabrescent, margin entire, apex acute. Male flowers: sessile or subsessile; bract and bracteole reniform, forming an involucel around perianth, sparsely hairy, apex cuspidate; stamens 3. Female flowers: bracts, perianth, and ovary hairy. Capsule black at maturity, long ellipsoid, thinly leathery, sparsely hairy.

Phenology: Aug.- Feb.

Parts used: Tuber.

Mode of use: Tubers are cooked and consumed as a vegetable.

Therapeutic value: Sliced tuber are soaked in running water and used as a common vegetable. It is also helpful in case of indigestion (Reddy *et al.,* 2009).

Diplocyclos palmatus (L.) Jeffrey in Kew Bull. 15: 352. 1962; Kumari in Nair and Henry, Fl. Tamil Nadu 1: 172. 1983. *Bryonia palmata* L. Sp. Pl. 1012. 1753. *B. laciniosa* L. Sp. Pl. 1013. 1753 p.p. excl. typus; Wight, Ic. t. 500. 1841; Clarke in Hook. f. Fl. Brit. India 2: 622. 1879. *Bryonopsis lanciniosa* sensu Naud. in Ann. Sci. Nat. Bot. ser. 4. 12: 141. 1859; Gamble, Fl. Pres. Madras 1: 377. 1957 (repr. ed.).

Climber. Leaves palmately 5-lobed, tendrils 2-fid. Flowers unisexual, in clusters. Male flowers: Corolla greenish yellow, broadly campanulate, petals ovate; Stamens 3, filaments free, inserted in the calyx tube, anthers ovate, pistil 0. Female flowers: Ovary globose, stigma 3-fid. Fruit globose, reddish when ripe.

Phenology: Nov.- Feb.

Parts used: Young twigs.

Mode of use: Prepared as curries and consumed.

Therapeutic value: It is used for the treatment of joints pain (Samuel and Andrews, 2008).

Drymaria cordata (L.) Willd. *ex* Roem. et Schultes subsp. **diandra** (Blume) Duke. Ann. Missouri Bot. Gard. 48: 253. 1861; Ramamurthy in Nair and Henry, Fl. Tamil Nadu 1: 23. 1983. *D. diandra* Blume, Bijdr. 62. 1825. *D. cordata* sensu Edgew. and Hook.f. in Hook. f. Fl. Brit. India 1: 244. 1874 p.p.; Gamble, Fl. Pres. Madras 1: 46. 1957 (repr. ed.).

Herbs. Leaves opposite, blade orbiculate to reniform, base cordate to rounded, apex rounded to acute. Flowers in axillary and terminal, mostly open, 5-15+-flowered cymes. Sepals with weak veins arching outward at midsection and confluent apically, lanceolate to ovate, 2-4(-5) mm, equal, apex acute to acuminate, not hooded, glabrous or with sessile glands. Petals 2-fid lobes 1-veined, vein unbranched, linear, trunk absent, base abruptly tapered, apex rounded to acute. Seeds dark reddish brown.

Phenology: Throughout the year.

Parts used: Leaves.

Mode of use: Leaves are eaten as leafy vegetables.

Therapeutic value: It is used for treating Wounds (Samuel and Andrews, 2008).

Eclipta prostrata (L.) L. Mant. Pl. 286. 1771; Koyama in Taxon 30: 505. 1981; Chandrasekaran in Henry *et al.,* Fl. Tamil Nadu 2: 37. 1987. *Verbesina prostrata* L. Sp. Pl. 902. 1753. *Eclipta alba* (L.) Hassk. Pl. Jav. Rar. 528. 1848; Hook. f. Fl. Brit. India 3: 304. 1881; Gamble, Fl. Pres. Madras 3: 304. 1957 (repr. ed.). *Verbesina alba* L. Sp. Pl. 902. 1753.

Herbs. Leaves lanceolate, papery, densely strigose-pubescent on both surfaces, base narrowed, sessile or shortly petiolate, margin serrulate, apex gradually acuminate. Flowers capitula wide, peduncle slender, involucre globose-campanulate, enlarging to 11 mm wide in fruit. Phyllaries 5 or 6, 2-seriate, oblong, acute, outer longer; Ray florets 2-seriate; lamina bifid or entire; Disk florets many, corolla 4-lobed. Achenes margin ribbed.

Phenology: Jun.-Sep.

Parts used: Leaves.

Mode of use: Leaves are cooked and eaten.

Therapeutic value: It is taken for curing fever, cough, cold, eye troubles and hair growth (Balakrishnan *et al.,* 2009).

Elaeagnus conferta Roxb. Fl. Ind. 1: 460. 1820; Gamble, Fl. Pres. Madras 2: 873. 1957 (repr. ed.); Kumari in Henry *et al.,* Fl. Tamil Nadu 2: 214. 1987. *E. latifolia* auct. non L. 1753; Bedd. Fl. Sylv. T. 180. 1872; Hook. f. Fl. Brit. India 5: 202. 1886 p.p.

Shrub. Leaf blade elliptic to elliptic-oblong, abaxially with scales dense, scattered, brownish, adaxially glabrescent, lateral veins 5-11 per side of midrib, base rounded or cuneate, margin obscurely undulate, apex shortly acuminate or subacute. Flowers 8-flowered. Pedicel 1-3 mm. Calyx tube tubular, 4-angled, constricted above ovary; lobes ovate. Stamens inserted in throat or on narrow annulus, exserted; Filaments and anthers small; Style hairy. Drupe ellipsoid.

Phenology: Oct.- Mar.

Parts used: Fruits.

Mode of use: Fruits are taken as raw.

Therapeutic value: It is used for the treatment of indigestion (Wu, 2011)

Elaeagnus indica Servettaz in Bull. Herb. Boissier ser. 2. 7: 393. 1908; Gamble, Fl. Pres. Madras 2: 872. 1957 (repr. ed.); Kumari in Henry *et al.,* Fl. Tamil Nadu 2: 214. 1987.

Straggler. Leaf blade elliptic to elliptic-oblanceolate, abaxially with scales dense, white silvery, adaxially glabrescent, base rounded or cuneate, margin obscurely undulate, apex shortly acuminate or subacute. Flowers 8-flowered. Pedicel 1-3 mm. Calyx tube tubular, 4-angled, constricted above ovary; lobes ovate, stellate pubescent within. Stamens inserted in throat or on narrow annulus, exserted; Filaments and anthers small; Style hairy. Drupe ellipsoid.

Phenology: Mar.- Dec.

Parts used: Fruits.

Mode of use: Fruits are taken as raw.

Elaeagnus kologa Schlecht. in DC. Prodr. 14: 611. 1857; Gamble, Fl. Pres. Madras 2: 873. 1957 (repr. ed.); Kumari in Henry *et al.,* Fl. Tamil Nadu 2: 215. 1987. *E. latifolia* L. Sp. Pl. 121. 1753 p.p.; Wight, Ic. t. 1856. 1852; Hook. f. Fl. Brit. India 5: 202. 1886 p.p.

Straggler. Leaf blade elliptic to elliptic-oblong, with scales brown, dense, base rounded or cuneate, margin obscurely undulate, apex shortly acuminate. Flowers 8-flowered. Pedicel 1-3 mm. Calyx tube tubular, 4-angled, constricted above ovary;

lobes ovate, sparsely scaly. Stamens inserted in throat or on narrow annulus, exserted; Filaments and anthers small; Style hairy. Drupe ellipsoid.

Phenology: Jan.- Mar.

Parts used: Fruits.

Mode of use: Fruits are taken as raw.

Elaeocarpus tectorius (Lour.) Poir. in Lam. Encycl. Suppl. 2: 704. 1812; Chithra in Nair and Henry, Fl. Tamil Nadu 1: 46. 1983. *Craspedum tectorium* Lour. Fl. Cochinch. 336. 1790. *Elaeocarpus oblongus* auct. non Gaertn. 1788; Mast. in Hook. f. Fl. Brit. India 1: 403. 1874; Gamble, Fl. Pres. Madras 1: 88. 1957 (repr. ed.). *E. robustus* Roxb. Fl. Ind. 2: 597. 1832; Mast. in Hook. f. Fl. Brit. India 1: 402. 1874.

Tree. Branchlets terete, glabrous. Leaves oblong, obovate or elliptic, apex acuminate with blunt ends, clustered at twig ends. Stipules small, lanceolate. Flowers cream coloured in racemose inflorescence. Sepals 5, free. Petals obovate, laciniate, ciliate. Stamens numerous, inserted on the disc. Filaments free, anther oblong. Ovules 2, style columnar. Torus glandular, woolly. Fruit drupe, 1- seeded.

Phenology: Apr.- May.

Parts used: Fruits.

Mode of use: Ripened fruits are taken as raw.

Elaeocarpus tuberculatus Roxb. Fl. Ind. 2: 594. 1832; Bedd. Fl. Sylv. t. 113. 1871; Mast. in Hook. f. Fl. Brit. India 1: 404. 1874; Gamble, Fl. Pres. Madras 1: 88. 1957 (repr. ed.); Chithra in Nair and Henry, Fl. Tamil Nadu 1: 46. 1983.

Tree. Branchlets terete, glabrous. Leaves alternate, oblong, elliptic, apex acuminate with blunt ends, clustered at twig ends. Stipules small, lanceolate. Flowers bisexual, white coloured in racemose inflorescence. Sepals 5, free. Petals obovate, laciniate, ciliate. Stamens numerous, inserted on the disc. Filaments free, anther oblong. Ovules 2, style columnar. Torus glandular, woolly. Fruit drupe, 1- seeded.

Phenology: Apr.- Jun.

Parts used: Fruit.

Mode of use: Ripened fruits are taken as raw.

Embelia ribes Burm. f. Fl. Ind. 62. t. 23. 1768; Clarke in Hook. f. Fl. Brit. India 3: 513. 1882; Gamble, Fl. Pres. Madras 2: 529. 1957 (repr. ed.); Chithra in Henry *et al.,* Fl. Tamil Nadu 2: 61. 1987. *Embelia glandulifera* Wight, Ic. t. 1207. 1848.

Shrub. Leaf blade elliptic, oblong, or narrowly obovate, papery, leathery or rarely cartilaginous, glabrous, pellucid or black punctate, base cuneate or rounded, margin revolute, entire, apex acuminate, rarely subacute; lateral veins inconspicuous. Flowers terminal, pinnately pyramidal paniculate, bracts subulate or lanceolate, sparsely puberulent abaxially, black punctate-lineate, margin entire, sparsely glandular puberulent, apex subulate. Flowers greenish or white, (4 or) 5-merous, papery. Sepals deltate, pellucid-punctate, glandular papillose adaxially and on margin, apex acuminate. Petals free, elliptic or oblong, papery, pellucid punctate, densely glandular granulose inside, margin entire, long glandular papillose, apex acute to obtuse. Stamens subequalling petals or slightly exserted, anthers ovate or oblong, base sagittate, apex rounded, connective black punctuate. Ovary glabrous; stigma capitate or 2- or 3-lobed. Fruit red or scarlet, globose or ovoid.

Phenology: Mar.- Apr.

Parts used: Fruit.

Mode of use: Ripened fruits are eaten as raw.

Therapeutic value: Fruit is a good appetizer, alterative, carminative, stomachic, anthelmintic, alexiteric, analgesic, purgative and tonic; useful in ascites, bronchitis, jaundice, hemicrania, dyspepsia and skin diseases. It is effective against intestinal worms specially tapeworms (http://www.mpbd.info/plants/embelia-ribes.php).

Eryngium foetidum L. Sp. Pl. 1: 232. 1753.

Plants basal rosette. Basal leaves numerous; blade lanceolate or oblanceolate, entire, base cuneate to decurrent, apex obtuse, callous-margined, crenate to finely spinulose-serrate. Upper leaves sessile, opposite deeply spinulose-serrate to parted. Flowers divaricately trifurcate; lateral branches often continuing to form a monochasium, heads numerous, short-pedunculate. Flower heads cylindrical. Bracts foliaceous, lanceolate, spreading to reflexed, margin 1–3-spinulose-serrate; bracteoles lanceolate, brightly scarious-margined. Calyx teeth ovate-lanceolate, equaling petals. Petals white or pale yellow. Styles erect, exceeding calyx teeth. Fruit ovoid-globose covered with tubercles.

Phenology: Apr.- Dec.

Parts used: Leaves.

Mode of use: Leaves are added in curries for flavours.

Therapeutic value: It is used treatment for burns, earache, fevers, hypertension, constipation, fits, asthma, stomach ache, worms, infertility complications, snake bites, diarrhea and malaria (Paul *et al.,* 2011).

Eugenia cotinifolia Jacq. Obs. 3: 2. 1764, subsb. **codyensis** (Munro *ex* Wight) Ashton in Dassanayake, Rev. Handb. Fl. Ceylon 2: 412. 1981; Chithra in Nair and Henry, Fl. Tamil Nadu 1: 153. 1983. *E. codyensis* Munro ex Wight, Ill. 2: 13. 1850; Duthie in Hook. f. Fl. Brit. India 2: 501. 1879; Gamble, Fl. Pres. Madras 1: 342. 1957 (repr. ed.).

Tree. Leaves decussate, ovate or oblong-lanceolate. Flowers in cymose inflorescence. Bracteoles persistent. Calyx tube globose, lobes 4, persistent, Petals 4, white, glandular. Stamens numerous, incurved. Ovary globose, 2-celled. Ovules numerous. Fruits berry, globose, orange when ripe, crowned with calyx lobes.

Phenology: Apr.- Dec.

Parts used: Fruit.

Mode of use: Ripened fruits are edible.

Euphorbia hirta L. Sp. Pl. 454. 1753; Gamble, Fl. Pres. Madras 2: 892. 1957 (repr. ed.); Chandrabose in Henry *et al.,* Fl. Tamil Nadu 2: 228. 1987. *E. pilulifera* sensu Hook. f. Fl. Brit. India 5: 250. 1887, non L. 1753.

Herbs. Leaves opposite. Stipules membranous, triangular, caduceus, leaf blade lanceolate-oblong, long elliptic, or ovate-lanceolate, adaxially green to red, sometimes with purple blotch along midrib, abaxially gray-green, both surfaces pilose, denser abaxially, base slightly oblique, margin entire or few serrulate below middle, finely serrulate above middle, apex acuminate or obtuse. Cyathia in dense, often headlike, pedunculate cymes at upper nodes, peduncle to 25 mm, all parts very hairy; involucre campanulate, marginal lobes 5, triangular-ovate. Male flowers 4 or 5; anthers red. Female flower: styles free; stigma slightly 2-lobed. Capsule 3-angular, seeds subglobose-tetragonal, reddish.

Phenology: Jun.-Dec.

Parts used: Fruits.

Mode of use: Inflorescence is edible as raw.

Therapeutic value: It helps to cure the fissures in lips.

Ficus racemosa L. Sp. Pl. 1060. 1753. *F. glomerata* Roxb. Pl. Cor. t. 123. 1799; Wight, Ic. t. 667. 1840; King in Ann. Roy. Bot. Gard. (Calcutta) 1: 173. t. 218 A. 1888; Gamble, Fl. Pres. Madras 3: 954. 1957 (repr. ed.); Chithra in Henry *et al.,* Fl. Tamil Nadu 2: 255. 1987.

Trees. Leaves alternate, leaf blade elliptic-obovate, elliptic, or narrowly elliptic, leathery, abaxially pale green, pubescent when young, glabrescent, adaxially dark green and glabrous, base cuneate to obtuse, margin entire, apex acuminate to obtuse. Figs in a tumorlike aggregate on short branchlets of old stem, occasionally axillary on leafy shoot or on older leafless branchlets, paired, reddish orange when mature, pear-shaped, basally attenuated into a stalk, apical pore navel-like, flat, involucral bracts triangular-ovate. Male and female flowers within same fig. Male flowers: near apical pore, sessile; calyx lobes 3 or 4; stamens 2 Gall and female flowers: calyx lobes linear, apex 3- or 4-toothed, style lateral, stigma clavate.

Phenology: May-Jul.

Parts used: Fruits.

Mode of use: Unripened fruits are used as vegetables.

Therapeutic value: The fruits are considered astringent, stomachic and carminative; given in menorrhagia, haemoptysis, bronchitis, dry cough, diseases of kidney and spleen. The unripe fruit is astringent to the bowels, tonic and styptic; allays thirst, useful in leucorrhoea. The ripe fruit is acrid and cooling; useful in biliousness, burning sensation, fatigue, urinary discharges, thirst, leprosy, menorrhagia and nose bleeding. The fresh juice of the ripe fruit is used as an adjunct to a metallic preparation, which is given in diabetes. Fruits are used for rheumatic pain (http://www.mpbd.info/plants/ficus-racemosa.php).

Gaultheria fragrantissima Wall. in Asiat. Res. 13: 397. 1820; Clarke in Hook. f. Fl. Brit. India 3: 457. 1882; Gamble, Fl. Pres. Madras 2: 522. 1957 (repr. ed.); Rajan in Henry *et al.,* Fl. Tamil Nadu 2: 58. 1987. *G. leschenaultii* DC. Prodr. 7: 593. 1839; Wight, Ic. t. 1195. 1848.

Shrub. Leaves scattered, leaf blade elliptic, oblong-elliptic, ovate, obovate, or narrowly obovate-lanceolate, leathery, base cuneate to broadly cuneate, sometimes obtuse-rounded, margin obtusely serrulate, apex acute, sometimes acuminate. Flowers axillary, racemose, many flowered, densely pubescent. Bracts ovate, abaxially glabrous or puberulous, margin ciliolate. Bracteoles apical, rarely distant, broadly ovate. Calyx glabrous; lobes triangular, margin ciliolate. Corolla white, tubular-urceolate, abaxially glabrous, lobes very short. Filaments dilated, puberulous; anthers 2-awned. Ovary pubescent. Calyx at fruiting blue-purple, fleshy, capsule globose, pubescent.

Phenology: Jan.- Aug.

Parts used: Fruits.

Mode of use: Ripened raw fruits are edible.

Therapeutic value: Unripened fruits increases appetite.

Glycosmis pentaphylla (Retz.) DC. Prodr. 1: 538. 1824, quoad basionym; Tanaka in J. Indian Bot. Soc. 16: 229. 1937; Hook. f. Fl. Brit. India 4: 499. 1785. p.p.; Vajravelu in Nair and Henry, Fl. Tamil Nadu 1: 61. 1983. *Limonia pentaphylla* Retz. Obs. 5: 24. 1789. *L. arborea* Roxb. Pl. Cor. t. 85. 1798. *Glycomis cochinchinensis* sensu Gamble. Fl. Pres. Madras 153. 1915 and 1: 109. 1957 (repr. ed.) p.p.

Phenology: Feb.- Apr.

Parts used: Fruits.

Mode of use: Ripened fruits are edible as raw.

Therapeutic value: Juice administered orally twice a day to cure chronic stomach pain (Karuppusamy, 2007).

Grewia hirsuta Vahl. Symb. Bot. 1: 34. 1790; Wight, Ic. t. 76. 1838; Mast. in Hook. f. Fl. Brit. India 1: 391. 1874; Gamble, Fl. Pres. Madras 1: 85. 1957 (repr. ed.); Vajravelu in Nair and Henry, Fl. Tamil Nadu 2: 44. 1983.

Shrubs. Leaf blade lanceolate, leathery, black-brown when dried, abaxially yellow-brown stellate tomentose, adaxially sparsely coarsely stellate, base narrow, shallowly cordate, margin serrulate, apex acuminate or rarely obtuse. Flowers in cymes 1-5 per leaf axil, 3- or 4-flowered. Bracts lanceolate. Sepals narrowly lanceolate, abaxially tomentose, adaxially glabrous. Petals narrowly ovate. Stamens small. Style longer than stamens, glabrous, stigma 4-lobed. Fruits drupe, globose or 2-lobed.

Phenology: June- Jul.

Parts used: Fruit.

Mode of use: Ripened fruits are edible as raw.

Therapeutic value: Fruits and roots are used to treat acute diarrhoea and dysentry (Senthilkumar *et al.,* 2006).

Grewia obtusa Wall. *ex* Dunn in Gamble, Fl. Pres. Madras 117. 1915 and 1: 83. 1957 (repr. ed.); Vajravelu in Nair and Henry, Fl. Tamil Nadu 2: 44. 1983. *G. bracteata* auct. non Heyne *ex* Roth 1821; Wight and Arn. Prodr. 76. 1834; Mast. in Hook. f. Fl. Brit. India 1: 389. 1874 p.p.

Shrubs. Leaf blade rhomboid, round or ovate, obscurely toothed. Flowers in cymose inflorescence. Bracts lanceolate. Sepals narrowly lanceolate, tomentose. Petals white, fading yellow, oblong, densely ciliate. Stamens sparsely puberulous. Style longer and stigma laciniate. Fruits globose, obscurely 4-lobed.

Phenology: Mar.- Jun.

Parts used: Fruits.

Mode of use: Ripened fruits are edible.

Grewia tiliaefolia Vahl, Symb. Bot. 1: 35. 1790; Bedd. Fl. Sylv. t. 108. 1871; Mast. in Hook. f. Fl. Brit. India 1: 386. 1874 p.p., Gamble, Fl. Pres. Madras 1: 84. 1957 (repr. ed); Vajravelu in Nair and Henry, Fl. Tamil Nadu 2: 45. 1983.

Trees. Leaf blade nearly orbicular or broadly ovate-orbicular, papery, black-brown when dry, sparsely hairy or glabrous, base obliquely cordate, margin serrulate, apex shortly acute. Flowers in cymes, gray-brown tomentose. Sepals oblong lanceolate, softly gray-yellow tomentose on both surfaces. Petals yellow, shorter and narrower than sepals. Stamens many, in 5 fascicles, slightly connate at base, slightly shorter than sepals. Ovary 2-loculed, hairy; style slightly longer than stamens. Drupe subglobose, drupe.

Phenology: Apr.- Aug.

Parts used: Fruits.

Mode of use: Ripened fruits are edible.

Grewia villosa Willd. in Ges. Naturf. Freunde Berlin Neue Schrifted 4: 205. 1803; Mast. in Hook. f. Fl. Brit. India 1: 388. 1874; Gamble, Fl. Pres. Madras 1: 85. 1957 (repr. ed.); Vajravelu in Nair and Henry, Fl. Tamil Nadu 2: 45. 1983.

Shrubs. Leaves alternate, petiolate, entire or serrate, rarely lobed, 3-7-costate, somewhat coriaceous. Flowers axillary or leaf-opposed, solitary or clustered, pedunculate, umbellate cymes or panicles. Flowers bisexual, pedicellate, yellow, orange, pink or white, pentamerous, bracteates. Sepals 5, free, leathery, mostly coloured and almost glabrous within, valvate. Petals 5, very rarely absent, free, mostly shorter than the sepals, clawed, claw thick, pitted or glandular near the point of insertion at the base of often glandular-hairy torus. Stamens numerous, free, anthers somewhat orbicular. Style subulate, with 2-4-lobed stigma. Drupe yellow, red or violet, globose, entire or up to 4-lobed.

Phenology: Sep.- Dec.

Parts used: Fruits.

Mode of use: Ripened fruits are edible.

Lantana camara L. var. **aculeata** (L.) Mold. in Torreya 34: 9. 1934; Srinivasan in Henry *et al.,* Fl. Tamil Nadu 2: 165. 1987. *L. aculeata* L. Sp. Pl. 627. 1753; Clarke in Hook. f. Fl. Brit. India 4: 562. 1885; Gamble, Fl. Pres. Madras 2: 761. 1957 (repr. ed.).

Shrub. Leaves opposite, decussate, ovate to ovate-oblong, crenate-serrate, acute to shortly acuminate, scabrid. Flowering heads axillary, peduncled, umbellate in flower, shorter to exceeding the subtending leaves, 2-3 cm across. Bracts lanceolate to linear, 5-7 mm broad, acute to subulate, rarely a few larger ones also present. Flowers mostly orange or yellow, turning to red or scarlet later. Calyx thin, pubescent. Corolla-tube pubescent, slightly enlarged and curved above the middle, limb 4 lobed with spreading. Fruit drupe globose, black, shining, 2-seeded.

Phenology: Throughout the year.

Parts used: Fruits.

Mode of use: Raw ripened fruits are edible.

Maesa indica (Roxb.) DC. in Trans. Linn. Soc. London 17: 134. 1834; Ramam. in Saldanha and Nicolson, Fl. Hassan Dist. 203. 1976; Chithra in Henry *et al.,* Fl. Tamil Nadu 2: 62. 1987. *Baeobotrys indica* Roxb. Fl. Ind. 2: 230. 1824. *Maesa dubia* (Wall.) DC. in Trans. Linn. Soc. London 17: 134. 1834; Clarke in Hook. f. fl. Brit. India 3: 510. 1882; Gamble, Fl. Pres. Madras 2: 527. 1957 (repr. ed.). *B. dubia* Wall. in Roxb. Fl. Ind. 2: 235. 1824. *Maesa perrottetiana* A. DC. in DC. Prodr. 8: 80. 1844; Gamble, Fl. Pres. Madras 2: 527. 1957 (repr. ed.). *Maesa indica* (Roxb.) DC. var. *perrottetiana* (A. DC.) Clarke in Hook. f. Fl. Brit. India 3: 509. 1882.

Shrubs. Leaf blade broadly ovate to oblong, papery, punctate-lineate, base obtuse or subrounded, margin serrate-dentate or -denticulate, teeth not callose, apex acute or acuminate. Flowers axillary or subterminal, racemose or paniculate, glabrescent or glandular granulose. Bracteoles broadly ovate, obtuse to rounded apically. Flowers white or light yellow-green. Calyx lobes broadly ovate, pellucid punctate, margin entire, sparsely ciliate, apex obtuse to rounded. Corolla campanulate, orange punctate-lineate; lobes broadly ovate, as long as or longer than tube, margin subundulate, apex acute. Stamens inserted at middle of corolla tube, included, filaments as long as or longer than anthers; anthers rounded. Style short; stigma lobed. Fruit globose, persistent calyx lobes closed at fruit apex.

Phenology: Apr.- Nov.

Parts used: Fruits.

Mode of use: Ripened fruits are edible.

Mahonia leschenaultii (Wall. ex Wight et Arn.) Takeda in Notes Roy. Bot. Gard. Edinburg 6: 223. 1917; Gamble, Dl. Pres. Madras 1: 23. 1957 (repr. ed.); Chithra in Nair and Henry, Fl. Tmail Nadu 1: 9. 1983. *Berberis leschenaultii* Wall. ex Wight et Arn. Prodr. 16. 1834; Wight, Ic. t. 940. 1845. *B. nepalensis* Spreng. var. *leschenaultii* (Wall *ex* Wight et Arn.) Hook. f. and Thoms. in Hook. f. Fl. Brit. India 1: 109. 1872.

Shrub to 4 m; branchlets with persistent leaf-bases. Leaves even-pinnate, 30 x 10 cm; leaflets ca. 6 pairs, opposite, inequilateral, oblong, ovate-lanceolate, 8.5 x 4.5 cm, coriaceous, glabrous, glossy above, base obtuse, auriculate, margin serrately spiny, apex acute; stipules in pairs, subulate. Racemes terminal, ca. 20 cm; peduncle to 2.5 cm; pedicel to 1 cm. Sepals 3+ 3, ovate, 5-7 x 4 mm, petaloid. Petals yellow, 3+ 3, subequal, 6 x 1.5 mm, oblong, notched at apex and shortly 2-lobed, with 2 basal glands within. Stamens 6, adnate to the base of the petal; filaments 2 mm, flat, thick; anthers 2 mm, dehiscing by recurved valves. Ovary oblong, to 4 mm, 1-celled; ovules ca. 4, ventral, ascending; stigma discoid or peltate, subsessile. Berry globose, to 8 mm, glaucous, 1-seeded.

Phenology: Mar.- May.

Parts used: Fruits.

Mode of use: Ripened fruits are edible.

Therapeutic value: Helps to cure fever and cold (Kumaravelu, 2008).

Mangifera indica L. Sp. Pl. 200. 1753; Bedd. Fl. Sylv. t. 162. 1871; Hook. f. Fl. Brit. India 2: 13. 1876; Gamble, Fl. Pres. Madras 1: 185. 1957 (repr. ed.); Chithra in Nair and Henry, Fl. Tamil Nadu 1: 88. 1983.

Trees. Leaf blade oblong to oblong-lanceolate, leathery, deep green adaxially, light green abaxially, glabrous on both sides, base cuneate to obtuse, margin entire, undulate, apex acute to long acuminate. Flowers paniculate, terminal, glabrous to tomentose-pilose. Bracts lanceolate, pubescent. Sepals ovate-lanceolate, glabrous to pubescent, acuminate. Petals light yellow with prominent red tree-shaped pattern adaxially, oblong or oblong-lanceolate, glabrous, recurved at anthesis. Fertile stamen 1, with ovate anther; staminodes 4. Ovary oblique, style and stigma small. Fruit drupe, bright yellow, oblong to subreniform, greenish yellow to red.

Phenology: Mar.- Jul.

Parts used: Fruits.

Mode of use: Ripened fruits are edible raw. Unripened fruits are used to make pickles and cooked as a vegetable.

Therapeutic value: Green fruits are considered anticholeric, antidysmenorrheic, antiscorbutic, astringent, and diaphoretic. Roasted green fruits are dissolved in sugar water and taken internally to prevent sunstroke and they may be just rubbed on the body. Ripe fruits are considered diuretic, laxative, and unguent, and the gum is used to treat scabies (http://www.rain-tree.com/mango.htm).

Moringa concanensis Nimmo *ex* Gibs. in Dalz. and Gibs. Bombay Fl. 311. 1861; Hook. f. Fl. Brit. India 2: 45. 1876; Hooker's Icon. Pl. t. 2596. 1899; Gamble, Fl. Pres. Madras 1: 192. 1957 (repr. ed.); Srinivasan in Nair and Henry, Fl. Tamil Nadu 1: 90. 1983.

Trees. Leaflets in 4-6 pairs, ovate, elliptic, or oblong, puberulous when young but glabrous at maturity, base rounded to cuneate, apex rounded to emarginated. Flowers widely spreading panicle. Bracts linear. Flowers white to cream, fragrant.

Sepals lanceolate to linear-lanceolate, usually puberulent. Petals spatulate, glabrous or puberulent at base. Stamens hairy at base. Ovary stipitate. Capsule 3-valved, dehiscent. Seeds subglobose, 3-angled.

Phenology: Throughout the year

Parts used: Leaves, flowers and fruits.

Mode of use: All the above are cooked and eaten.

Therapeutic value: The leaves are anthelmintic, aphrodisiac, cures hallucinations, dry tumours and high cough (Bhoomika *et al.,* 2007). Flowers are anti-inflammatory, analgesic and antipyretic (Jayabharathi, 2011). Dried seeds are used in ophthalmic preparation, venereal affection, in goitre, glycosuria and lipid disorders (Ayurvedic Pharmacopoeia).

Morus alba L. Sp. Pl. 986. 1753; Hook. f. Fl. Brit. India 5: 492. 1888; Gamble, Fl. Pres. Madras 3: 958. 1957 (repr. ed.); Chithra in Henry *et al.,* Fl. Tamil Nadu 2: 256. 1987.

Shrubs. Leaf blade ovate to broadly ovate, irregularly lobed, abaxially sparsely pubescent along midvein or in tufts in axil of midvein and primary lateral veins, adaxially bright green and glabrous, base rounded to cordate, margin coarsely serrate to crenate, apex acute, acuminate. Male flowers: calyx lobes pale green, broadly elliptic; filaments inflexed in bud; anthers 2-loculed, globose to reniform. Female flowers: sessile; calyx lobes ovoid, with marginal hairs; ovary sessile, style absent; stigmas with mastoid like protuberance. Fruit syncarp, purple, ovoid.

Phenology: Apr.- Aug.

Parts used: Fruit.

Mode of use: Ripened fruits are edible as raw. It is also used for the jams, jellys, cooling beverages and wines.

Therapeutic value: It is good for brain, heart, spleen and helps curing diarrhea and intestinal ulcers. It is considered as a laxative and oral juice administration good drink during convalescence after a febrile illness (Soni *et al.,* 2009).

Murraya koenigii (L.) Spreng. Syst. Veg. 2: 315. 1826; Hook. f. Fl. Brit. India 1: 503. 1875; Gamble, Fl. Pres. Madras 1: 111. 1957 (repr. ed.). *Bergera koenigii* L. Mant. Pl. 565. 1771; Wight, Ic. t. 13. 1838; Vajravelu in Nair and Henry, Fl. Tamil Nadu 1: 61. 1983.

Tree. Leaves pinnate, leaflets simple, entire, imparipinnate oblique, oblong-lanceolate. Flowers in panicled corymbose inflorescence. Flowers small, white, fragrant. Sepals short, 5 lobed. Petals white, linear or oblanceolate. Stamens 10, subequal. Ovary 2- celled, ovoid, ovules 2. Fruit globose, purplish black when ripe.

Phenology: Mar.- Dec.

Parts used: Leaves and fruits.

Mode of use: Leaves are added in curries and also used to make chutneys. Ripened fruits are edible.

Therapeutic value: The leaves of plant are used as tonic, stomachic, carminative, internally in dysentery also checking vomiting, anthelmintic, analgesic, cures piles, allays heat of the body, thirst, inflammation and itching (Bonde *et al.,* 2011). Fruits are also considered as astringent.

Opuntia dillenii (Ker- Gawl.) Haw. Suppl. Pl. Succ. 79. 1819; Clarke in Hook. f. Fl. Brit. India 2: 657. 1879; Gamble, Fl. Pres. Madras 1: 387. 1957 (repr. ed.); Ramamurthy in Nair and Henry, Fl. Tamil Nadu 1: 176. 1983. *Cactus dillenii* Ker-Gawl. In Bot. Reg. 3: t. 255. 1818.

Shrubs. Larger, terminal joints green to gray-green, obovate or elliptic-obovate to suborbicular; Spines 1-12(-20) per areole on most areoles, spreading, yellow, brown banded or mottled, subulate, straight or curved, 1.2-4(-6) cm, basally flattened; glochids yellow. Leaves subulate, 4.5-6 mm, deciduous. Flowers: Sepaloids greenish with yellow margin, broadly deltoid-obovate to obovate, margin entire or slightly crisped, apex mucronate. Petaloids spreading, bright yellow, obovate or cuneate-obovate, margin entire or slightly undulate, apex rounded, truncate, or emarginated. Filaments yellow, anthers small. Style yellow or yellowish, stigmas 5, pale green. Fruit purple, seeds orbicular.

Phenology: Jun.- Dec.

Parts used: Fruit.

Mode of use: Ripened fruits are consumed raw. Jams and jellies are produced from the fruit, which resemble strawberries and figs in color and flavour.

Therapeutic value: Decoctions of the fruit are taken as purgatives (Bean and Saubel 1972).

Oxalis corniculata L. Sp. Pl. 435. 1753; Wight, Ic. t. 18. 1838; Edgew. and Hook. f. in Hook. f. Fl. Brit. India 1: 436. 1874; Gamble, Fl. Pres. Madras 1: 94. 1957 (repr. ed.); Vajravelu in Nair and Henry, Fl. Tamil Nadu 1: 51. 1983.

Herbs. Leaves alternate or pseudoverticillate, leaflet blades obcordate, green or suffused purplish red, variably adaxially and abaxially pubescent, apex deeply emarginated. Flowers umbellate, 2-7-flowered; peduncle usually slightly longer than petioles. Bracts linear-lanceolate. Sepals oblong-lanceolate, margin ciliate especially at apex. Petals bright yellow, oblong-obovate. Capsule long cylindric, 5-sided, strigose with abundant simple hairs and a few septate hairs on dehiscence sutures. Seeds brown to brownish red, ovoid-oblong.

Phenology: Feb.- Oct.

Parts used: Leaves and Fruits.

Mode of use: Leaves and fruits are cooked and consumed.

Medicinal uses: Leaves as such eaten or leaf juice is taken orally to check dysentery (Senthilkumar *et al.,* 2006).

Oxalis latifolia H. B. K. Nov. Gen. Sp. 5: 237. T. 467. 1822; Fyson, Fl. South Indian Hill St. 2: t. 51. 1932; Matthew in Rec. Bot. Surv. India 20 (1): 57. 1969; Vajravelu in Nair and Henry, Fl. Tamil Nadu 1: 51. 1983.

Herbs. Leaves alternate or whorled, digitately or pinnately compound. Flowers axillary and/or terminal in (pseudo) umbels. Flowers 5-merous, bisexual, regular. Sepals 5, free, imbricate. Petals 5, connate at base. Stamens 5+5; outer ones short, opposite the petals; inner ones long, opposite the sepals; filaments fused at base. Ovary superior, 5-celled; ovules numerous per cell on axile placentae; styles 5; stigmas capitate, entire or 2-fid. Capsule oblong or ovoid, loculicidal.

Phenology: Feb.- Nov.

Parts used: Tubers.

Mode of use: Tubers are cooked and eaten.

Passiflora edulis Sims in Bot. Mag. t. 1989. 1818; Fyson, Fl. South Indian Hill St. 2: t. 187. 1932; Gamble, Fl. Pres. Madras 1: 370. 1957 (repr. ed.); Srinivasan in Nair and Henry, Fl. Tamil Nadu 1: 169. 1983.

Climber. Leaves membranous, base cuneate or cordate, deeply 3-lobed, middle lobe ovate, lateral lobes ovate-oblong, glabrous. Flowers in reduced cyme, central flower not developed, one lateral branch converted to a tendril, flower opposite tendril. Bracts green, broadly ovate or rhombic, margin irregularly serrulate. Flowers hypanthium. Sepals green outside, light green or white inside. Corona in 4 or 5 series; outer 2 series ligulate with filiform distal half, base light green, middle purple, apex white; inner 2 or 3 series filiform, green and purple; operculum recurved, margin entire or irregularly lacerate apically, androgynophore tall. Filaments flat, coherent at base; anthers light yellow-green, oblong. Styles flat, stigma reniform. Fruit purple at maturity, ovoid, glabrous; Seeds many, ovoid.

Phenology: June- Nov.

Parts used: Fruits.

Mode of use: Ripened fruits are edible raw.

Medicinal uses: Juice of ripe fruit is taken to cure jaundice (Senthilkumar *et al.*, 2006).

Passiflora foetida L. Sp. Pl. 959. 1753; Gamble, Fl. Pres. Madras 1: 570. 1957 (repr. ed.); Srinivasan in Nair and Henry, Fl. Tamil Nadu 1: 169. 1983.

Climber. Leaf blade broadly ovate to oblong-ovate, membranous, glandular ciliate, abaxially pubescent with small nectaries in distal part, adaxially sericeous, mixed with occasional glandular hairs, base cordate, 3-lobed or unlobed, margin irregularly undulate, apex acute. Flowers reduced to a single flower, opposite tendril. Bracts 3-4, pinnatifid; lobes threadlike, glandular pilose apically. Flowers white or light purple with white spots or lines. Sepals awned. Corona 3-5-seriate, filamentous, two outer series, inner three series. Androgynophore tall. Stamens coherent at base, flat, anthers oblong. Styles small and stigmas capitate. Berry orange or orange-red, ovoid-globose, glabrous.

Phenology: July- May.

Parts used: Fruits.

Mode of use: Ripened fruits are edible.

Passiflora leschenaultii DC. Prodr. 3: 326. 1828; Wight, Ic. t. 39. 1838; Mast. in Hook. f. Fl. Brit. India 2: 599. 1879; Gamble, Fl. Pres. Madras 1: 530. 1957 (repr. ed.); Srinivasan in Nair and Henry, Fl. Tamil Nadu 1: 170. 1983.

Climbers. Leaves alternate, palmately lobed, semi-orbicular, rounded at base, broadly truncate. Stipules minute to large, sometimes caducous. Flowers in cymose. Bracts and bracteoles small to large, forming a conspicuous involucre or 0. Flowers white, 4- or 5-merous, bisexual or unisexual. Calyx-tube saucer shaped or cupular; lobes 4 or 5, imbricate, sometimes connate, persistent. Petals 4 or 5, imbricate. Corona extrastaminal, inserted on the calyx-tube, annular or thread-like. Male: Stamens 5, inserted on androgynophore or on calyx-tube; anthers oblong, apiculate. Bisexual, female: Ovary superior, sessile or on a gynophore, 1-celled; ovules numerous on 3 parietal placentae; styles 3, short, prominent; stigmas capitate to subglobose, sometimes laciniate. Fruit yellow when ripened, indehiscent berry.

Phenology: July- May.

Parts used: Fruits.

Mode of use: Ripened fruits are edible. Tender fruits are used as a vegetable.

Therapeutic uses: It is used for pitta, burning sensation, constipation, flatulence, dyspepsia, urinary retentions, calculus, and skin diseases (http://ayurvedicmedicinalplants.com).

Passiflora mollissima (H. B. K.) Bailey in Rhodora 18: 156. 1916. *Tacsonia mollissima* H. B. K. Nov. Gen. Sp. 2: 144. 1817; Fyson, Fl. South Indian Hill St. 2: t. 188. 1932; Srinivasan in Nair and Henry, Fl. Tamil Nadu 1: 170. 1983.

Climbers. Leaves alternate, often simple, entire or lobed; lobes elliptic to oblong, palmately nerved; foliar glands on petiole or at leaf-base. Stipules minute to large, sometimes caducous. Flowers in cymose, with basal flowers transformed into tendrils. Bracts and bracteoles small to large, forming a conspicuous involucre or 0. Flowers 4- or 5-merous, bisexual or unisexual. Calyx-tube saucer shaped or cupular; lobes 4 or 5, imbricate, sometimes connate, persistent. Petals 4 or 5, imbricate. Corona extrastaminal, inserted on the calyx-tube, annular or thread-like. Male: Stamens 5, inserted on androgynophore or on calyx-tube; anthers oblong, apiculate. Bisexual, female: Ovary superior, sessile or on a gynophore, 1-celled; ovules numerous on 3 parietal placentae; styles 3, short, prominent; stigmas capitate to subglobose, sometimes laciniate. Fruit an indehiscent berry.

Phenology: Mar.- Apr.

Parts used: Fruits.

Mode of use: Ripened fruits are edible as raw.

Therapeutic uses: It is used as antifeedant, antifungal, and antibacterial (Edwin *et al.,* 2007).

Persea americana Mill. Gard. Dict. Ed. 8. 1768; Chandrasekaran in Henry *et al.,* Fl. Tamil Nadu 2: 212. 1987. *Laurus persea* L. Sp. Pl. 370. 1753. *Persea gratissima* Gaertn. f. Fruct. 3: 222. 1807; Curtis, Bot. Mag. T. 4580. 1851; Gamble, Fl. Pres. Madras 2: 869. 1957 (repr. ed.); Matthew in Rec. Bot. Surv. India 20 (1): 191. 1969.

Tree. Bark gray-green, longitudinally fissured. Leaves alternate; petiole 2-5 cm, slightly sulcate adaxially, sparsely pubescent; leaf blade usually somewhat glaucous abaxially, green adaxially, narrowly elliptic, elliptic, ovate, or obovate. Flowers in cymose panicles, most of them inserted on lower part of branchlet, pedunculate. Pedicels densely yellowish brown pubescent. Flowers yellow-greenish, perianth densely yellowish brown pubescent outside and inside. Perianth tube obconical, perianth lobes 6, oblong, obtuse, outer 3 smaller, all dilated after anthesis and caducous. Fertile stamens 9; filaments filiform, complanate, densely pilose, those of 3rd whorl each with 2 complanate ovate and orange glands at base, others glandless; anthers 4-celled. Staminodes 3, of innermost whorl, sagittate-cordate, glabrous, stalked. Ovary ovoid; Style densely pilose; stigma slightly dilated, discoid. Fruit yellow-green or reddish brown, ovoid or globose.

Phenology: Feb.- Sep.

Parts used: Fruits.

Mode of use: Ripened fruits are edible as raw.

Therapeutic uses: It is used as an ethnomedicine for hypertension, jaundice and diabetes (Lans, 2006).

Phoenix loureirii Kunth. Enum. Pl. 3: 257. 1841; Bhargavan in Henry *et al.,* Fl. Tamil Nadu 3: 53. 1989. *Phoenix humilis* Royle var. *pedunculata* Beccari, Malesia 3: 379 and 387. t. 44. ff. 13- 15, 18- 21, 25- 27. 1890; Beccari and Hook. f. Fl. Brit. India 6: 427. 1892; Gamble, Fl. Pres. Madras 3: 1088. 1957 (repr. ed.).

Tree. Stems often clustered, usually clothed with remains of petiole. Leaves subglaucous, leaflets interruptedly fascicled along the rachis, not prominently scaly beneath but with deciduous pale scales and persistent linear brown scales on midrib beneath. Peduncle elongating in fruits. Male flowers, petals blunt. Female flowers with calyx half as long as corolla, lobes very small. Petals with short central lobe. Fruit oblong, orange.

Phenology: Jan.- Apr.

Parts used: Fruits

Mode of use: Ripened fruits are eaten raw.

Phyllanthus emblica L. Sp. Pl. 982. 1753; Hook. f. Fl. Brit. India 5: 289. 1887; Chandrabose in Henry *et al.,* Fl. Tamil Nadu 2: 236. 1987. *Emblica officinalis* Gaertn. Fruct. 2: 122. 1791; Wight, Ic. t. 1896. 1852; Gamble, Fl. Pres. Madras 2: 906. 1957 (repr. ed.).

Trees. Stems terete, sparsely lenticellate, with very reduced short shoots producing groups of leafy shoots. Leafy shoots angular, tawny pubescent. Stipules triangular-ovate, margins entire or denticulate, ciliate. Leaf blade oblong or linear-oblong, papery to leathery, paler abaxially, green adaxially, drying reddish or brownish, base shallowly cordate and slightly oblique, margin narrowly revolute, apex truncate. Fascicles with many male flowers and sometimes 1 or 2 larger female flowers. Male flowers: Sepals 6, membranous, yellow, obovate or spatulate, subequal, apex obtuse or rounded, margin entire or shallowly denticulate; Stamens 3; filaments coherent into column, anthers erect. Female flowers: Sepals 6, oblong or spatulate, apex obtuse or rounded, thicker, margin membranous, ovary ovoid, styles 3, connate at base. Fruit drupe, globose.

Phenology: Sep.- Apr.

Parts used: Fruits.

Mode of use: Ripe and unripened fruits are edible. It used to make pickles and jams.

Therapeutic uses: The fruit juice is consumed to regulate the body temperature (Britto and Mahesh, 2007).

Phyllanthus indofischeri Bennet in Indian Forester 109: 221. 1983; Chandrabose in Henry *et al.,* Fl. Tamil Nadu 2: 237.1987. *Emblica fischeri* Gamble in Kew Bull. 1925: 330. 1925 and Fl. Pres. Madras 2: 906. 1957 (repr. ed.). *Phyllanthus fischeri* sensu Ellis in Bull. Bot. Surv. India 22: 193. 1980 (1982), non Pax 1895.

Tree. Leaves compound, leaflets numerous, arranged in the branchlets, leaflets simple, opposite, glabrous, entire, elliptic-oblong at base and retuse at apex. Stipules yellowish green, minute, scarious. Flowers small, monoecious in axillary fasicled inflorescence, male flowers in upper axils and female flowers in the lower; Male Flower: Calyx green, sepals oblong, 5- 6-lobed; Corolla absent; Stamens 3, filament attached in the column, anther small; Female Flower: Calyx green, sepals oblong, 5- 6 lobed; Corolla absent; Style connate at base, recurved; Stigma bifid. Fruit green when unripe, yellowish green when ripe, seeds 6, green, stony.

Phenology: Dec.- Mar.

Parts used: Fruits.

Mode of use: Ripe and unripened fruits are edible. Fruits are used for making beverages.

Physalis minima L. Sp. Pl. 183. 1753; Clarke in Hook. f. Fl. Brit. India 4: 238. 1883; Gamble. Fl. Pres. Madras 2: 659. 1957 (repr. ed.); Chithra in Henry *et al.,* Fl. Tamil Nadu 2: 114. 1987. *P. pubescens* sensu Wight, Ill. 2: t. 166 B. f. 6. 1850, non L. 1753.

Herb. Stems prostrate or erect, pubescent with long many-celled hairs. Leaf blade ovate or ovate-lanceolate, pubescent along veins, base cuneate, often oblique, margin entire, sinuate, or with a few coarse teeth, apex acuminate. Calyx campanulate, pubescent, lobes deltate, short acuminate, densely ciliate. Corolla yellow. Anthers light yellow. Fruiting pendulous. Fruiting calyx green, subglobose or ovoid. Fruit berry globose.

Phenology: Throughout the year.

Parts used: Fruits.

Mode of use: Ripened fruits are edible raw.

Therapeutic uses: It is used as a remedy for abscesses, coughs, fevers and sore throats, among others (Duke and Ayensu, 1985).

Physalis peruviana L. Sp. Pl. ed. 2. 2: 1670. 1763; Clarke in Hook. f. Fl. Brit. India 4: 238. 1883; Fyson, Fl. South Indian Hill St. 2: t. 361. 1932; Gamble, Fl. Pres. Madras 2: 659. 1957 (repr. ed.); Chithra in Henry *et al.,* Fl. Tamil Nadu 2: 114. 1987.

Herb. Stems erect, sparingly branched, densely pubescent. Leaf blade broadly ovate to cordate, densely pubescent, base cordate, margin entire or with a few indistinct teeth, apex short acuminate. Calyx broadly campanulate. Corolla yellow, spotted in throat. Filaments and anthers blue-purple; anthers 3-4 mm long. Fruiting calyx green, ovoid. Fruit berry yellow.

Phenology: Throughout the year.

Parts used: Fruits.

Mode of use: Ripened fruits are consumed raw.

Therapeutic uses: It acts as a febrifuge, a diuretic and has anti-inflammatory, antirheumatic properties (http://food-nutrition.factoidz.com).

Picris hieracioides L. Sp. Pl. 792. 1753; Wight, Ic. t. 1143. 1846; Hook. f. Fl. Brit. India 3: 393. 1881; Gamble, Fl. Pres. Madras 2: 512. 1957 (repr. ed.); Chandrasekaran in Henry *et al.,* Fl. Tamil Nadu 45. 1987.

Herb. Stem erect, branched especially in apical third, hirsute. Basal leaves and lower stem leaves narrowly elliptic to broadly lanceolate, base attenuate into a winged petiole-like portion, margin entire to coarsely and sharply dentate, apex acuminate, acute, or obtuse. Middle and upper stem leaves sessile, lanceolate to linear, smaller, base semiamplexicaul to clasping. Synflorescence corymbiform to corymbosely paniculiform. Involucre cylindric-campanulate to ovoid. Phyllaries whitish, rigid; outer and inner phyllaries linear-lanceolate, apex acute to acuminate. Florets yellow. Achene brown, fusiform.

Phenology: June- Oct.

Parts used: Fruits.

Mode of use: Ripened raw fruits are edible.

Piper nigrum L. Sp. Pl. 28. 1753; Hook. f. Fl. Brit. India 5: 90. 1886; Gamble, Fl. Pres. Madras 2: 845. 1957 (repr. ed.); Kumari in Henry *et al.,* Fl. Tamil Nadu 2: 204. 1987.

Climber. Leaf blade ovate to ovate-oblong, rarely suborbicular, thick, leathery, glabrous, base rounded, usually slightly oblique, apex acute, base, alternate, others basal; reticulate veins prominent. Flowers polygamous, usually monoecious. Spikes leaf-opposed, to as long as leaves; peduncle nearly as long as petioles, glabrous. Bracts spatulate-oblong, adaxially adnate to rachis. Stamens 2, 1 on each side of ovary; filaments thick, short; anthers reniform. Ovary globose; stigmas 3 or 4, rarely 5. Fruit drupe red when ripe, drying black when unripe, globose.

Phenology: June- Oct.

Parts used: Fruits.

Mode of use: Ripened dry fruits are added in curries and edible. Raw dry fruits are also eaten.

Therapeutic uses: Fruits are carminative, antiperiodic, rubefacient, stimulant and aphrodisiac; used for the treatment of cough and cold. It is also useful in the treatment of cholera, obstinate fevers, dyspepsia, constipation, piles, colic, gastric troubles, ascites, anaemia, worms, asthma, sore throats, toothache and skin diseases (http://www.mpbd.info/plants/piper-nigrum.php).

Pithecellobium dulce (Roxb.) Benth. in London J. Bot. 3: 199. 1844; Bedd. Fl. Sylv. t. 188. 1872; Baker in Hook. f. Fl. Brit. India 2: 302. 1878; Gamble, Fl. Pres. Madras 1: 308. 1957 (redpr. ed.); Vajravelu in Nair and Henry, Fl. Tamil Nadu 1: 139. 1983. *Mimosa dulcis* Roxb. Pl. Cor. t. 99. 1798.

Trees. Leaflets sessile, 1 pair per pinna, elliptic or obovate-elliptic, both surfaces glabrous, reticulate veins raised abaxially, base slightly oblique, apex obtuse or emarginate. Flowers pedunculate heads, aggregated in terminal panicles. Calyx funnel-shaped, tomentose. Corolla small, white. Stamens numerous, connate into a tube at base. Legume blackish brown, curved. Seeds dark brown, shiny, ovoid-ellipsoid.

Phenology: Mar.- Jul.

Parts used: Aril.

Mode of use: Ripened fruit aril is eaten raw.

Portulaca oleracea L. Sp. Pl. 1753, var. **oleracea**; Dyer in Hook. f. Fl. Brit. India 1: 246. 1874; Gamble, Fl. Pres. Madras 1: 47. 1957 (repr. ed.); Chandrasekaran in Nair and Henry, Fl. Tamil Nadu 1: 24. 1983.

Herb. Leaf blade flat, obovate, base cuneate, apex obtuse, rounded, truncate, or retuse. Flowers in clusters, surrounded by involucre of 2-6 bracts. Sepals green, helmeted, ca. 4 mm, apex acute, keeled. Petals 5, yellow, obovate, slightly connate at base, apex retuse. Stamens 7-12, anthers yellow. Ovary glabrous. Stigma 4-6-lobed. Capsule ovoid. Seeds glossy black.

Phenology: May- Sep.

Parts used: Leaves.

Mode of use: Leaves are cooked and taken as leafy vegetable.

Therapeutic uses: Its leaves are used to treat insect or snake bites on the skin (Bensky and Dan, 2004) oils, sores, pain from bee stings, bacillary dysentery, diarrhoea, hemorrhoids, postpartum bleeding, and intestinal bleeding (Tierra and Michael, 1988).

Potentilla indica (Andr.) Wolf in Asch. and Graebn. Syn. Mitt. Fl. 6 (1): 661. 1904. *Fragaria indica* Andr. Bot. Repos. 7: t, 479. 1807; Wight, Ic. t. 989. 1845; Hook. f. Fl. Brit. India 2: 343. 1878; Gamble, Fl. Pres. Madras (repr. ed.); Chithra in Nair and Henry, Fl. Tamil Nadu 1: 141. 1983.

Herb; creeping stolons. Leaves digitately 3-5-foliate, leaflets obovate, usually crenate. Stipules adnate to the petiole. Flowers yellow in solitary cymose inflorescence. Bracts leafy. Bracteoles 5 close under the calyx. Petals 5, ovate to rounded. Stamens many, persistent; filaments filiform, anther didymous. Ovules 2, style lateral. Fruits achenes, reddish when ripened.

Phenology: July- Sep.

Parts used: Fruits.

Mode of use: Ripened raw fruits are edible.

Therapeutic uses: Anticoagulant, antiphlogistic, antiseptic, depurative, febrifuge etc. (http://www.pfaf.org/user/Plant.aspx?LatinName=Duchesnea+indica).

Prosopis cineraria (L.) Druce in Bot. Exch. Club Soc. Brit. Isles 3: 422. 1914; Vajravelu in Nair and Henry, Fl. Tamil Nadu 1: 139. 1983. *Mimosa cineraria* L. Sp. Pl. 517. 1752. "*cinérea*". *Prosopis spicigera* L. Mant. Pl. 68. 1767; Bedd. Fl. Sylv. t. 56. 1870; Baker in Hook. f. Fl. Brit. India 2: 288. 1878; Gamble, Fl. Pres. Madras 1: 297. 1957 (repr. ed.).

Shrub. Leaves alternate, bipinnate, with 1-3 pairs of pinnae, rachis prolonged beyond the last pinnae as a soft bristle. Leaflets 10-20 pairs, 7-17 mm long, 2-3 mm broad, entire, oblong, obtuse, sometimes mucronate. Stipules spiny, in pair. Flowers in dense axillary pedunculate spikes. Flowers greenish yellow. Calyx cup-shaped, 5 toothed, teeth small. Petals 5, free, tip and margin hairy. Stamens 10, free, exserted, anthers tipped with deciduous glands. Pod pedicellate, light yellow, glabrous.

Phenology: Mar.- June.

Parts used: Aril.

Mode of use: Ripened arils of the fruits are eaten raw.

Psidium gujava L. Sp. Pl. 470. 1753; Duthie in Hook. f. Fl. Brit. India 2: 468. 1878, "*guyava*"; Gamble, Fl. Pres. Madras 1: 334. 1957 (repr. ed.); Chithra in Nair and Henry, Fl. Tamil Nadu 1: 154. 1983.

Tree. Leaf blade oblong to elliptic, leathery, pubescent. Flowers solitary, hypanthium campanulate. Calyx cap near rounded, irregularly opening. Petals white. Stamens numerous. Ovary adnate to hypanthium. Style as long as stamens. Fruit globose, ovoid with persistant calyx lobes.

Parts used: Fruits.

Mode of use: Ripened raw fruits are consumed.

Therapeutic uses: Fruits are tonic, cooling and laxative; good for colic and bleeding gums. Fruits and its conserve are also astringent and used in diarrhoea and dysentery (http://www.mpbd.info/plants/psidium-guajava.php).

Rhodomyrtus tomentosa (Ait.) Hassk. in Flora 25 (2): 35. 1842; Duthie in Hook. f. Fl. Brit. India 2: 469. 1878; Gamble, Fl. Pres. Madras 1: 333. 1957 (repr. ed.); Chithra in Nair and Henry, Fl. Tamil Nadu 1: 154. 1983. *Myrtus tomentosa* Ait. Hort. Kew. ed. 1. 2: 159. 1789; Wight, Ic. t. 522. 1842.

Shrub. Leaves opposite, leaf blade elliptic to obovate, leathery, pubescent when young but glabrescent, base broadly cuneate, apex rounded to obtuse and often slightly emarginated. Flowers 1-3, stipitate, hypanthium obovoid, grey tomentose. Calyx lobes 5, sub-rounded. Persistent. Petals 5, pink, obovate. Stamens numerous, red. Ovary 3-4-loculed; Style short, stigma simple. Berry purplish black when mature.

Phenology: Apr.- May.

Parts used: Fruits.

Mode of use: Ripened fruits are edible raw. The fruits have been reported to be used to make jams or jellies.

Therapeutic uses: Unripe fruits have green skin and taste slightly astringent. The ripened fruits are used to treat dysentery and diarrhoea in Malaysia (http://www.greenculturesg.com).

Ricinus communis L. Sp. Pl. 1007. 1753; Hook. f. Fl. Brit. India 5: 457. 1887; Gamble, Fl. Pres. Madras 2: 933. 1957 (repr. ed.); Chandrabose in Henry *et al.,* Fl. Tamil Nadu 2: 239. 1987.

Herb. Stem single, but sometimes bushlike, younger parts glaucous, whole plant often purplish. Leaves alternate, palmatifid, peltate, lobes lanceolate, margin serrate. Flowers in terminal panicle. Male flowers: Perianth cupular, spiliting into 3- 5 lobes; Stamens numerous, filaments connate and repeatedly branched, anther cells divergent. Female flowers: Tepals 5, subequal, ovary 3- locular, ovules 3 pendulous, style 3, papillose, stigmatiferous. Fruit capsule, ovoid, 3- lobed, prickly.

Phenology: June- Dec.

Parts used: Seeds.

Mode of use: Matured fruits are added in food and taken.

Therapeutic uses: Seed oil is a strong purgative; used externally as a massage for rheumatic pains, joint pain, paralysis and internally for the treatment of constipation. Seeds paste is used as counter irritant (http://www.mpbd.info/plants/ricinus-communis.php).

Rivea hypocrateriformis (Desr.) Choisy, Convolv. Orient. 26. 1834; Clarke in Hook. f. Fl. Brit. India 4: 184. 1883; Gamble, Fl. Pres. Madras 2: 635. 1957 (repr. ed.); Chandrabose in Henry *et al.,* Fl. Tamil Nadu 2: 111. 1987. *Convolvulus hypocrateriformis* Desr. in Lam. Encycl. 3: 561. 1792.

Lianas. The stems becoming woody with age, pubescent. Leaves rounded-cordate, obtuse apically, densely appressed pubescent below. Flowers usually solitary, occasionally subspicate. Sepals unequal, ovate, obtuse apically, densely short villose. Corolla white, salverform. Stamens villous at base. Ovary 4-locular, stigma linearly 2-fid, oblong. Fruit indehiscent or tardily dehiscent, dry-baccate. Seeds brown, glabrous.

Phenology: Feb.- June.

Parts used: Leaves.

Mode of use: Leaves are made into curries and taken as leafy vegetables.

Therapeutic uses: The Juice of leaves is used for skin disease of hair scalp (Brahmbhatt, 2010).

Rubus ellipticus Smith in Rees, Cyclop. 30: n. 16. 1815; Hook. f. Fl. Brit. India 2: 336. 1878; Gamble, Fl. Pres. Madras 1: 313. 1957 (repr. ed.); Chithra in Nair and Henry, Fl. Tamil Nadu 1: 143. 1983. *R. gowreephul* Roxb. Fl. Ind. 2: 517. 1832; Wight, Ic. t. 230. 1839.

Shrub. Branchlets purplish brown with sparse curved prickles and dense purplish brown bristles. Leaves imparipinnate, 3-foliate, lateral leaflets subsessile, leaflets elliptic-obovate, abaxially dense tomentose with purplish red bristles, sharply serrare, apex acute, abruptly pointed, shallowly cordate. Flowers in terminal, dense glomerate racemes, flowers several in clusters. Bracts linear, pubescent. Calyx pubescent, yellowish tomentsoe, sparsely bristly; Sepals erect, ovate, apex acute and abruptly pointed. Petals white, densely pubescent, base clawed. Stamens numerous, shorter than petals. Filaments broadened and base flattened. Ovary pubescent; Styles glabrous, stigma small. Fruit subglobose, golden yellow, glabrous.

Phenology: Dec.- May.

Parts used: Fruits.

Mode of use: Ripened fruits are edible raw and also it is used for making jams.

Therapeutic value: It is used for indigestion (Kala, 2005).

Rubus leucocarpus Arn. Nova Acta Phys. Med. Acad. Caes. Leop. Carol. Nat. Cur. 18: 16. 1834; Chithra in Nair and Henry, Fl. Tamil Nadu 1: 143. 1983. *R. lasiocarpus* Smith var. *subglaber* Thw. Enum. Pl. Zeyl. 101. 1859; Hook. f. Fl. Brit. India 2: 339. 1878. *R. niveus* Thunb. var. *subglaber* (Thw.) Gamble, Fl. Pres. Madras 442. 1919 and 1: 313. 1957 (repr. ed.).

Shrub. Branchlets purplish brown with sparse curved. Leaves imparipinnate, 3-4 pairs, leaflets ovate-lanceolate, abaxially dense tomentose with purplish red bristles, sharply serrare, apex acute, abruptly pointed, shallowly cordate. Flowers in terminal, dense glomerate racemes, flowers several in clusters. Bracts linear, pubescent. Calyx pubescent, yellowish tomentose, sparsely bristly; Sepals erect, ovate, apex acute and abruptly pointed. Petals pink, densely pubescent, base clawed. Stamens numerous, shorter than petals. Filaments broadened and base flattened. Ovary pubescent; Styles glabrous, stigma small. Fruit globose, golden yellow, glabrous.

Phenology: Dec- Apr.

Parts used: Fruits.

Mode of use: Ripened fruits are edible raw and also it is used for making jams.

Rubus rugosus Smith in Rees, Cyclop. 30: n. 34. 1815, var. **rugosus;** Gamble, Fl. Pres. Madras 1: 312. 1957 (repr. ed.); Chithra in Nair and Henry, Fl. Tamil Nadu 1: 143. 1983. *R. moluccanus* sensu Hook. f. Fl. Brit. India 2: 330. 1878 p.p., non L. 1753.

Shrub. Leaves imparipinnate, leaflets 3-4 pairs, orbicular, the lobes rounded, scarsely lobulate. Bracts broadly ovate, lacinate at apex. Flowers in terminal, dense glomerate racemes, flowers several in clusters. Bracts linear, pubescent. Calyx pubescent, yellowish tomentose, sparsely bristly; Sepals erect, ovate, apex acute and abruptly pointed. Petals pink, densely pubescent, base clawed. Stamens numerous, shorter than petals. Filaments broadened and base flattened. Ovary pubescent; Styles glabrous, stigma small. Fruit globose, purplish black, glabrous.

Phenology: Dec.- May.

Parts used: Fruits.

Mode of use: Ripened fruits are edible raw and also it is used for making jams.

Therapeutic value: Crushed fruits applied on skin diseases; fruits eaten as appetizer (Shankar and Ramesh Babu, 2012).

Rumex vesicarius L. Sp. Pl. 336. 1753; Hook. f. Fl. Brit. India 5: 61. 1886; Gamble, Fl. Pres. Madras 2: 835. 1957 (repr. ed.); Chandrasekaran in Henry *et al.,* Fl. Tamil Nadu 2: 200. 1987.

Herb. Branching from the base, somewhat succulent, glaucous, fragile. Leaves with petiole, triangular-hastate, entire. Flowers in racemes simple or branched. Pedicels solitary, bearing twin or sometimes ternate flower. Valves of fruit-bearing primary flower suborbicular and broad, without a marginal nerve, entire, reticulate, membranous with parallel basal lobes, two of the valves with a small basal grain, the third without a grain; the second - and if present the third - fruit somewhat or hardly smaller, folded together with the primary one and hidden by it. Fruit nuts, primary flower, greyish-brown, that of secondary flower more dark brown.

Phenology: Mar.- May.

Parts used: Leaves.

Mode of use: Young leaves are used as leafy vegetables.

Therapeutic value: Leaf juice is astringent, cooling, aperient and a strong diuretic; relieves toothache, checks nausea and promotes appetite (http://www.mpbd.info/plants/rumex-vesicarius.php).

Scutia myrtina (Burm. f.) Kurz in J. Asiat. Soc. Bengal 44: 168. 1875; Gamble, Fl. Pres. Madras 1: 160. 1957 (repr. ed.); Chithra in Nair and Henry, Fl. Tamil Nadu 1: 77. 1983. *Rhamnus myrtinus* Burm. f. Fl. Ind. 60. 1768. *Scutia indica* Brongn. Mam. Fam. Rhamn. 56. 1826; Wight, Ill. 1: t. 73. 1840; Lawson in Hook.f. Fl. Brit. India 1: 640. 1875.

Shrubs. Branches opposite to subopposite; young branches puberulent; older branches brown or red-brown, striate, glabrous. Spines mostly 2 per node, recurved. Leaves opposite or subopposite. Stipules lanceolate. Leaf blade abaxially pale green, adaxially shiny, deep green, brown when dry, elliptic, leathery, both surfaces glabrous, minutely serrulate, apex shortly acuminate or acute. Flowers yellow-green, few in axillary fascicles or shortly pedunculate in axillary condensed cymes, glabrous. Petals 4 -5, deeply emarginate to deeply bilobed. Stamens 4 - 5, surrounded by and equaling petals. Ovary globose, base filling calyx tube, but not immersed in disk, 2-loculed; style short and stout. Stigma undivided or inconspicuously 2- or 3-lobed. Drupe obovoid-globose, base with persistent calyx tube.

Phenology: Sep.- Dec.

Parts used: Fruits.

Mode of use: Ripened fruits are consumed raw.

Securinega leucopyrus (Willd.) Muell.- Arg. in DC. Prodr. 15 (2): 451. 1866; Chandrabose in Henry *et al.,* Fl. Tamil Nadu 2: 240. 1987. *Flueggea leucopyrus* Willd. Sp. Pl. 4: 757. 1805; Wight, Ic. t. 1875; Hook. f. Fl. Brit. India 5: 328. 1887; Gamble, Fl. Pres. Madras 2: 907. 1957 (repr. ed.).

Shrub. Leaf blade elliptic, obovate, or rotund, papery to thinly leathery, base cuneate to obtuse, margin entire, apex rounded; lateral veins 5-7 pairs. Flowers axillary, appearing with expanding leaves or at leafless nodes, cymose. Male flowers: Sepals 5, ovate or rotund, margins entire or denticulate; disk segments 5, angular; stamens 5; filaments and anthers small; rudimentary ovary 2- or 3-lobed, lobes erect or recurved. Female flowers: Sepals 5, elliptic or ovate, disk annular, subentire at apex. Berry subglobose, whitish when ripe, seeds brownish.

Phenology: Apr.- Oct.

Parts used: Fruits.

Mode of use: Ripened fruits are eaten raw.

Solanum nigrum L. Sp. Pl. 186. 1753; Clarke in Hook. f. Fl. Brit. India 2: 229. 1883; Gamble, Fl. Pres. Madras 2: 657. 1957 (repr. ed.); Chithra in Henry *et al.,* Fl. Tamil Nadu 2: 116. 1987. *S. rubrum* Mill. Gard. Dict. ed. 8. n. 4. 1768, non L. 1767; Wight, Ic. t. 344. 1840.

Herb. Stems often angular, sparsely pubescent. Leafblade ovate, pubescent or glabrescent, base cuneate, entire or coarsely dentate, apex obtuse. Flowers in axillary umbles. Calyx cup shaped, lobes subdeltate, pubescent, abaxially, ciliate. Corolla white, lobes ovate- oblong, pubescent, abaxially, ciliate, spreading. Filaments short, anther oblong. Style short and stigma simple. Fruit berry, dull black, globose.

Phenology: May- Nov.

Parts used: Leaves and fresh and dry fruits.

Mode of use: Leaves are used as green leaves. Fresh fruits are edible raw. Dried fruits are used to make curries.

Therapeutic value: Decoction and also aqueous extract of the leaves is very useful in dropsy for its diuretic action. Young shoots are given in skin diseases, and are very useful in psoriasis. The Chinese employ the leaf juice to alleviate the pain in inflammation of the kidneys and bladder, and in virulent gonorrhoea. Fruits are tonic, diuretic and aphrodisiac; useful in anasarca, heart diseases, hiccup, asthma, fever, diarrhoea and bronchitis. Paste of the green fruit is applied to ringworm. Syrup of the fruit is expectorant and diaphoretic; used as a cooling drink in fevers. Seeds are laxative; useful in giddiness, gonorrhoea, thirst and inflammations (http://www.mpbd.info/plants/solanum-nigrum.php).

Solanum sisymbriifolium Lam. Tabl. Encycl. 2: 25. 1792; Fyson, Fl. South Indian Hill St. 2: 4. 359. 1932; Gamble, Fl. Pres. Madras 2: 658. 1957 (repr. ed.); Chithra in Henry *et al.,* Fl. Tamil Nadu 2: 112. 1987.

Herb. Stems often angular, sparsely pubescent. Leaf blade oblong-lanceolate, deeply pinnatifid, often stellate pubescent, base cuneate, entire or coarsely dentate, apex obtuse. Flowers in axillary umbles. Calyx cup shaped, lobes inflated in fruit pubescent, abaxially, ciliate. Corolla white, lobes ovate-oblong, pubescent, abaxially, ciliate, spreading. Filaments short, anther oblong. Style short and stigma simple. Fruit berry, globose, yellowish when ripe.

Phenology: May- Dec.

Parts used: Fruits.

Mode of use: Fruits are used as vegetable.

Solanum surattense Burm. f. Fl. Ind. 57. 1768. *S. xanthocarpum* Schrader et Wendl. Sert. Hannov. 1: 8. t. 2. 1795; Clarke in Hook. f. Fl. Brit. India 4: 236. 1883; Gamble, Fl. Pres. Madras 2: 658. 1957 (repr. ed.); Chithra in Henry *et al.,* Fl. Tamil Nadu 2: 117. 1987. *S. jacquinii* Willd. Sp. Pl. 1: 1041. 1798; Wight, Ic. t. 1401. 1848.

Subshrub. Stems often angular, prickly. Leafblade lacerate, pubescent, base cuneate, entire or coarsely dentate, apex obtuse, prickly along midnerve. Flowers in axillary cymes. Calyx cup shaped, lobed, prickly, acuminate. Corolla violet, lobes ovate-oblong, pubescent abaxially, ciliate, spreading. Filaments short, anther oblong. Style short and stigma simple. Fruit berry, green with white stripes, yellow when ripened.

Phenology: Throughout the year.

Parts used: Fruits.

Mode of use: Unripened fruits are used as vegetable.

Therapeutic value: Fresh or dried fruits are kept in fire and the smoke is inhaled through mouth once a week to treat toothache (Poongodi *et al.,* 2011). It is used for urinary troubles and skin diseases (Biswas *et al.,* 2010)

Solanum torvum Sw. Prodr. 47. 1788; Wight, Ic. t. 1840; Clarke in Hook. f. Fl. Brit. India 4: 234. 1883; Gamble, Fl. Pres. Madras 2: 658. 1957 (repr. ed.); Chithra in Henry *et al.,* Fl. Tamil Nadu 2: 117. 1987.

Shrub. Stems with stout, recurved reddish or pale-yellow prickles. Leaves solitary or paired, leafblade ovate to elliptic, many- branched, stellate hairs, base cordate or cuneate, margin sinuate, apex acute. Flowers in racemose panicle inflorescence. Calyx cup-shaped, pubescent, lobes ovate-lanceolate. Corolla white, lobes ovate-lanceolate, stellate pubescent. Filaments short, anther small. Style short and stigma simple. Berry globose, yellowish when mature.

Phenology: Throughout the year.

Parts used: Fruits.

Mode of use: Unripened fruits are used as vegetable.

Therapeutic value: Good in enlarged spleen. The boiled fruits with rice at night for 2-3 days to expell thread-worms (http://www.mpbd.info/plants/solanum-torvum.php).

Solanum trilobatum L. Sp. Pl. 188. 1753; Wight, Ic. t. 854. 1844- 1845; Clarke in Hook. f. Fl. Brit. India 4: 236. 1883; Gamble, Fl. Pres. Madras 2: 659. 1957 (repr. ed.); Chithra in Henry *et al.,* Fl. Tamil Nadu 2: 117. 1987.

Phenology: Dec.- Feb.

Parts used: Leaves and fruits.

Mode of use: Leaves are made into chutneys and eaten. Dried fruits are cooked and eaten.

Therapeutic value: The bruised fresh leaves, used externally, are said to ease pain and abate inflammation, and the Arabs apply them to burns and ulcers. The juice of the fresh herb is sometimes used for fever and to allay pain. The fruit is used as a cosmetic; rubbing the seeds on the cheeks to remove freckles (http://www.evergreenenterprises.co.in/solanum-trilobatum-herbs.htm).

Syzygium calophyllifolium Walp. Rep. 2: 180. 1843; Gamble, Fl. Pres. Madras 1: 339. 1957 (repr. ed.); Chithra in Nair and Henry, Fl. Tamil Nadu 1: 155. 1983. *Eugenia calophyllifolia* (Walp.) Wight, Ic. t. 1000. 1845; Duthie in Hook. f. Fl. Brit. India 2: 494. 1878.

Tree. Leaves simple, opposite, entire, sub-orbicular, emarginated, coriaceous. Stipules absent. Flowers white in terminal corymbose inflorescence. Bracteoles green, small, deciduous; Bracts green, small, linear. Calyx green, tubular, funnel-shaped, sepals 4-lobed, ovoid. Corolla white, petals 4, united, forms a calyptra; Stamens numerous, filaments short and anther minute; Style 1, slender, thin and stigma simple. Fruit oblong, purplish black when ripe, 1-seeded.

Phenology: Mar.- Jun.

Parts used: Fruits.

Mode of use: Ripened fruits are taken raw.

Syzygium cumini (L.) Skeels. in U.S. D. A. Bur. Pl. Industr. Bull. 248: 2. 1912; Chithra in Nair and Henry, Fl. Tamil Nadu 1: 156. 1983. *Myrtus cumini* L. Sp. Pl. 471. 1753. *Eugenia jambolanum* Lam. Encycl. 3: 198. 1789; Wight, Ic. t. 535; Duthie in Hook. f. Fl. Brit. India 2: 499. 1879. *Syzygium jambolanum* (Lam.) DC. Prodr. 3: 259. 1828; Gamble, Fl. Pres. Madras 1: 340. 1957 (repr. ed.).

Trees. Branchlets grayish white when dry, terete. Leaf blade broadly elliptic to narrowly elliptic, leathery, abaxially slightly pale when dry, adaxially brownish green to blackish brown and slightly glossy when dry, base broadly cuneate to rarely rounded, apex rounded to obtuse and with a short cusp. Flowers axillary on flowering branches or occasionally terminal, paniculate cymes. Hypanthium obconic or long pyriform. Calyx lobes inconspicuous. Petals 4, white or light purple, coherent, ovate and slightly rounded. Stamens 3-4 mm. Style as long as stamens. Fruit red to black, ellipsoid to pot-shaped, with persistent calyx tube.

Phenology: Feb.- Sep.

Parts used: Fruits.

Mode of use: Ripened fruits are taken raw.

Therapeutic value: The fruit is acrid, sweet, cooling and astringent to the bowels and removes bad smell form mouth, biliousness, stomachic, astringent, diuretic and antidiabetic (Nadkarni, 1976). Extracts of fruit and seeds were found to have antidiabetic, antiinflammatory, hepatoprotective, antihyperlipidemic, diuretic and antibacterial activities (Modi *et al.,* 2010).

Syzygium densiflorum Wall. ex Wight et Arn. Prodr. 329. 1834; Chithra in Nair and Henry, Fl. Tamil Nadu 1: 156. 1983. *S. arnottianum* Walp. Rep. 2: 180. 1843; Gamble, Fl. Pres. Madras 1: 338. 1957 (repr. ed.). *Eugenia arnottiana* (Walp.) Wight, Ic. t. 999. 1845; Duthie in Hook. f. Fl. Brit. India 2: 483. 1878.

Tree. Leaves simple, opposite, elliptic, caudate- acuminate at apex, acute at base. Stipules absent. Flowers cream coloured in dense cymes of umbel inflorescence. Bracteoles green, small, deciduous; Bracts green, small, linear. Calyx green, tubular, funnel-shaped, sepals 4, lobed, ovoid. Corolla white, petals 4, united, forms a calyptra; Stamens numerous, filaments short and anther minute; Style 1, slender, thin and stigma simple. Fruit oblong, purplish black when ripe, 1 seeded.

Phenology: Mar.- Jul.

Parts used: Fruits.

Mode of use: Ripened fruits are edible raw.

Talinum triangulare Willd. Sp. Pl. 162. 1799; Subramaniam and Chandrabose in Madras Agri. J. 40. 51; Chandrasekaran in Nair and Henry, Fl. Tamil Nadu 1: 25. 1983.

Herbs. Branchlets succulent. Leaves alternate, decussate, oblong or obovate. Flowers in terminal clusters, surrounded by whorl of involucre. Bracteoles hairy. Sepals 2, keeled, deciduous, base connate. Petals 5, pink. Stamens numerous. Ovules numerous, style 5-fid. Capsule globose, valvular, seeds smooth, numerous.

Phenology: Throughout the year.

Parts used: Leaves.

Mode of use: Leaves are consumed as vegetables.

Therapeutic value: It is used for curing asthma (Samuel *et al.,* 2010)

Tamarindus indica L. Sp. Pl. 34. 1753; Bedd. Fl. Sylv. t. 184. 1872; Baker in Hook. f. Fl. Brit. India 2: 273. 1878; Gamble, Fl. Pres. Madras 1: 290. 1957 (repr. ed.); Vajravelu in Nair and Henry, Fl. Tamil Nadu 1: 133. 1983.

Trees. Bark dark ashy, irregularly longitudinally splitting. Leaflets oblong, small, glabrous, base obliquely rounded, apex rounded or emarginate. Flowers few, yellowish tinged with purplish red stripes; peduncles and pedicels yellowish green puberulent. Bracteoles enclosing flower bud before anthesis. Calyx tubular, lobes, lanceolate-oblong, reflexed after anthesis. Petals obovate, sub-equal to calyx lobes, margin repand, curled. Stamens pubescent near base, free parts of filaments short, anthers elliptic. Ovary slightly incurved, terete. Legume brownish, straight or arcuate, terete-oblong.

Phenology: May- Dec.

Parts used: Aril and seeds

Mode of use: Ripened fruits are added for flavours in curries and also unripened and ripened fruits are edible raw. Seeds are fried and eaten.

Therapeutic value: Pulp of the ripe fruit is refrigerant, carminative, digestive, astringent, laxative and tonic to the heart; it is a household remedy for fever, dyspepsia, gastritis, dysentery and diarrhoea; also useful in diseases supposed to cause by deranged bile, such as burning of the body and costiveness. Washing of hairs with pulp-soaked water removes dandruff. Drink of pulp lowers blood cholesterol. Gargle with tamarind water is useful in sore throat. Seeds are astringent; used for diarrhoea (http://www.mpbd.info/plants/tamarindus-indica.php).

Toddalia asiatica (L.) Lam. Tabl. Encycl. 2: 116. 1797; Gamble, Fl. Pres. Madras 1: 107. 1957 (repr. ed.) incl. vars. *Paullinia asiatica* L. Sp. Pl. 365. 1753; Vajravelu Nair and Henry, Fl. Tamil Nadu 1: 62. 1983. *Toddalia aculeata* Pers. Syn. 1: 249. 1805; Wight, Ic. t. 66. 1840; Hook. f. Fl. Brit. India 1: 479. 1875.

Shrubs or woody climbers, usually armed. Leaflet blades usually sessile or subsessile, elliptic or narrowly elliptic to obovate to oblanceolate, base narrowly cuneate to attenuate, apex acuminate or rarely acute to obtuse or rounded. Flowers simple; Sepals campanulate, lobed. Petals cream-white, ovate to elliptic. Male flowers: Petals 5, stamens 5, free, prominent. Female flowers: ligulate; Ovary oblong; Style short and stigma capitate. Fruit berry, reddish when ripe.

Phenology: Throughout the year.

Parts used: Fruits.

Mode of use: Ripened fruits are edible raw.

Therapeutic value: Unripe fruit and root are rubbed down with oil to make a stimulant liniment for rheumatism (http://www.mpbd.info/plants/todalia-asiatica.php).

Tribulus terrestris L. Sp. Pl. 387. 1753; Edgew. and Hook. f. Fl. Brit. India 1: 423. 1874; Gamble, Fl. Pres. Madras 1: 92. 1957 (repr. ed.); Chithra in Nair and Henry, Fl. Tamil Nadu 1: 49. 1983. *T. lanuginosus* L. Sp. Pl. 387. 1753; Wight, Ic. t. 98. 1838.

Herbs. Stem prostrate, glabrous, pubescent, or hispid. Leaves opposite, even-pinnate, with 6-16 leaflets; leaflet blades oblong to obliquely oblong, base slightly oblique, margin entire, apex acute to obtuse. Flower solitary. Sepals 5, persistent. Petals golden yellow, obovate. Stamens inserted on base of disk, with scale-form appendages. Ovary 5-angled, 5-locular, with 3 or 4 ovules per locule; stigma 5-parted. Fruit schizocarp, pubescent or glabrous, with 5 carpels with 2 hardened 4-6 mm spines at mid margin, surface spiny or prickly.

Phenology: May- Sep.

Parts used: Leaves.

Mode of use: Young leaves are used as leafy vegetables.

Therapeutic value: Jaundice, Urinary troubles (Rajendran *et al.,* 2008; Rajadurai *et al.,* 2009)

Vaccinium leschenaultii Wight, Ic. t. 1188. 1848, var. **leschenaultii**; Clarke in Hook. f. Fl. Brit. India 3: 455. 1882; Gamble, Fl. Pres. Madras 2: 522. 1957 (repr. ed.); Rajan in Henry *et al.,* Fl. Tamil Nadu 2: 57. 1987.

Tree. Leaves ovate-lanceolate or orbicular, minutely serrate. Flowers in racemose inflorescence. Calyx 5, tube truncate, teethed, adnate to the ovary. Corolla pinkish, 5-teethed, urceolate, deciduous. Stamens 10. Filaments short and anther small. Ovary 10-locular; Style columnar and stigma small. Fruits berry, purplish when ripe; seeds numerous.

Phenology: Feb.- May.

Parts used: Fruits.

Mode of use: Ripened and unripened fruits are edible raw. It also used for making pickles.

Vaccinium neilgherrense Wight in Calcutta J. Nat. Hist. 8: 173. 1847 and Ic. t. 1189. 1848; Clarke in Hook. f. Fl. Brit. India 3: 454. 1882, "*nilgherrense*"; Gamble, Fl. Pres. Madras 2: 521. 1957 (repr. ed.); Rajan in Henry *et al.,* Fl. Tamil Nadu 2: 58. 1987.

Tree. Leaves lanceolate, acuminate, crenate-serrate. Flowers in cymose inflorescence. Calyx 5, tube truncate, teethed, adnate to the ovary. Corolla white, 5-teethed, urceolate, deciduous. Stamens 10. Filaments short and anther small. Ovary 10-locular; Style columnar and stigma small. Fruits berry, purplish when ripe; seeds numerous.

Phenology: Feb.- Apr.

Parts used: Fruits.

Mode of use: Ripened and unripened fruits are edible raw. It also used for making pickles.

Zingiber officinale Roscoe in Trans. Linn. Soc. London. 8: 348. 1807; Baker in Hook. f. Fl. Brit. India 6: 246. 1892; Gamble, Fl. Pres. Madras 3: 1040. 1957 (repr. ed.); Ramamurthy in Henry *et al.,* Fl. Tamil nadu 3: 29. 1989.

Herb with pseudostems. Leaves sessile; ligule slightly 2-lobed, membranous. Leaf blade lanceolate or linear-lanceolate, glabrescent. Inflorescences arising from rhizomes, ovoid. Bracts pale green, sometimes yellowish at margin, ovate, apex mucronate. Bracteoles equaling bracts. Calyx small. Corolla yellowish green; lobes lanceolate; Central lobe of labellum with purple stripe and cream blotches, oblong-obovate, shorter than corolla lobes; lateral lobes ovate, free nearly to base. Stamen dark purple; anther connective appendage curved.

Phenology: Oct.

Parts used: Rhizome.

Mode of use: It is used as a flavor for curries.

Therapeutic value: The rhizome juice is mixed with honey is prescribed for cold fever (Choudhury *et al.,* 2010), debility, digestive aid. Rhizomes are mixed with salt and taken after food (Rahmatullah *et al.,* 2009).

Ziziphus mauritiana Lam. Encycl. 3: 319. 1789, var. **mauritiana.** *Z. jujuba* (L.) Gaertn. Fruct. 1: 203. 1788, non Mill. 1768; Wight, Ic. t. 99. 1838; Lawson in Hook. f. Fl. Brit. India 1: 632. 1875; Gamble, Fl. Pres. Madras 1: 157. 1957 (repr. ed.); Chithra in Nair and Henry, Fl. Tamil nadu 1: 78. 1983. *Rhamnus jujuba* L. Sp. Pl. 194. 1753.

Tree. Leaves orbicular to rotund, rusty tomentose below, apex rounded. Flowers in cymes in dense fascicles. Sepals tubular, ovate- triangular, spreading. Petals ovate, greenish, grooved, obovate- spathulate. Stamens 5, enclosed by petals. Ovule 1, erect; Styles 2, connate to middle; Stigma bifid. Drupe oblong- globose. Pale brownish when ripe.

Phenology: July- Dec.

Parts used: Fruits.

Mode of use: Ripened fruits are edible raw.

Ziziphus oenoplia (L.) Mill. Gard. Dict. ed. 8. n. 3. 1768; Lawson in Hook. f. Fl. Brit. India 1: 634. 1875; Gamble, Fl. Pres. Madras 1: 158. 1957 (repr. ed.); Chithra in Nair and Henry, Fl. Tamil Nadu 1: 78. 1983. *Rhamnus oenoplia* L. Sp. Pl. 194. 1753.

Tree. Thorns recurved, blackish galls on the stem. Leaves ovate-lanceolate. Flowers in cymes in dense fascicles, 5-8-fid. Sepals tubular, ovate-triangular, spreading. Petals ovate, greenish, grooved, obovate-spathulate. Stamens 5, enclosed by petals. Ovule 1 erect; Styles 2, connate to middle; Stigma bifid. Drupe globose, rugose. Pale brownish when ripe.

Phenology: June- Nov.

Parts used: Fruits.

Mode of use: Ripened fruits are edible raw.

Therapeutic value: Fruit is used as an ingredient in the pills used for stomachache (http://www.mpbd.info/plants/zizyphus-oenoplia.php).

Ziziphus rugosa Lam. Encycl. 3: 319. 1789; Wight, Ic. t. 339. 1840; Lawson in Hook. f. Fl. Brit. India 1: 636. 1875; Gamble, Fl. Pres. Madras 1: 158. 1957 (repr. ed.); Chithra in Nair and Henry, Fl. Tamil Nadu 1: 78. 1983.

Straggler. Thorns stout, recurved. Leaves orbicular- rotund. Flowers in cymes in dense fascicles, 5-8-fid. Sepals tubular, ovate-triangular, spreading. Petals 0, Disc 5-lobed. Stamens 5, enclosed by petals. Ovule 1 erect; Styles 2, connate below the middle; Stigma bifid. Drupe globose. Pale brownish when ripe.

Phenology: Jan.- Jun.

Parts used: Fruits.

Mode of use: Ripened fruits are eaten raw.

Therapeutic value: Fruits are used in the treatment of carbuncle, menorrhoea, syphilis and ulcer, tongue (Tyagi, 2005.). The fruit is described as demulcent and enters into the treatment of throat and broncho-pulmonic irritations (Yadav and Singh, 2010).

Results and Discussion

Acquiring bioresources is a common activity of the indigenous people in Nilgiri District. A total of 111 species of wild plants belonging to 79 genera listed in 47 families were identified in the present study. Out of 47 families identified the most widely used species belonged to the family Euphorbiaceae with 8 species; Solanaceae with 7 species; Myrtaceae with 6 speices; Amaranthaceae and Tiliaceae with 5 species; Moraceae, Passifloraceae, Rosaceae and Rhamnaceae, Rutaceae with 4 species; Dioscoreaceae, Eleagnaceae and Mimosaceae with 3 species; Liliaceae, Convolvulaceae, Caryophyllaceae, Berberidaceae, Arecaceae, Sapindaceae, Apiaceae, Cucurbitaceae, Zingiberaceae, Asteraceae, Elaeocarpaceae, Myrsinaceae, Oxalidaceae, Vaccinaceae with 2 species and the rest of the families were represented by each one species.

It provides a database regarding the ethnobotanical facts of these lesser-known under-utilized wild edible species through frequent interactions and discussions were made with the local people in the study area. The present study indicates that the tribal people possess a thorough knowledge on the usage of wild plants used by them in their day to day's life. In addition, the wild edibles would not only be the supplement for the food but also it contributes therapeutically.

Wild edible plants are gathered in the form of fruits, leaves, roots, tubers, flowers etc., and these plants still share a good proportion of tribal dishes all over the world (Anon., 1970-1988; Duke and Atchley, 1986; Neog and Mohan, 1994; Samant and Dhar, 1997). Fruit plays a major role in all above the stuffs, which is followed by the leafy vegetables. Regular consumption of fruit is associated with reduced risks of cancer, cardiovascular disease (especially coronary heart disease), stroke, Alzheimer disease, cataracts, and some of the functional declines associated with aging (Liu, 2003). Ambe (2001) in overview of their nutritional value, the preservation of these fruits also has economical advantages, as there is a significant trade in some of these wild edible fruits. Tubers contain carbohydrates in the form of starch but they are easily digestable when compared to those of grains, cereals and pulses. They also supply minerals, vitamins and fibers etc. (Rajendran and Manian, 2009).

Traditional wild edible species have been meeting the protein, carbohydrate, fat, vitamin and mineral requirements of the local residents to a greater extent (Sebastian and Bhandari, 1990; Ohiokpehai, 2003).

Plants like *Elaeagnus conferta, Embelia ribes, Ficus racemosa, Glycosmis pentaphylla, Grewia hirsuta, Morus alba, Passiflora leschenaultia, Physalis peruviana, Rhodomyrtus tomentosa, Ricinus communis, Rubus ellipticus, Solanum nigrum, Syzygium cumini, Tamarindus indica, Ziziphus oenoplia, Ziziphus rugosa* are used for curing stomach problems. The species like *Alangium salvifolium, Artocarpus heterophyllus, Artocarpus hirsutus, Psidium gujava, Syzygium cumini* are used for body cooling.

Artocarpus heterophyllus, Artocarpus hirsutus, Azadirachta indica, Piper nigrum and *Solanum nigrum* are used as aphrodisiac. *Alangium salvifolium, Azadirachta indica, Embelia ribes, Opuntia dillenii, Ricinus communis* are used as purgative. *Artocarpus heterophyllus, Ficus racemosa, Mangifera indica, Murraya koenigii, Rhodomyrtus tomentosa, Syzygium cumini, Tamarindus indica* are used as astringents. *Capsicum annuum, Ficus racemosa, Mahonia leschenaultia, Physalis peruviana* and *Piper nigrum* are used for curing cold.

Endemic species such as *Artocarpus hirsutus, Elaeagnus indica, Elaeagnus kologa, Mahonia leschenaultii, Phyllanthus indofischeri, Syzygium arnottianum, Syzygium calophyllifolium, Vaccinium leschenaultii* and *Vaccinium neilgherrensis* also used by tribes for edible and medicinal purposes. It was also found that many endemic edible plants are still exploited in the study area. Due to over exploitation and depletion of biodiversity, there is a need to collect and conserve those species before the threat of extinction. Plants used frequently such as *Amaranthus* spp., *Centella asiatica, Colocasia esculenta, Dioscorea bulbifera, Murraya koenigii* etc., are easily propagated and maintained in the backyards of houses. The dietary forms of these wild edibles are mostly the leafy vegetables, fruits and tubers. The nutritional value of traditional leafy vegetables is higher (Nordeide *et al.,* 1996; Orech *et al.,* 2007; Sundriyal *et al.,* 2001) than several common vegetables. They also contain antioxidants which offer protection against many chronic diseases like Heart disease and certain type of cancers (Saxena, 1999).

The outcome of the present study reveals the diversity of wild edible plant wealth of the Nilgiri District. Consequently, the present documentation is important to preserve their valuable indigenous knowledge, for sustainable utilization and conservation for future generation. Otherwise, they may be lost in the near future and another need of this study is domestication of this wild species. Eco-tourism and urbanization have totally changing this life style. Hence, the oral transmission of traditional knowledge is declining rapidly, so there is an urgent need to document the specific indigenous knowledge for serving future generation.

References

Akah, P.A., Ezike, A.C., Nwafor, S.V., Okoli, C.O. Enwerem, N.M. 2003. Evaluation of the anti-asthmatic property of *Asystasia gangetica* leaf extracts. J. Ethnopharmacol. **89(1):**25-36.

Ambe, G.A. 2001. Less fruits sauvages comestibles des savanes guinéennes de Cote-d'Ivoire: etat de la connaissance par une population locale, les Malinké. Biotechnol. Agron. Soc. Environ. **5:** 43-58.

Anonymous, 1994. Ethnobiology in India-A Study Report. Ministry of Environment and Forests, Government of India, Delhi.

Anonymous, 1970-1988. Wealth of India: raw materials; Council of Scientific and Industrial Research, Delhi. 1-12 (Reprinted).

Ashok Kumar, B.S., Lakshman, K., Jayaveera K.N., Nandeesh, R, Mani Tripathi, S.N., Vamshi Krishn,a N., Manjunath, M. and Suresh, M.V. 2009. Estimation of Rutin and Quercetin in *Amaranthus viridis* L. by High Performance Layer chromatography (HPLC). Ethno. Leaflets **13:** 437-42.

Balakrishnan, V., Prema, P., Ravindran, K.C and Philip Robinson, J. 2009. Ethnobotanical studies among Villagers from Dharapuram Taluk, Tamil Nadu, India. Global J. Pharma., **3(1):** 08-14.

Bean, J. L. and Saubel, K.S. 1972. Temalpakh: Cahuilla Indian Knowledge and Usage of Plants. Malki Museum Press, Morongo Indian Reservation, pp. 225.

Bensky and Dan, 2004. Chinese Herbal Medicine, Materia Medica. China: Eastland Press Inc.

Bhowmik, D. Chiranjib, Jitender Yadav, Tripathi K. K. and Sampath Kumar, K. P. 2010. Herbal remedies of *Azadirachta indica* and its medicinal application. J. Chem. Pharm. Res. **2(1):** 62-72.

Biswas, A. Bari, M.A., Roy, M. and Bhadra, S.K. 2010. Inherited folk pharmaceutical knowledge of tribal people in the Chittagong Hill tracts, Bangladesh. Indian J. Trad. Knowl. 9(1): 77-89.

Bonde, S.D., Nemade, L S., Patel, M.R. and Patel, A.A. 2011. *Murraya koenigii* (Curry leaf): Ethnobotany, Phytochemistry and Pharmacology - A Review. Int. J. Pharm. Phytopharmacol. Res. **1(1):** 23-27.

Brahmbhatt, M. R., Patel, J. M., Patel, V. B. and A. K. Saluja. 2010. Analgesic and Anti-inflammatory activity of leaves of *Rivea hypocrateriformis*. J. Pharmaco. Phytother. **1:** 1.

Britto, J.D. and Mahesh, R. 2007. Exploration of Kani Tribal Botanical Knowledge in Agasthiayamalai Biosphere Reserve - South India. Ethno. Leaflets, **11:** 258-265.

Choudhury, M.D., Bawari, M. and Singha, L.S. 2010. Some antipyretic Ethno-medicinal Plants of Manipuri community of Barak Valley, Assam, India, Ethno. Leaflets **14:** 21-28.

Duke, J.A. and Atchley, A. 1986. Handbook of proximate analysis tables of higher plants. CRC Press, Inc., Boca Raton, Florida.

Duke, J.A. and Ayensu. E.S. 1985. Medicinal Plants of China. Reference Publications.

Edwin, E., Sheeja, E., Dhanabal, B. and Suresh. 2007. Antihyperglycemic activity of *Passiflora mollissima* Bailey. Ind. J. Pharma. Sci. **69 (4):** 570- 571.

Ganesan, S., Ponnuchamy, M., Kesavan, L. and Selvaraj, A. 2009. Floristic composition and practices on the selected sacred groves of Pallapatty village (Reserved Forest), Tamil Nadu. Indian J. Trad. Knowledge 8(2): 154-162.

Goyal, B.R., Agarwal, B.B., Goyal, R.K. and Mehta, A.A., 2007. Phyto- pharmacology of *Moringa oleifera* Lam. An overview. Nat. Prod. Rad. **6(4):** 347- 353.

Grivetti L.E. and Ogle, B.M. 2000. Value of traditional foods in meeting macro- and micronutrient needs: The wild plant connection. Natl. Res. Rev.**13:**31-46.

Hemborn, P.P. 1996. Contact therapy practiced by Mundas Chotanagar (Bihar). Ethno. **8:** 36-39.

Herzog, F., Gautier- Begun, D. and Muller, K. 2012 Uncultivated plants for human nutrition in Cote d Ivoire (online, accessed on 13th May, 2012).

Hosagoudar, V.B. and Henry, A.N. 1996. Ethnobotany of tribes Irular, Kurumban and Paniyan of Nilgiris in Tamil Nadu, Southern India. J. Econ. Taxon. Bot. **112:** 272-283.

Jain, S.K. and Srivastava, S. 2005. Traditional uses of some Indian plants among islanders of the Indian Ocean. Indian J. Trad. Knowl. **4(4):** 354- 357.

Jayabharathi, M. and Chitra, M. 2011. Evaluation of anti-inflammatory, analgesic and antipyretic activity of *Moringa concanensis* Nimmo. J. Chem. Pharm. Res. **3(2):**802-806.

Kala, C.P. 2005. Ethnomedicinal botany of the Apatani in the Eastern Himalayan region of India. J. Ethno. and Ethnomed. **1**(11): 1-8.

Karuppusamy, 2007. Medicinal plants used by Paliyan tribes of Sirumalai hills of Southern India. Nat. Prod. Rad. **6 (5):** 436- 442.

Khan, M., Kumar, S. and Hamal, I.A. 2009. Medicinal plants of sewa river catchment area in the Northwest Himalaya and its implication for conservation. Ethno. Leaflets, **13:** 1113- 39.

Khare, C.P. 200. Indian Medicinal Plants. An illustrated Dictionary. Springer.

Kumaravelu, 2008. Ethnobotanical knowledge of the Toda tribes of Nilgiris. Eco. news, **14 (2):** 17- 18.

Lans, CA. 2006. Ethnomedicines used in Trinidad and Tobago for urinary problems and diabetes mellitus. J. Ethnobiol. Ethnomed. **2:** 45.

Liu. 2003. Health benefits of fruits and vegetables are from additive and synergistic combinations of phytochemicals, Amer. J. Clin.Nut. **78(3):**517S.

Modi, D.C., Patel, J.K., Shah, N.N. and Nayak, B.S. 2010. Pharmacognostic studies of the seed of *Syzygium cumini* Linn. Pharma Sci. Monitor **1(1):** 20-26.

Myers, N., Mittermeier, R.A., Mittermeier, C.G., da Fonseca, G.A.B. and Kent, J. 2000. Biodiversity hotspots for conservation priorities. Nature, **403:** 853- 858.

Neog, M. and Mohan, N.K. 1994. Minor and less known fruits of Assam. Indian Horticulture.

Nordeide, M.B., Hatloy, A., Folling, M., Lied, E. and A. Oshoug. 1996. Nutrient composition and nutritional importance of green leaves and wild foods in an agricultural district, Koutiala, in Southern.

Ogle, B.M. 2001. Wild vegetables and micronutrient nutrition- studies on the significance of wild vegetables in women's Diets in Vietnam, (Comprehensive summaries of Uppsala, Dissertations from the Faculty of Medicine).

Ogle, B.M., Nguyen Nhut Xuan Dung, Do Tran Thanh and Hambraeus L. 2001. The contribution of Wild Vegetables to micronutrient intakes among women: An example from the Mekong Delta, Vietnam, Ecol. Food Nutr. **40:**159-184.

Ogle, B.M., Ho Thi Tuyet, Hoang Nghia Duyet and Nguyen Nhut Xuan Dung. 2003. Food, Feed or Medicine: The multiple functions of edible wild plants in Vietnam. Econ. Bot. **57(1):**103-117.

Ohiokpehai, O.2003. Nutritional aspects on street foods in Botswana. Pak. J. Nutri. **2(2):** 76- 81.

Orech, F.O., Aagaard- Hansen, J. and H. Friis. 2007. Ethnoecology of traditional leafy vegetables of the Luo people of Bondo District, Western Kenya. Int. J. Food. Sci. Nutr. **58(7):**522-530.

Paul J. H. A., Seaforth, C.E. and T. Tikasingh. 2011. *Eryngium foetidum* L.: A review, Fitoterapia **82(3):** 302- 308.

Poongodi, A., Thilagavathi, S., Aravindhan, V. and Rajendran, A. 2011. Observations on some ethnomedicinal plants in Sathyamangalam forests of Erode district, Tamil Nadu, India. J. Med. Pl. Res. **5(19):** 4709-4714.

Rahmatullah, M., Ferdausi,D., Mollik, M.A.H., Azam, M.N.K., Taufiq-Ur-Rahman,M. and Jahan, R. 2009. Ethnomedicinal Survey of Bheramara Area in Kushtia District, Bangladesh. Amer.-Eur. J. Sustai. Agri., **3(3):** 534-541.

Rajadurai, M., Vidhya, V.G., Ramya, M. and Anusha Bhaskar, 2009. Ethno-medicinal plants used by the traditional healers of Pachamalai hills, Tamil Nadu, India. Ethno. Med., **3(1):** 39- 41.

Rajendran, A. and Manian, S. 2009. Wild edible plant resources of the Kolli hills, Eastern Ghats, Tamil Nadu, India. In proceedings of the National seminar Recent trends in the conservation and utilization of under utilized wild edible plants, (CUWEP). LAMBERT, Academic publishing. 117- 127.

Rajendran, K., Balaji, P. and Jothi Basu, M. 2008. Medicinal plants and their utilization by villagers in southern districts of Tamil Nadu. Ind. J. Trad. Knowl. **7(3):** 417-420.

Ravikumar, K. and Vijayasankar, R. 2003. Ethnobotany of Malayali tribes in Melpattu village, Javvadhu hills of Eastern Ghats, Tiruvannamalai district, Tamil Nadu. J. Econ. Taxon. Bot. **27:** 715-726.

Reddy, C.S., Reddy, K. N., Murthy, E. N. and Raju, V. S. 2009. Traditional medicinal plants in Seshachalam hills, Andhra Pradesh. Indian J. Med. Pl. Res. **3(5):** 408-412.

Reddy, K.N., Trimurthulu, G. and Sudhakar Reddy, C. 2010. Medicinal plants used by ethnic people of Medak district, Andhra Pradesh. Ind. J. Trad. Knowl. **9 (1):** 184- 190.

Rodgers, W.A. and Panwar, H.S. 1988. Planning a Wildlife Protected Area Network in India. Vol. I and II. Wildlife Institute of India, Dehradun.

Samant, S.S. and Dhar, U. 1997. Diveristy, Endemism and Economic Potential of Wild Edible Plants of Indian Himalaya. Inter. J. Sustain. Develop. World ecol. **4:**179-191.

Samuel, A.J.S.J., Kalusalingam, A., Chellappan,D.K., Gopinath, R., Radhamani,S., Husain, H.A., Muruganandham, V. and Promwichit, P. 2010. Ethnomedical survey of plants used by the Orang Asli in Kampung Bawong, Perak, West Malaysia. J. Ethno. Ethnomed. **6:** 5.

Samuel, J. K. and Andrews, B. 2008. Traditional medicinal plant wealth of Pachalur and Periyur hamlets Dindigul district, Tamil Nadu. Indian J. Tradi. Knowl. **9 (2):** 264- 270.

Saxena, R. 1999. How green is your diet? Nutrition, **33(3):** 9.

Sebastian, M.K. and Bhandari, M.K. 1990. Edible wild plants of the forest areas of Rajasthan. J. Econ. Taxon. Bot. **14(3):**689- 694.

Senthilkumar, M., Gurumoorthi, P. and Janardhanan, K. 2006. Some medicinal plants used by Irular, the tribal people of Marudhamalai hills, Coimbatore, Tamil Nadu. Nat. Pro. Rad. **5(5):** 382- 388.

Shaheen, M.E.K., Syef, M.A., Saha, S.S., Islam, M.S., Hossain,M.D.A., Sujan, M.A.I. and Rahmatullah, M. 2010. Medicinal plants used by the folk and tribal medicinal practitioners in two villages of Khakiachora and Khasia Palli in Sylhet District, Bangladesh. Advan. Nat. and Appl. Sci. **5(1):** 9-19.

Shankar, R. and Ramesh Babu, D. 2012. Conservation of folk healing practices and commercial medicinal plants with special reference to Nagaland, Int. J. Bio. Con. **4(3):**155-163.

Soni, S., Sahu, K.R., Dewagan, S., Yogita Soni and Sanjeev Katre. 2009. Medicinal value of [*Morus alba*] Mulberry Plant. Shodh, Samiksha aur Mulyankan, Inter. Res. Jour. 2 (5): 449-450.

Sundriyal, M. and Sundriyal, R.C. 2001. Wild Edible Plants of the Sikkim Himalaya: Nutritive values of selected species, Eco. Bot., **55:**377-390.

Sundriyal, M., Sundriyal, R., Sharma, C.E. and Purohit, A. N. 1998. Wild edibles and other useful plants from the Sikkim Himalaya, India. Oecol. Mont. **7:** 43- 54.

Tierra, C.A., and Michael, N.D. 1988. Planetary Herbology. Lotus Press. pp. 199.

Tyagi, D.K. 2005. Pharma Forestry, Field Guide to Medicinal Plants. Atlantic publishers and Distributors.

Wu, C. Dai, R., Bai, J., Chen,Y., Yu,Y., Meng, W. and Deng, Y 2011. Effect of *Elaeagnus conferta* Roxb. (Elaeagnaceae) dry fruit on the Activities of Hepatic Alcohol Dehydrogenase and Aldehyde Dehydrogenase in Mice. Tro. J. Pharma. Research **10 (6):** 761-766.

Yadav, A. and Singh, P. 2010. Analgesic and anti-inflammatory activities of *Zizyphus rugosa* root barks. J. Chem. Pharm. Res. **2(3):**255-259.

Yadav, N., Vasudeva, N., Singh, S. and Sharma, S.K. 2007. Medicinal properties of genus *Chenopodium* Linn. Nat. Prod. Rad. (**2**)**:** 131- 134.

http://www.herbnature. tk/2011/02/spinachamaranthus- tricolor-l.html.

http://ayurvedicmedicinalplants.com.

http://www.ecoindia.com/flora/trees/turmeric-plant.html.

http://www.rain-tree.com/mango.htm.

http://food-nutrition.factoidz.com.

http://www.pfaf.org/user/Plant.aspx?LatinName=Duchesnea+indica.

http://www.greenculturesg.com.

http://www.evergreenenterprises.co.in/solanum-trilobatum-herbs.htm.

http://www.mpbd.info/plants.php.

Chapter 7

Study on *Cymbopogon* Spreng. (Poaceae): An Aromatic Grass

T. Nirmala Mary[1]*, *D. Someswari*[2] and *N. Sarathchandra Bose*[3]

[1]*Department of Botany and Microbiology,*
[2]*Department of Botany, T.J.P.S. College, Guntur – 522 006*
[3]*Department of Biochemistry,*
Acharya Nagarjuna University, Guntur – 522 510

ABSTRACT

The morphological features pertaining to the inflorescence among four species of *Cymbopogon i.e., C. caesius* under series Rusae, *C. flexuosus* under series Citrati and *C. jwarancusa* under Schoenanthi series were studied in the present investigation. Lower glume of the sessile spikelet was found to be interesting variable in distinguishing the species, thereby considered key character under taxonomic series. Cytological observations revealed *C. coloratus* and *C. caesius* as diploids (2n=20), *C. flexuosus* and *C. jwarancusa* as tetraploids (2n=40). In addition, the presence of B. chromosomes were noted in *C. caesius*. In general meiosis was found normal with regular bivalents.

Introduction

The grasses that yield food and fodder are known from times immemorial and considerable amount of research has been carried out on them. But the grasses that

* Corresponding Author: E-mail: prof.tnmary@gmail.com

yield essential oil, aroma chemicals and raw material for pharmaceutical products are among the least studied in grass family. The present grass under the study *Cymbopogon* is an aromatic grass that belongs to the tribe Andropogoneae of Poaceae.

The genus *Cymbopogon* is versatile, as the essential oil obtained from various species, varieties is a source of terpenes used in perfumery, flavouring and pharmaceutical preparations. Many of them are used in treatment of various human ailments such as cough, fever, gout, leprosy, stomach disorders and has sedative properties (Kirtikar and Basu, 1935; Kokate *et al.*, 2001). The oils of some species of *Cymbopogon* have powerful germicidal and antibacterial properties (Dayal and Purohit, 1972). Thus species of *Cymbopogon* have got lot of commercial importance in Indian trade.

The present study pertains to the study of reproductive characters (Spike diversity) and their cytological behaviour in four species of *Cymbopogon* growing in and around areas of Guntur.

Materials and Methods

The four species of *Cymbopogon* under study were *C. coloratus* (Hook. f.) Stapf., *C. caesius* (Hook. et Arn.) Stapf., *C. flexuosus* (Nees ex Steud.) Wats. and *C. jwarancusa* (Jones) Schult. that belong to three different series *i.e.*, Rusae, Citrati and Schoenanthi. The spikelet features were studied under dissection microscope. For cytological studies florets of suitable size were fixed in acetic alcohol solution 1 : 3, slides were prepared using anther squash technique with 2 per cent acetocarmine and meiosis was studied.

Results and Discussion

Cymbopogon is a very complex genus with respect to its species allocation. The number of species belonging to this genus range from 102 to 140 and have varied combination of constituents in their essential oil. The genus is partitioned into five taxonomical series by Soenarko (1977) based on the chemical markers (Figure 7.1) and their geographical distribution is given in Figure 7.2 and Table 7.1.

Table 7.1: Broad Geographical Distribution of the Five Series of the Genus *Cymbopogon* in World

Sl.No.	*Series*	*Geographical Distribution*
1	Refracti	Eastern Australia, Pacific islands (Tahiti Hawaii, New Caledonia)
2	Proceri	Australia, New Guinea, Timor, Moluccas
3	Schoenanthi	Saharao-Sindian region, Africa to Himalayas
4	Rusae	Africa to Indochina
5	Citrati	India to South East Asia

The plants of the genus *Cymbopogon* are perennial tall tufted grasses with usually aromatic leaves. Leaf blades are commonly glabrous. The inflorescence is very complex, a panicle supported by a spathe. Racemes commonly occur in pairs, borne on a common peduncle enclosed by a spatheole. Each raceme consists of several

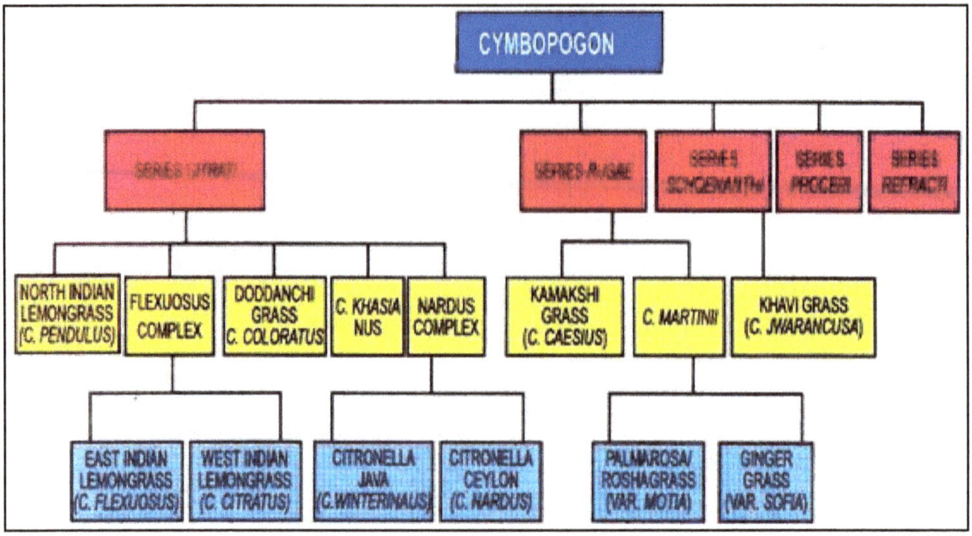

Figure 7.1: Cultivated and Cultivable Species of *Cymbopogon*

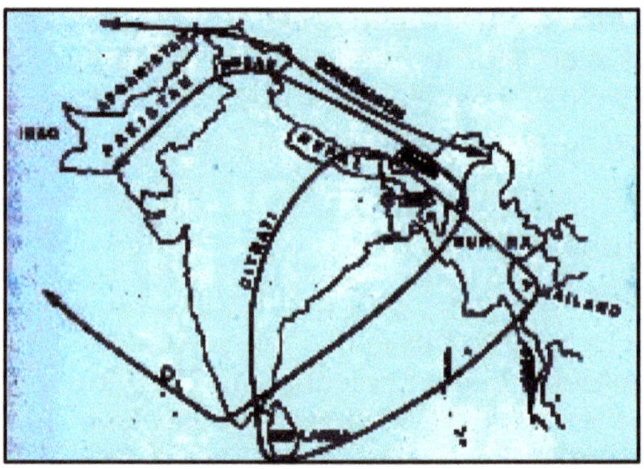

Figure 7.2: Distribution of Three Varieties of *Cymbopogon* in Indian Subcontinent (Gupta, 1970)

pairs of spikelets, in each pair one is sessile and other is pedicelled. The sessile spikelet is hermaphrodite and the pedicelled spikelet is male. Sessile spikelet contains two florets enclosed in two glumes. The structure of the lower glume is the most important character in *Cymbopogon* for differentiation of the series and the species (Figure 7.3).

Key to identification of different series: According to Soenarko (1977):

1. Lower glume of sessile spikelet with deep median groove from middle downwards, often corresponding to a keel inside, leaf blades rounded to cordate at the base .... Series: Rusae
2. Lower glume of sessile spikelet flat, often wrinkled on the back, with narrow to broad wings along the keels .... Series: Citrati

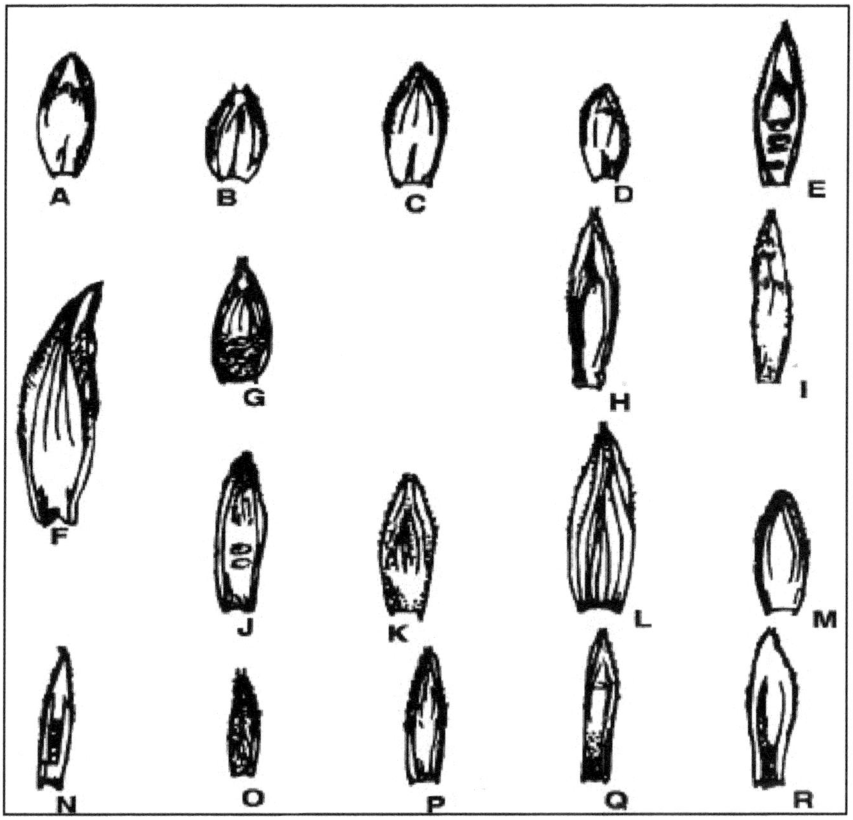

Figure 7.3: Lower Glume of Sessile Spikelet of *Cymbopogon* spp.

A: *C. martini*; B: *C. caesius;* C: *C. polyneuros*; D: *C. asmastonii;* E: *C. coloratus*; F: *C. distans*; G: *C. khasianus*; H: *C. pendules*; I: *C. citratus*; J: *C. flexuosus*; K: *C. nardus* var. *nardus*; L: *C. nardus* var. *confertiflorus*; H: *C. winterrianus*; N: *C. gidarba*, O: *C. microtheca*; P: *C. jwarancusa*; Q: *C. commutatus;* R: *C. pospischilli* (adopted from Bor, 1954).

3. Lower glume of sessile spikelet shallowly or deeply concave on the back, wingless to very narrowly wigned .... Series: Schoenanthi

Among the wild aromatic grasses of *Cymbopogon*, the species from plains *i.e.*, *C. coloratus* and *C. caesius* (Figure 7.4) were found difficult to distinguish from each other, since both of them have false panicles. The only distinctive feature between them is the presence (*C. caesius*) or absence (*C. coloratus*) of a groove in the glume of sessile spikelet.

The hilly polyploid races of *Cymbopogon viz.*, *C. flexuosus* and *C. jwarancusa* were found robust in their growth (Figure 7.5). The arrangement of spikelets was crowded in *C. jwarancusa* while of lax-type in *C. flexuosus*. *C. jwarancusa* is clearly identified by its awns in sessile spikelet. A trend towards elaboration of the inflorescence is followed by the condensation of the raceme segment in both the species.

Figure 7.4: *Cymbopogon caesius*

Figure 7.5: *Cymbopogon coloratus*

Figure 7.6: *Cymbopogon jwarancusa*

The chromosome number of 38 species including 27 Indian species/varieties and their cytotypes have been recorded so far (Table 7.2). The perusal of chromosome number reveals that the basic number is x = 10 as also deduced earlier by de Wet (1954), de Wet and Anderson (1956); although Celarier (1956) suspected the basic number to be 5 for the tribe Andropogonae. Another basic number x = 11 has also been proposed for the genus since 20 and 22 have been recorded in *C. caesius* (Singh, 1966), *C. gidarba* and *C. travancorensis* (Narayan and Jagadischandra 1966). However, Jagadischandra and Narayan (1971) have suggested the extra chromosomes encountered above the normal complement in such cases might be due to supernumerary chromosomes. Hence the accepted basic number is considered to be ten only.

Table 7.2: Wild and Cultivated Species of *Cymbopogon* and their Chromosome Numbers

Species	*2n (Chromosome No.)*
Series: Procari	
C. bombycinus (R. Br.) Domin.	20
C. procerus (R. Br.) Domin.	20
Series: Schoenanthi	
C. gidarba (Buch-Ham.) A. Camms	20,
	20^{+1*}
	39,
	9 40
	40^{+1}
C. jwarancusa subsp. *jwarancusa* (Jones) Schult.	20
	20^{+1*}
	40
C. jwarancusa subsp *olivieri* (Bloss) Bor	20
C. jwarancusa (Jones) Schult. var. *assamensis*	20
C. parkeri Stapf	20, 40
C parkeri (Stapf) var. *jammuensis*	40
C. schoanantus (L.) Spreng.	20
	20^{+1}
C. stracheyi (Hook.f.)	20, 40
C. plurinodis (Stapf) Stapf ex Burti-Davy	40
C. microtheca (Hook. f.) A. Camus	20
Series Rasae	
C. caesius (Nees) Stapf	20
	20^{+1*}
	22
	40
C. excavatus (Hochst.) Stapf	20
C. martinii (Roxb.) Wats.	20
	20^{+1}
	40
	40^{+1*}
C. martinii (Roxb.) Wats. var. *rafla* Gupta	40

Contd...

Table 7.2–*Contd...*

Species	*2n (Chromatic No.)*
C. nervatus (Hochst.) Chiov	24
	20
	20^{+1}
	40^{+2}
	40
C. coloratus (Hook.f.) Stapf	40
	40^{+1}
C. distans (Steud.) Wats.	20, 40
C. distans (Steud.) Wats. var. *mundensis* Gupta	20
C. flexuosus (Nees ex Steud.) Wats.	20
	20^{+1*}
	40, 16
	16, 39
C. flexuosus (Nees ex Steud.) Wats *var. coimbatorensis* Gupta	40
C. goeringii (Steud.) A. Camus	20
C. khasianus (Munro ex Hack.) Bor	60
C. ladakhensis Gupta	20
C. motia Gupta	40
C. nardus var. *nardus* (L.) Rendle	20
	40
	60
	15
C. nardus var. *confertiflorus* (Steud.) Bor	20
	40
	60
C. pendulus (Steud.) Wats.	20
	40
	60
C. proximus (Stapf) Phillps	40
C. ramnagarensis Gupta	20
C. totrilis (Preal) Hltche.	20
C. travancorensis Bor	20
	20^{+2}
	40, 40^{+2}
C. winterianus Jowitt	20
Hybrid (triploid) species of	30^{+1}
Cymbopogon	30

* The superscribed numbers over the 2n numbers indicates the presence of B-chromosomes.

(*See* Lavania, 2000).

The species *C. coloratus* and *C. caesius* were found to be diploid with chromosome number of 2n=20, only bivalents were observed at diakinesis and metaphase. However, among populations of *C. caesius,* PMC sowing ten + one B chromosome, eleven chromosomes were also noticed. B chromosomes were present near the A chromosomes. Even the morphological differences among the populations of *C. caesius* with regard to the presence of B′ were distinguished by the deep yellow colour of the culm in B populations. The presence of additional nucleoli at diakinesis was also noted in *C. coloratus*.

The aromatic species from hilly areas *viz., C. flexuosus* and *C. jwarancusa* showed tetraploid chromosome numbers 2n=40 at meiosis. Only bivalents were of frequent occurrence at diakinesis and metaphase and later stages were normal but poor seed set was observed among these populations.

From perusal of literature, it is understood that B chromosomes have been observed in 10 out of 12 species studied from south India and 2 out of 9 from north India (Jagadischandra 1970; Verma and Sobti 1985). They were noticed both in diploid and tetraploid forms of the species based on the detailed study of structure and behaviour of B choromosomes of *C. travancorensis*, Jagadischandra (1976) demonstrated that B chromosomes originate form A chromoomes.

Verma and Sobti (1985) extensively studied the effect of B chromosomes on chiasma frequnecy and chromatin length and suggested their effect is variable and species specific. It has been observed that individuals possessing B chromosomes exhibit some phenotypic variation as seen in *C. caesius, C. gidarba* and *C. travancorensis* (Jagadischandra 1975b) which support the present study.

There are various reports of natural hybridization and polyploidy in the existing species of *Cymbopogon* (Gupta, 1971; Soenarko, 1977). Since naturally occurring polyploid forms exhibit an enhanced amount of oil content, such proposition is more useful in species that do not flower or set viable seeds (*C. winterianus, C. citratus*). Moreover, the prevalent vegetative mode of reproduction in *Cymbopogon* could be exploited for oil yield from polyploid races for commercial applicability (Gupta, 1969a, 1969b; Jagadischandra, 1975b; Rao, 1991).

When spike characters of species of *Cymbopogon* are considered, they are very complex and are considered as advanced. In addition to certain morphological features of the tribe Andopogoneae, as reviewed by Stebbins (1986), the presence of bicellular hairs is primitive, while basal reduction of spikelets, 2 lodicules, 3 stamens, large embryos, kranz syndrome, dumbell-shaped siliceous cells are considered as advanced.

The trends of elaboration in Andropogoneae can be interpreted as adaptations to the very different climates in which the taxa are dominant, particularly with respect to reproductive characters. On the other hand, elaboration of complexity may promote greater phenotypic plasticity, which in warm climate may help to adapt the plant to irregularities in the onset and end of moist seasons (Stebbins, 1986).

Thus among Poaceae, as in higher plants and animals generally different characteristics have evolved at different rates within the same evolutionary line. Nearly every taxon is a mosaic of relatively primitive and more advanced character states (Stebbins, 1986). The degree of advancement of a particular taxon is expressed not in terms of the degree of advancement of every characters, but in proportion of primitive versus advanced states. With respect to advanced character states, the tribe Andropogoneae is distinctive in that advancement affects reproductive characters much unlike vegetative characters as in Bambusoideae. Many of such characters seen in *Cymbopogon* might have become established by chance in small populations during the early evolution.

From the present study, it is understood that the complex inflorescence structure and the nature of raceme pairs delimit the characters of the species of *Cymbopogon*. Since there is prevalence of natural hybridization among the species, the cytological variability is found to be common. The regular meiosis indicates that the species of *Cymbopogon* have become established due to allopolyploidy. There is need to extend the various studies to other species to precisely define the chromosome genetics both from the fundamental as well as applied angles. Evaluating the economic value of the aromatic grass, greater scope exists for further research on the genus *Cymbopogon*.

Acknowledgements

Authors are thankful to the Head, Department of Botany and Microbiology, Acharya Nagarjuna university, Nagarjunanagar for providing the necessary research facilities. The second author acknowledges the financial assistance given to her by University Grants Commission under Faculty Improvement Programme.

References

Celarier, R.P. 1956. Additional evidence for five as the basic number of the Andropogoneae. Rhdora. **58**: 690.

Dayal, A. and Purohit, 1971. Screening of some Indian essential oils for antifungal properties. Flavours Inds., **2**: 484-485.

De Wet, J.H.S. 1954. Chromosome number of a few South African grasses. Cytologia, **19:** 176-181.

De Wet, J.H.S. and Anderson, L.J. 1956. Chromosome number in Transvaal grasses. Cytologia, **21**: 1-10.

Gupta, B.K. 1969a. Studies in the genus *Cymbopogon.* Spreng. I. Chromosome studies in Indian Cymbopogons. Proc. Indian Acad. Sci. **B70**: 80-87.

Gupta, B.K. 1969b. Studies in the genus *Cymbopogon.* Spreng II chromotaxonomic studies in Indian Cymbopogons. Proc. Indian Acad. Sci. **B70**: 241-247.

Gupta, R.K. 1971. Ecology of pastoral areas in the arid zone of Rajasthan. Ann. Arid. Zone, **10**: 136-157.

Jagadishchandra, K.S. 1970. Polyploid drop. of degenerating chromosomes in *Cymbopogon martinii.* Proc. 57th Indian Sci. Congr. Part-II, p.286. Indian Sci. Congr. Assn., Calcutta.

Jagadishchandra, K.S. 1975a. Recent Studies on *Cymbopogon.* Spreng. In Advancing Frontiers in cytogenetics. pp. 377, P. Kachroo (ed.). Hindustan Publ. Corp., New Delhi.

Jagadishchandra, K.S. 1975b. Recent studies on *Cymbopogon* Spreng. (Aromatic grasses) with special reference to Indian taxa: Taxonomy, Cytogenetics, Chemistry and scope. J. Plant Crops. **3**: 43-57.

Jagadishchandra, K.S. 1976. Origin and behaviour of a pair of standard B chromosomes in *Cymbopogon travencorensis* Bor (Poaceae). Proc. Dunn Dobzh Symp. Genet. 301-311.

Jagadishchandra, K.S. and Narayan, K.N. 1971. Studies on supernumerary chromosomes in the genus *Cymbopogon* (Poaceae). Proc. First All India Congr. of Cytol. And Genetics. J. Cytol. And Genet. Congr. Suppl. 104-112.

Kirtikar, K.K. and Basu, B.D. 1935. Indian Medicinal Plants Part-IV 2nd Ed. 2675-2683 Lalitha Mohan Basu, Allahabad.

Kokate, C.K., Purohit, A.P., and Gokhale, S.B. 2001. Text book of Phamacognocy, 17th Edn. Nirali Prakashan, Pune, India, 361-362.

Lavania, U.C. 2000. Cytogenetics of *Cymbopogon.* In: *Cymbopogon* the aromatic grass Monograph. Director, Central Institute of Medicinal and Aromatic Plants, Lucknow, pp.145-157.

Narayan, K.N. and Jagadishchandra, K.S. 1966. Cytological studies in species of *Cymbopogon* from South India. J. Mysore University, **20:** 66-72.

Rao, B.L. and Setu, S.N. 1991. Ckp-25 an hybrid lemongrass. Indian Performer. **35**: 148-149.

Singh, F. 1966. Basic chromosome numbers in *Cymbopogon caesius* (Gramineae). Curr. Sci. **35**: 418-419.

Soenarko, S. 1977. The Genus *Cymbopogon* Sprengel (Gramineae). Reinwardtia **9**: 225-375.

Stebbins, G.L. 1986. Grass systematics and Evolution: Past, Present and Future. Edited by Thomas R. Soderstrom. K.W. Hilu, Christopher S. Cambell, and Mary E. Bark Worth.

Verma, V. and Sobti, S.N. 1985. Cytogenetical studies in the genus *Cymbopogon* Spreng. II. Meiotic studies in six species and their varieties. Cytologia. **50**: 351-359.

Chapter 8

Wild Edible Plants Used by the Tribals of East Godavari District, Andhra Pradesh

*J. Suneetha, J. Koteswara Rao, P. Prabhakara Rao and T.V.V. Seetharami Reddi**

Department of Botany, Andhra University, Visakhapatnam – 530 003, Andhra Pradesh, India

Keywords: *Food plants, Tribal people, East Godavari, Andhra Pradesh.*

The wild plants have provided an important source of food since time immemorial. Our ancestors collected wild grains, vegetables, fruits, tubers etc. for food. In general, plants provide 65 per cent of the global requirements of edible protein (Young and Pellett, 1994) and in particular about 80 per cent of the protein consumed by the humanity in developing countries is accrued from plants (Singh and Singh, 1992).

In India, studies on wild food plants have been carried out by several workers (Jain, 1963, 1964; Singh and Singh, 1985; Radhakrishnan *et al.,* 1996; Maliya, 2007; Kumar and Kumari, 2007; Pratap *et al.,* 2007; Arinathan *et al.,* 2009, 2011; Singh *et al.,* 2009; Binu *et al.,* 2010; Subramanian *et al.,* 2011). Studies on wild edible plants on limited scale were conducted in the Eastern Ghats of Andhra Pradesh (Krishna Prasad *et al.,* 1999; Sudhakar and Vedavathy 1999; Reddy *et al.,* 2007). Keeping these facts in

* Corresponding Author: E-mail: reddytvvs@rediffmail.com

view the wild edible plants especially tubers, rhizomes, corms, bulbs, root types and stems were surveyed and documented from the East Godavari district, Andhra Pradesh.

East Godavari district (Figure 8.1) is situated on the Northeast of Andhra Pradesh in the geographical condition of 16°-30' and 18°-20' of the northern Latitude and 81°-30' and 82°-36' of eastern Longitude occupying an area of 10,807 sq kms. It is bounded on the North by Visakhapatnam district and the state of Odisha, on the East and South by Bay of Bengal, on the West by West Godavari district and on the Northwest by Khammam district. The total population of the district is 49, 01,420 of which 1,91,561 are the tribal people comprising 3.91 per cent of the population. Konda reddi, Konda dora, Konda kammara, Konda kapu, Manne dora and Valmiki are the main tribal communities.

Materials and Methods

Several field trips were undertaken in tribal habitations of East Godavari district during 2005-2007. At each time of visit different tribal hamlets and forest pockets were chosen in different seasons to collect more information. The information was accrued after discussions with several tribal persons, village heads, elderly women and other local informants. Repeated interviews through questionnaires were made in different villages to authenticate the information. Plant specimens were collected and identified with regional floras. All the specimens were deposited in the Herbarium of the Department of Botany, Andhra University, Visakhapatnam. All the collected information on wild food plants was analyzed. The plants are enumerated alphabetically with their botanical name, family, vernacular name, English name and mode of uses.

Enumeration

***Abelmoschus esculentus* (L.) Moench. Malvaceae**
VN: Bendakaya E: Lady's finger

Tender fruits and leaves are used as vegetables and also sold in market.

***Abelmoschus manihot* (L.) Medic. Malvaceae**
VN: Peda benda

Leaves are used as vegetables.

***Acalypha indica* L. Euphorbiaceae**
VN: Kuppinta E: Indian acalypha

Leaves are used as vegetables.

***Achyranthes aspera* L. Amaranthaceae**
VN: Ucchulu chettu E: Prickly chaff flowers

Leaves are used as vegetables.

***Aegle marmelos* (L.) Correa Rutaceae**
VN: Maredu E: Bael tree

Fruits are edible and pulp is used in the preparation of jam and candy.

***Aerva lanata* (L.) Juss. Amaranthaceae**
VN: Pindikoora

Leaves are used as vegetables.

***Agaricus campestris* L. Agaricaceae**
VN: Kukkagodugu E: Mushroom

This mushroom and other types of edible mushrooms are collected from wild during rainy season and consumed as delicious dish.

Whole plant is used as vegetable.

***Alangium salvifolium* (L. *f.*) Wang. Alangiaceae**
VN: Udugu chettu E: Sage-leaved alangium

Fruits are edible.

***Albizia amara* (Roxb.) Boivin Fabaceae**
VN: Narlinga

Leaf paste is used for hair wash.

***Albizia lebbeck* (L.) Willd. Fabaceae**
VN: Dirishena E: Lebbeck tree

Tender twigs are used as fodder and fallen leaves as manure.

Leaves are used as vegetables.

Roasted seeds are edible.

***Allophyllus cobbe* (L.) Raeusch. Sapindaceae**
VN: Gaju chettu

Fruits are edible.

***Alstonia scholaris* (L.) R. Br. Apocynaceae**
VN: Edakulapala E: Devil tree

Leaves are used as vegetables.

***Alternanthera sessilis* (L.) R. Br. ex DC. Amaranthaceae**
VN: Ponnaganti koora

Leaves and shoots are used as vegetables.

***Amaranthus spinosus* L. Amaranthaceae**
VN: Mullathota koora E: Prickly amaranth

Tender leaves and shoots are used as vegetables.

***Amaranthus tricolor* L. Amaranthaceae**
VN: Thotakura E: Spinach

Leaves and tender stems are used as vegetables.

***Amaranthus viridis* L. Amaranthaceae**

VN: Kodijuttu koora

Leaves and tender stems are used as vegetables and sold in the market.

***Amorphophallus paeoniifolius* (Dennst.) Nicolson var. c*ampanulatus* (Decne.) Sivadasan Araceae**

VN: Siri kanda E: Elephant foot yam

Corms are boiled along with the leaves of tamarind and eaten as a vegetable mostly during winter season.

***Amorphophallus sylvaticus* (Roxb.) Kunth Araceae**

VN: Adavi Kanda

Corms are used as vegetables.

***Ampelocissus latifolia* (Roxb.) Planch. Vitaceae**

VN: Adavi draksha

Fruits are edible.

***Ampelocissus tomentosa* (Heyne ex Roth) Planch. Vitaceae**

VN: Atukula baddu

Fruits are edible.

***Anacardium occidentale* L. Anacardiaceae**

VN: Geedimamidi E: Cashew nut

Ripened hypocarp is edible and also used as vegetable.

Kernels are edible.

***Ananas comosus* (L.) Merr. Bromeliaceae**

VN: Anas pandu E: Pineapple

Fruits are edible.

***Angiopteris evecta* (Forst.) Hoff. Angiopteridaceae**

VN: Kalojikanda

Stems are edible.

***Annona reticulata* L. Annonaceae (Figure 8.2)**

VN: Ramapalam E: Bullocks heart of India

Fruits are edible and used in religious rites and also sold in the market.

***Antidesma ghaesembilla* Gaertn. Euphorbiaceae**

VN: Giridi

Fruits are eaten to reduce thirstiness.

Fruits are edible.

Arachis hypogaea **L. Fabaceae**
VN: Verusengalu E: Groundnut

Seeds are edible and also sold in the market.

Ardisia solanacea **Roxb. Myrsinaceae**
VN: Kunti

Leaves are used as vegetables.

Fruits are edible.

Areca catechu **L. Arecaceae**
VN: Poka-vakka E: Areca nut tree

Dried fruits are edible and sold in the market.

Argyreia nervosa **(Burm. *f.*) Boj. Convolvulaceae**
VN: Samudra paala E: Elephant creeper

Climbing stems are used for making ropes.

Leaves are used as fodder.

Aristolochia indica **L. Aristolochiaceae (Figure 8.3)**
VN: Nalla eswari E: Indian birthwort

Fruits are edible and sold in the market.

Artocarpus heterophyllus **Lam. Moraceae**
VN: Panasa E: Jack fruit

Ripened perianth and boiled seeds are edible.

Dried seeds are used as vegetables.

Asparagus racemosus **Willd. Liliaceae**
VN: Pilli peechera, Pilli teegalu E: Climbing asparagus.

Tender shoots are cooked and eaten as vegetables.

Tuberous roots are boiled and eaten.

Atalantia monophylla **(L.) Correa Rutaceae**
VN: Gaja nimma

Fruits are edible.

Atylosia scarabaeoides **(L.) Benth. Fabaceae**
VN: Adavi ulava teega

Fruits are edible.

Azadirachta indica **A. Juss. Meliaceae**
VN: Yepa chettu E: Margosa tree

Fruits are edible.

***Balanites aegyptiaca* (L.) Del. Balanitaceae**
VN: Gara

Fruit pulp is edible.

***Bambusa arundinacea* (Retz.) Roxb. Bambusaceae**
VN: Bongu Veduru E: Thorny bamboo

Boiled grains are cooked and eaten.

Newly born shoots are used as pickles.

***Basella alba* L. var. *rubra* (L.) Stewart Basellaceae**
VN: Bacchali koora E: Indian spinach

Leaves are used as vegetables.

***Bauhinia purpurea* L. Caesalpiniaceae (Figure 8.4)**
VN: Pedda are E: Pink bauhinia

Tender leaves and inflorescence are used as vegetables.

Fried seeds are eaten.

***Bauhinia racemosa* Lam. Caesalpiniaceae (Figure 8.5)**
VN: Arechettu E: Sweta kanchanara

Tender leaves and fruits are used as vegetables.

Seeds are used as fish poison.

***Bauhinia vahlii* Wight et Arn. Caesalpiniaceae (Figure 8.6)**
VN: Add, Adda chettu E: Camel's foot climber

Leaves after drying are stitched together nicely as plates and used them for eating food and for food packing throughout Andhra Pradesh and other adjoining areas. As such they are in good demand in the local markets and thus form a good source of income for tribals.

Tender fruits and flower buds are used as vegetables.

Seeds are used at the time of Pongal festival and also eaten to enhance health.

***Bauhinia variegata* L. Caesalpiniaceae**
VN: Deva kanchanum E: Variegated bauhinia

Roasted seeds are edible.

***Benincasa hispida* (Thunb.) Cogn. Cucurbitaceae**
VN: Budida gummadi E: Ash gourd

Fruits are used as vegetables.

***Benkera malabarica* (Lam.) Tirveng. Rubiaceae**
VN: Pedimi

Fruits are edible.

Beta vulgaris **L. Chenopodiaceae**
VN: Bitrutu E: Beet root

Tuberous roots are edible and used as vegetables and sold in the market.

Boerhavia diffusa **L. Nyctaginaceae**
VN: Atuka mamidi E: Spreading hog weed.

Leaves are used as vegetables.

Bombax ceiba **L. Bombacaceae (Figure 8.7)**
VN: Buruga E: Red silk cotton tree

Tender seedlings and calyx are highly nutritive and used as vegetables.

Borassus flabellifer **L. Arecaceae**
VN: Tadi E: Great fan palm

Boiled primary root, terminal bud, tender kernel, fruit pulp and ripened fruit is edible.

Borreria articularis **(L. f.) Willd. Rubiaceae**
VN: Madanaku

Whole plant except roots are used as vegetables.

Brassica nigra **(L.) Koch. Brassicaceae**
VN: Aavalu E: True mustard

Seeds are used as spice and also sold in the market.

Seed powder is specially used in the preparation of pickles.

Oil is extracted from seeds.

Bridelia montana **(Roxb.) Willd. Euphorbiaceae**
VN: Panchuka E: Feather foil

Fruits are edible.

Bridelia retusa **(L.) Spreng. Euphorbiaceae**
VN: Koramanu E: Spinous kinotree

Ripened fruits are edible.

Buchanania axillaris **(Desr.) Ramam. Anacardiaceae**
VN: Sarpappu, Sara

Fruits are edible.

Buchanania lanzan **Spreng. Anacardiaceae**
VN: Chinna morlu E: Cheronjee

Fruits are edible and sold in the market.

Butea monosperma **(Lam.) Taub. Fabaceae (Figure 8.8)**
VN: Moduga E: Flame of the forest

Flowers are used for decorative purposes.

Cajanus cajan **(L.) Millsp. Fabaceae**
VN: Kandulu E: Red gram

Seeds are used as pulse and sold in the market.

Seeds and seed flour are used in culinary preparations and also in the preparation of sweets.

Calamus rotang **L. Arecaceae**
VN: Pemu E: Cane palm

Fruits are edible.

Canavalia gladiata **(Jacq.) DC. Fabaceae**
VN: Thamba teega E: Jack bean

Seeds are edible.

Young pods and seeds are used as vegetables.

Canavalia virosa **(Roxb.) Wight et Arn. Fabaceae**
VN: Adavi chemba kaya

Pods are used as vegetables.

Canna indica **L. Cannaceae**
VN: Metta tamara E: Indian reed

Rhizome yields starch which is used as food.

Canthium parviflorum **Lam. Rubiaceae**
VN: Balusu E: Carray cheddie

Leaves are used as vegetables.

Ripened fruits are edible.

Capparis zeylanica **L. Capparidaceae**
VN: Adonda, Aredonda E: Caper plant

Fruits are edible.

Leaves are used as fodder.

Capsicum annuum **L. Solanaceae**
VN: Mirapakaya E: Chillies

Fruits are used in the preparation of chutney and pickles.

Fruit is used as spice due to pungent nature.

Capsicum frutescens **L. Solanaceae**
VN: Seema mirapa E: Golkonda chillies

Fruits are used for chutney and pickles.

Fruit is used as spice due to pungent nature.

***Caralluma adscendens* (Roxb.) Haw. Asclepiadaceae**

VN: Kundeti kommulu

Shoots are cooked and eaten as vegetables.

***Cardiospermum halicacabum* L. Sapindaceae**

VN: Budda teega S: Indravalli H: Kanphuti E: Blister creeper

Leaves are used as vegetables.

***Careya arborea* Roxb. Barringtoniaceae**

VN: Dud chettu S: Kumbini H: Kamba E: Patana oak

Leaves are used as meal plates.

***Carissa carandas* L. Apocynaceae**

VN: Vaka E: Jasmine flowered carissa

Ripened fruit has cooling effect and acidic and used as vegetable dish.

Fruits are used in culinary preparations along with dal.

Fruits are edible and sold in the market.

Ripened fruits are used in the preparation of pickles.

***Carissa spinarum* L. Apocynaceae**

VN: Kalivi E: Karanda

Ripened fruits are edible and used in the preparation of pickles and chutney.

***Cassia occidentalis* L. Caesalpiniaceae**

VN: Eddukommu Koora mokka E: Stinking weed

Tender twigs are used as tooth brushes.

Leaves are fried and eaten.

***Cassia sophera* L. Caesalpiniaceae**

VN: Kasunda E: Senna sophera

Leaves are used as vegetables.

***Cassia tora* L. Caesalpiniaceae**

VN: Tantepu E: Foetid cassia

Tender leaves are used as vegetables.

***Celosia argentea* L. Amaranthaceae**

VN: Gunaka koora E: Cox comb plant

Leaves and young branches are used as vegetables.

***Centella asiatica* (L.) Urban Apiaceae**

VN: Saraswathi aku E: Indian pennywort

Leaves are cooked and used as vegetables.

Chenopodium album **L. Chenopodiaceae**
VN: Aku koora

Leaves and tender twigs are cooked and eaten as vegetables.

Chlorophytum arundinaceum **Baker Liliaceae**
VN: Tella nelatadi.

Tuberous roots are edible.

Cicer arietinum **L. Fabaceae**
VN: Chanakamula E: Bengal gram

Leaves are used as vegetables. Seeds are used as pulse and also sold in the market. Boiled seeds are eaten and also used in culinary purposes. Seed flour is used in the preparation of sweets etc. Both green plants and dry straw are used as fodder.

Cissus quadrangularis **L. Vitaceae**
VN: Nalleru E: Adamant creeper

Tender stems and leaves are used as vegetables.

Citrus aurantifolia **(Christm.) Swingle Rutaceae**
VN: Narinja E: Sour orange.

Fruits are edible and also sold in the market.

Citrus limon **(L.) Burm. Rutaceae**
VN: Nimma E: Lemon

Fruits are edible and also sold in the market.

Citrus medica **L. Rutaceae**
VN: Maadiphalam E: Lemon

Fruits are edible and sold in the market.

Citrus sinensis **(L.) Osbeck Rutaceae**
VN: Battayi E: Sweet orange

Fruits are edible and sold in the market.

Cleome gynandra **L. Cleomaceae**
VN: Vaminta

Leaves are used as vegetables and also used in the preparation of chutnies.

Cleome monophylla **L. Cleomaceae**
VN: Vaminta

Tender leaves are used as vegetables.

Cleome viscosa **L. Cleomaceae**
VN: Kukka vaminta E: Wild mustard

Leaves are used as vegetables.

Figure 8.1: Study area; **Figure 8.2:** *Annona reticulata*; **Figure 8.3:** *Aristolochia indica*; **Figure 8.4:** *Bauhinia purpurea*; **Figure 8.5:** *B. racemosa*; **Figure 8.6:** *B. vahlii*

Figure 8.7: *Bombax ceiba*; Figure 8.8: *Butea monosperma*;
Figure 8.9: *Costus speciosus*; Figure 8.10: *Dillenia pentagyna*;
Figure 8.11: *Dioscorea bulbifera*; Figure 8.12: *D. hispida*

Clerodendrum serratum **(L.) Moon Verbenaceae**
VN: Bommalamarri E: Beetle killer

Tender leaves and flowers are used as vegetables.

Coccinia grandis **(L.) Voigt. Cucurbitaceae**
VN: Kakidonda E: Ivy gourd

Fruits are used as vegetables and also sold in the market.

Cocculus hirsutus **(L.) Diels Menispermaceae**
VN: Cheepuru teega E: Broom creeper

Tender leaves are used as vegetables.

Cocos nucifera **L. Arecaceae**
VN: Kobbari chettu E: Coconut

Toddy is extracted from the young inflorescence.

Water in the cavity of green unripened fruit is used as beverage.

Empty coconut shells are used as containers of toddy, water and coconut oil.

Commelina benghalensis **L. Commelinaceae**
VN: Vennaddu laaku

Tender stems and leaves are used as vegetables.

Commiphora caudata **(Wight et Arn.) Engl. Burseraceae**
VN: Konda mamidi E: Hill mango

Fruits are used in the preparation of pickles.

Fruits are edible.

Cordia dichotoma **Forst. f. Boraginaceae**
VN: Eriki E: Sebesten plum

Fruits are edible.

Unripe fruits are used as vegetables and also used in the preparation of pickles

Cordia monoica **Roxb. Boraginaceae**
VN: Bottuku koora

Fruits are edible.

Coriandrum sativum **L. Apiaceae**
VN: Dhaniyalu E: Coriander

Leaves are used as vegetables.

Seeds are used as condiment and forms an important ingredient in the preparation of curry powder.

Costus speciosus **(Koen.) Sm. Zingiberaceae (Figure 8.9)**
VN: Chengalva kostu E: Spiral ginger

Rhizomes are used as vegetables and boiled and eaten.

Crotalaria juncea **L. Fabaceae**
VN: Janumu E: Sunhemp

Floral buds are used as vegetables.

Cucumis melo **L. Cucurbitaceae**
VN: Dosa E: Musk melon

Tender fruits are used as vegetables and also sold in the market.

Fruits are edible.

Cucumis sativus **L. Cucurbitaceae**
VN: Dosakaya E: Common cucumber

Unripe fruits are used as vegetables and fruits are edible.

Curculigo orchioides **Gaertn. Hypoxidaceae**
VN: Nela tadi E: Black musali

Tubers are boiled and eaten.

Daucus carota **L. Apiaceae**
VN: Gajjaragadda E: Carrot

Leaves and tuberous roots are used as vegetables and also sold in the market.

Tuberous roots are edible and used in the preparation of sweets and jams.

Debregeasia longifolia **(Burm. f.) Wedd. Urticaceae**
VN: Karangi E: Capsy wild rhea

Fruits are edible.

Decalepis hamiltonii **Wight et Arn. Asclepiadaceae**
VN: Maredu geddalu

Roots are made into pickle.

Delonix regia **(Boj. ex Hook.) Rafin. Caesalpiniaceae**
VN: Agnipoolu E: Gul mohur.

Leaves are used as vegetables.

Dendrocalamus strictus **(Roxb.) Nees Poaceae**
VN: Veduru karra E: Solid bamboo

Tender shoots are used as vegetables and also used in the preparation of pickles.

Grains are boiled, cooked and eaten.

Derris scandens **(Roxb.) Benth. Fabaceae**
VN: Nalla teega

Leaves are used as vegetables.

Digera muricata **(L.) Mart. Amaranthaceae**
VN: Chenchalikura

Leaves are used as vegetables.

Dillenia indica **L. Dilleniaceae**

VN: Chalta E: Elephant apple

Fleshy sepals are eaten and also used in the preparation of pickles.

Fruit are used as a substitute for tamarind.

Fruits are edible and sold in the market.

Dillenia pentagyna **Roxb. Dilleniaceae (Figure 8.10)**

VN: Rai E: Dog teak

Ripened fruits are edible.

Dioscorea alata **L. Dioscoreaceae**

VN: Pendalamu E: Greater yam

Tubers are boiled and eaten.

Dioscorea bulbifera **L. Dioscoreaceae (Figure 8.11)**

VN: Adavi dumpa E: Bulb bearing yam

Tubers and bulbils are cooked and used as vegetables.

Dioscorea hispida **Dennst. Dioscoreaceae (Figure 8.12)**

VN: Konda dumpa E: Tiger root yam

Tubers are boiled and eaten.

Dioscorea oppositifolia **L. Dioscoreaceae**

VN: Gentica dumpa E: Betel yam

Tubers are directly put on fire or boiled and eaten. It is one of the major foods during lien season.

Dioscorea pentaphylla **L. Dioscoreaceae**

VN: Dukka pendalam E: Prickly yam

Flower buds are cooked and eaten as vegetables.

Root tubers are edible.

Dioscorea tomentosa **Koen. ex Spreng. Dioscoreaceae**

VN: Niluvu dumpa

Tubers are used as vegetables.

Diospyros chloroxylon **Roxb. Ebenaceae**

VN: Ulimiri

Tender leaves are used as vegetables.

Fruits are edible.

Diospyros melanoxylon **Roxb. Ebenaceae**

VN: Tumki, Beediaku

Tender leaves are used as vegetables.

Diospyros montana **Roxb. Ebenaceae**

VN: Erragada

Fruits are edible.

Dolichos trilobus **L. Fabaceae**

VN: Adavichikkudu

Young pods are used as vegetables.

Elettaria cardamomum **Maton. Zingiberaceae**

VN: Elakulu E: Cardamom

Fruit is used as spice.

Eleusine coracana **(L.) Gaertn. Poaceae**

VN: Ragulu E: Finger millet

Tribals prepare country liquor from the grains by fermentation.

Various delicious dishes such as cakes, puddings are prepared from flour of the grain.

Light food (malt) is cooked with the flour of grains and extensively eaten.

Malt is said to be better food for digestion and good health.

Grains are sold in the market.

Entada rheedii **Spreng. Mimosaceae**

VN: Gilla teega E: Giants rattle

Seeds are edible.

Erythroxylum monogynum **Roxb. Erythroxylaceae**

VN: Devadharu E: Bastard sandal

Leaves are used as vegetables.

Fruits are edible.

Ficus benghalensis **L. Moraceae**

VN: Marri S: Bahupada H: Bor E: Banyan tree

Tender leaves are used as vegetables.

Ficus carica **L. Moraceae**

VN: Anjooramu E: Common fig.

Fruits are edible.

Ficus hispida **L. Moraceae**

VN: Bemmadu E: Wild fig

Fruits are used as vegetables.

Ficus racemosa **L. Moraceae**

VN: Medi E: Cluster fig

Tender fruits are used as vegetables.

Stem sap and ripened fruits are fermented for preparing arrack.

Ripened fruits are edible.

***Ficus religiosa* L. Moraceae**
VN: Raavi chettu E: Sacred fig tree

Tender leaves are cooked and used as vegetables.

***Gardenia gummifera* L. *f.* Rubiaceae**
VN: Bikki E: Cumbi gum

Fruits are edible.

***Gardenia latifolia* Ait. Rubiaceae (Figure 8.13)**
VN: Konda jama

Fruits are edible.

***Garuga pinnata* Roxb. Burseraceae**
VN: Garuga E: Garuga

Fruits are edible and made into pickles.

***Glycosmis pentaphylla* (Retz.) DC. Rutaceae**
VN: Konda gogulu

Fruit is edible.

***Gnetum ula* Brogn. Gnetaceae**
VN: Laluga teega

Seeds are roasted and eaten.

***Grewia hirsuta* Vahl Tiliaceae**
VN: Chirijana

Fruit is edible.

***Grewia rhamnifolia* Heyne ex Roth Tiliaceae**
VN: Tegalli

Fruits are edible.

***Guazuma ulmifolia* Lam. Sterculiaceae**
VN: Rudraksha S: Rudraksha H: Rudraksha E: Musket tree

Fruits are edible.

***Guizotia abyssinica* (L. *f.*) Cass. Asteraceae**
VN: Valiselu

Oil is extracted from seeds.

***Hemidesmus indicus* (L.) R. Br. Periplocaceae**
VN: Sugandhapala E: Indian sarsaparilla

Root decoction is used as a stimulant drink.

Hibiscus cannabinus L. **Malvaceae**
VN: Gogu E: Bimlipatam-Jute

Leaves are used as vegetables and also used in the preparation of pickles.

Hibiscus sabdariffa **L. Malvaceae**
VN: Erra gongura

Leaves and flowers are used as vegetables.

Roasted seeds are edible.

Seed kernels are used in the preparation of jams and pickles.

Hibiscus surattensis **L. Malvaceae**
VN: Adavi gogu

Leaves are used as vegetables.

Holoptelea integrifolia **(Roxb.) Planch. Ulmaceae**
VN: Nemali chettu E: South Indian elm

Seeds are edible.

Ipomoea aquatica **Forsk. Convolvulaceae**
VN: Tuti kaada E: Water spinach

Whole plant is cooked and used as vegetable.

Ipomoea batatas **(L.) Lam. Convolvulaceae**
VN: Silkada dumpa E: Sweet potato

Tuberous roots are edible and also used for culinary purposes.

Leaves are used as vegetables.

Ipomoea muricata **(L.) Jacq. Convolvulaceae**
VN: Thutikada

Leaves and young shoots are cooked and used as vegetables.

Justicia glauca **Rottl. Acanthaceae**
VN: Konda pindi

Tender leaves are used as vegetables.

Lablab purpureus **(L.) Sweet Fabaceae**
VN: Chikkudu E: Indian butter bean

Tender pods are used as vegetables and also sold in the market.

Pods are made into pickles.

Lagenaria siceraria **(Molina) Standl. Cucurbitaceae**
VN: Sorakaya, E: Bottle-gourd

Tender fruits and leaves are used as vegetables.

Lantana camara **L. Verbenaceae**
VN: Gabbuseeki E: Wild sage

Fruits are edible.

Leucas aspera **(Willd.) Link. Lamiaceae**
VN: Thunika kaya mokka E: Thumb

Leaves are used as vegetables.

Luffa acutangula **(L.) Roxb. Cucurbitaceae**
VN: Beera kaya E: Ribbed gourd

Fruits are used as vegetables and also sold in the market.

Fruits are made into chutney.

Luffa cylindrica **(L.) Roem. Cucurbitaceae**
VN: Netibirakaya E: Dishcloth gourd

Fruits are used as vegetables and also sold in the market.

Fruits are made into chutney.

Lycopersicon esculentum **Mill. Solanaceae**
VN: Tomato E: Tomato

Raw fruits are eaten and also used in salads, soups, chutnies and pickles.

Fruits are used as vegetables and also sold in the market.

Macrotyloma uniflorum **(Lam.) Verdc. Fabaceae**
VN: Ulavalu E: Horse gram

A liquid delicacy is prepared from seeds and used as curry.

Seeds are cooked and eaten and also sold in the market.

Madhuca indica **J. F. Gmel. Sapotaceae (Figure 8.14)**
VN: Ippa chettu E: Indian butter tree

Dry flowers, fresh and roasted corolla are edible and also sold in the market.

Liquor is prepared from corolla and sometimes tribals survive with this liquor for days together without food.

Flowers are fermented by using indigenous distilleries.

Flowers are used in the preparation of delicious cake.

The tree is well protected by tribals for its flowers which are used for making liquor. There is good market for this liquor.

Oil extracted from seeds is used for culinary purposes and also for hair oil.

Fruits are edible.

Mangifera indica **L. Anacardiaceae**
VN: Maavichettu E: Mango

Fruits edible and both ripened and unripened fruits are sold in the market.

Country liquor is prepared from fruits.

Juice is prepared from fruits.

Mango jelly is prepared from fruits and eaten during lien season and also sold in the market. Bread made from seed kernel is eaten.

Seeds collected in summer season are dried and stored. The seed kernels of stored seeds are crushed and made into gruel. The gruel is used as food at the time of scarcity of food particularly during rainy season.

Dried cotyledons are cut into small pieces, kept in running water for a day, then ground into paste for preparing cakes and porridge.

***Manihot esculenta* Crantz. Euphorbiaceae**
VN: Karrapendalamu E: Cassava

Leaves are used as vegetables

***Manilkara hexandra* (Roxb.) Dubard Sapotaceae**
VN: Pedda pala E: Milk tree

Fruits are edible and best source of pickle and sauce.

***Manilkara zapota* (L.) P. Royen. Sapotaceae**
VN: Sapota E: Sapota

Fruits are edible and also sold in the market.

***Marsilea quadrifolia* L. Marsiliaceae**
VN: Ciklintakura

Leaves are used as vegetables.

***Martynia annua* L. Martyniaceae (Figure 8.15)**
VN: Telukondi kaya E: Devils claw

Oil is extracted from fruits and seeds.

***Memecylon edule* Roxb. Melastomaceae**
VN: Alli chettu, Kukkalli

Fruits are edible.

***Mentha spicata* L. Lamiaceae**
VN: Pudina E: Spearmint

Leaves are used as flavouring agent and in the preparation of chutney and also used as vegetable.

***Merremia vitifolia* (Burm. *f.*) Hall. *f.* Convolvulaceae**
VN: Katti kura E: Bombay naval

Tender leaves and flowers are used as vegetables.

***Mimusops elengi* L. Sapotaceae**
VN: Pogada E: Spanish cherry

Fruits are edible.

Mirabilis jalapa **L. Nyctaginaceae**
VN: Badrakshi E: Four-O'clock plant

Tubers and leaves are used as vegetables.

Momordica charantia **L. Cucurbitaceae**
VN: Kakara E: Bitter gourd

Tender fruits are used as vegetables.

Momordica dioica **Roxb. Cucurbitaceae**
VN: Goddu kakara teega E: Small bitter gourd

Tender fruits are used as vegetables and sold in the market.

Morinda pubescens **Sm. Rubiaceae**
VN: Chekka chettu, E: Morinda tree

Fruits are edible.

Moringa oleifera **Lam. Moringaceae**
VN: Munaga chettu E: Indian horse raddish

Fruits and leaves are used as vegetables and also sold in the market.

Murraya koenigii **(L.) Spreng. Rutaceae**
VN: Karivepaku E: Curry leaf

Leaves are used as vegetables, chutney; also used as condiment to flavour curries etc.

Musa paradisiaca **L. Musaceae**
VN: Arati E: Plantain

Fruits are edible and also sold in the market.

Musa rosacea **Jacq. Musaceae**
VN: Adavi arati E: Wild banana

Fruits are edible.

Nelumbo nucifera **Gaertn. Nelumbonaceae**
VN: Tamara E: Sacred

Rhizome, tender petioles and seeds are used as vegetables.

Nymphaea nouchali **Burm. *f.* Nymphaeaceae**
VN: Kaluva

Tender leaves, petioles and stems are eaten as vegetables.

Seeds are cooked and used as vegetables.

Ocimum americanum **L. Lamiaceae**
VN: Kukkatulasi E: Hoary basil

Leaves are used as vegetables.

Olax scandens **Roxb. Olacaceae**

VN: Mekabanda

Fruits are edible.

Opuntia dillenii **(Ker-Gawl.) Haw. Cactaceae (Figure 8.16)**

VN: Bonthajamudu E: Prickly pear

Floral buds are used as vegetables

Fruits are edible.

Oxalis corniculata **L. Oxalidaceae**

VN: Pulicinatku E: Indian sorrel

Leaves are used in the preparation of sauce, salad and pickles and also used as vegetables.

Panicum sumatrense **Roth ex Roem. et Schult. Poaceae**

VN: Chamalu E: Little millet

Grains are cooked and eaten.

Pedalium murex **L. Pedaliaceae**

VN: Palleru chettu H: Bada gokhru

Leaves are used as vegetables.

Pennisetum typhoides **(Burm.*f.*) Stapf ex Hubb. Poaceae**

VN: Gantelu E: Pearl millet

Grains are cooked and eaten.

Pergularia daemia **(Forsk.) Chiov. Asclepiadaceae**

VN: Dishtiveru E: Whitlow plant

Leaves and floral buds are used as vegetables.

Perilla frutescens **(L.) Brit. Lamiaceae**

VN: Pappukura

Seed flour is used for flavouring eatables.

Phoenix loureirii **Kunth Arecaceae**

VN: Konda eetha E: Hill date palm

Fruits are edible.

Phoenix pusilla **Gaertn. Arecaceae**

VN: Chitti eetha

Fruits are edible.

Phoenix sylvestris **(L.) Roxb. Arecaceae**

VN: Eatha E: Date sugar palm

Tender fruits are used for making pickles.

Tender buds and fruits are edible.

Figure 8.13: *Gardenia latifolia*; Figure 8.14: *Madhuca indica*;
Figure 8.15: *Martynia annua*; Figure 8.16: *Opuntia dillenii*;
Figure 8.17: *Pueraria tuberosa*; Figure 8.18: *Woodfordia fruticosa*

Phyllanthus debilis **Klein ex Willd. Euphorbiaceae**
VN: Gunne chinta kura

Leaves are used as substitute for tamarind.

Phyllanthus emblica **L. Euphorbiaceae**
VN: Usirica E: Indian gooseberry

Raw fruits are edible and made into pickles.

Physalis minima **L. Solanaceae**
VN: Kuppante E: Country gooseberry

Fruits are edible.

Piper betle **L. Piperaceae**
VN: Tamalapaku E: Betel leaf

Leaves with lime is chewed by most of the tribals.

Piper nigrum **L. Piperaceae**
VN: Miriyalu E: Black pepper

Fruits are used as spice and as an ingredient in various medicines and also sold in the market.

Polyalthia cerasoides **(Roxb.) Bedd. Annonaceae**
VN: Chilaka duddigara

Fruits are edible.

Polyalthia longifolia **(Sonner) Thw. Annonaceae**
VN: Naramamidi E: Cemetery tree

Fruits are edible.

Polygonum chinense **L. Polygonaceae**
VN: Sarpakshi

Fruits are edible; leaves cooked and eaten.

Pongamia pinnata **(L.) Pierre Fabaceae**
VN: Ganuga, Kagu, Kanugu E: Indian beech tree

Tender pods are used as vegetables and also sold in the market.

Portulaca oleracea **L. Portulacaceae**
VN: Pappukooraku E: Common purslane

Tender branches and leaves are used as vegetables.

Premna tomentosa **Willd. Verbenaceae**
VN: Naguru, Kokkita

Ripened fruits are edible.

Psidium guajava **L. Myrtaceae**
VN: Jama chettu E: Guava

Fruits are edible.

Pterocarpus marsupium **Roxb. Fabaceae**
VN: Yegisa E: Kino tree

Flowers and seeds are edible.

Pueraria tuberosa **(Willd.) DC. Fabaceae (Figure 8.17)**
VN: Nela gummadi E: Indian kudzu

Tubers are boiled and eaten and also used as vegetable.

Punica granatum **L. Punicaceae**
VN: Dhanimma E: Pomegranate

Fruits are edible and also sold in the market.

Sapindus emarginatus **Vahl Sapindaceae**
VN: Kunkudu E: Soapnut

Seed kernel is edible.

Sarcolobus carinatus **Wall. Asclepiadaceae**
VN: Pala boddy teega

Fleshy fruits are edible.

Sauropus quadrangularis **(Willd.) Muell.- Arg. Euphorbiaceae**
VN: Multi vitamin plant E: Triplinerved lizard foot

Leaves are used as vegetables.

Schleichera oleosa **(Lour.) Oken. Sapindaceae**
VN: Busi chettu E: Lac tree; honey tree

Fruits are very tasty, edible, pickled and also sold in the market.

Sesamum indicum **L. Pedaliaceae**
VN: Nuvvulu E: Sesame

Seeds are edible and used in preparation of sweet with jaggery.

Seed oil is used for culinary purposes and also used as hair oil.

Sesbania grandiflora **(L.) Poir. Fabaceae**
VN: Avisa E: Sesbania

Young leaves, flowers and pods are used as vegetables.

Setaria italica **(L.) Beauv. Poaceae**
VN: Korralu E: Indian millet

Grains are cooked and eaten.

Smilax zeylanica **L. Smilacaceae**

VN: Konda thambera gadda E: Rough bind weed

Young leaves are used as vegetables and meal plates.

Cooked tuberous roots are edible.

Solanum americanum **Mill. Solanaceae**

VN: Jaji koora mokka E: Black night shade

Leaves and tender fruits are used as vegetables.

Solanum anguivi **Lam. Solanaceae**

VN: Chitti mulaga E: Indian night shade

Fruits are used as vegetables.

Solanum melongena **L. Solanaceae**

VN: Vankaya E: Brinjal

Fruits are used as vegetables and also sold in the market.

Fruits are made into pickles.

Solanum surattense **Burm. *f.* Solanaceae**

VN: Tikka vankaya E: Dwarf lid brinjal

Fruits are used as vegetables.

Solanum torvum **Sw. Solanaceae**

VN: Kondusti

Fruits are used as vegetables.

Solanum tuberosum **L. Solanaceae**

VN: Bangaladumpa E: Potato

Tubers are used as vegetables and also sold in the market.

Tubers are used as main winter vegetables.

Solena heterophylla **Lour. Cucurbitaceae**

VN: Nalla dhonda

Tuberous roots and tender fruits are used as vegetables.

Sphaeranthus indicus **L. Asteraceae**

VN: Bedasaramu E: East Indian globe thistle

Seedlings are used as vegetables.

Spondias pinnata **(L. *f.*) Kurz Anacardiaceae**

VN: Konda mamidi E: Wild mango

Fruits are edible and also used in the preparation of pickles.

Stachytarpheta urticaefolia **(Sal.) Sims. Verbenaceae**

VN: Yerri tulsi E: Brazilian tree

Leaves are used as vegetables.

Sterculia foetida **L. Sterculiaceae**
VN: Gurrapu badam E: Pinari

Seeds are edible and also sold in the market.

Sterculia urens **Roxb. Sterculiaceae**
VN: Kovila chettu, E: Gum karaya

Seeds are roasted and eaten.

Sterculia villosa **Roxb. ex DC. Sterculiaceae**
VN: Gudilam E: Elephant rope tree

Seeds are edible.

Syzygium cumini **(L.) Skeels Myrtaceae**
VN: Neredu E: Indian cherry

Fruits are edible and also sold in the market.

Talinum portulacifolium **(Forsk.) Asch. et Schweinf. Portulacaceae**
VN: Sima bachchali

Leaves are used as vegetables.

Fruits are edible and also sold in the market.

Terminalia catappa **L. Combretaceae**
VN: Badam E: Indian almond tree

Fruits are edible and also sold in the market.

Trema orientalis **(L.) Bl. Ulmaceae**
VN: Boggu E: Indian nettle tree

Fruits are edible.

Trianthema portulacastrum **L. Aizoaceae**
VN: Yerra galijereu E: Horse-purslane

Leaves are used as vegetables.

Tribulus terrestris **L. Zygophyllaceae**
VN: Palleru E: Land caltrops

Leaves are used as vegetables.

Trichosanthes anguina **L. Cucurbitaceae**
VN: Potla kaya E: Snake gourd

Fruits are used as vegetables and sold in the market.

Tridax procumbens **L. Asteraceae**
VN: Gayapaaku E: Coat-buttons

Leaves are used as vegetables.

***Trigonella foenum-graecum* L. Fabaceae**
VN: Menthulu E: Fenugreek

Leaves are used as vegetables. Seeds are used as condiment and also sold in the market.

***Vernonia cinerea* (L.) Less. Asteraceae**
VN: Picchi donka, E: Flebane-purple

Leaves are used as vegetables.

***Vigna mungo* (L.) Hepper Fabaceae**
VN: Minumulu E: Black gram

Seeds are edible and sold in the market.

Seeds are used as pulse and also for culinary purposes.

***Vigna radiata* (L.) Wilczek Fabaceae**
VN: Pesara E: Green gram

Pods are used as vegetables.

Seeds are also used for culinary purposes.

***Woodfordia fruticosa* (L.) Kurz Lythraceae (Figure 8.18)**
VN: Arepuvvu E: Fire flame brush

Uses: Flowers are used as vegetables.

***Wrightia tinctoria* R. Br. Apocynaceae**
VN: Pedda pala E: Sweet indraja

Flowers are used as vegetables.

***Xanthium pungens* L. Asteraceae**
VN: Peddatithiram E: Bur weed

Tender leaves are used as vegetables.

***Ximenia americana* L. Olacaceae**
VN: Nakkera E: False sandal wood

Ripened fruit pulp is edible.

***Xylia xylocarpa* (Roxb.) Taub. Mimosaceae**
VN: Konda tangedu E: Iron wood tree

Fruits are edible.

***Ziziphus mauritiana* Lam. Rhamnaceae**
VN: Regu chettu E: Indian jujube

Different types of pickles are prepared from fruits.

Fruits are edible and sold in the market.

Ziziphus nummularia **(Burm. f.) Wight et Arn. Rhamnaceae**

VN: Chittiregu

Fruits are edible.

Ziziphus oenoplia **(L.) Mill. Rhamnaceae**

VN: Parimi E: Jackal jujube

Fruits are edible and also a favourite food of bears.

Ziziphus rugosa **Lam. Rhamnaceae**

VN: Pindi parimiki

Fruits are edible.

Ziziphus xylopyrus **(Retz.) Willd. Rhamnaceae**

VN: Gotte chettu

Fruits are edible and also used for making dyes.

Results and Discussion

A total of 242 plant species were recorded that are used for edible purposes by the tribals of East Godavari district and a few species are sold in local markets. These 242 species belonged to 180 genera and 82 families. Most of the edible parts are fruits (118) followed by leaves (84), seeds (37), tubers (22), flowers (11), tender stems and branches (9 each) and root (1). Tubers are collected and consumed from time to time by the tribal people over a period of 4-5 months from June-October. The ethnobotanical observations revealed that the tribals are using edible plants in different forms. Most of the tribals go into the forest regularly early in the morning for collection of wild edibles. The tribal people for their diet mostly use fruits (118) and leaves (84). Leaves are collected in different seasons, cooked and eaten with their staple food. Most of the tribal people are using leafy vegetables as part of their food. These are either collected from forest areas or found as weeds in moist areas of cultivated and open fields. Women are found to play a major role in the collection and preparation of wild leafy vegetables. Some plants are also used in the preparation of curries. Fruits are eaten as raw as well as in ripen form. Fruits of *Buchanania lanzan, Diospyros chloroxylon, D. melanoxylon, D. perigrina, Ehretia laevis, Phoenix acaulis* and *Ziziphus mauritiana* are eaten depending upon their ripening in different seasons. Fruits of *Benincasa hispida, Coccinia grandis, Cordia dichotoma, Cucumis sativus, Ficus hispida, Momordica charantia, Solanum americanum, S. anguivi, S. surattense* and *Trichosanthes anguina* are used in the preparation of curries. Flowers of *Bauhinia vahlii, Butea monosperma, Clerodendrum serratum, Crotalaria juncea, Dioscorea pentaphylla, Merremia vitifolia, Pergularia daemia, Pterocarpus marsupium* and *Sesbania grandiflora* are used as vegetables. Tubers of *Dioscorea bulbifera, D. oppositifolia, D. pentaphylla* and *D. tomentosa* have good nutritional value. Fruits of *Madhuca indica* are edible. Although the detailed nutritional value of these wild plants is not known, the species are consumed very commonly by all the tribal people.

The study revealed that the tribal communities of East Godavari district continue to have and use knowledge about wild food species, including their habitat, collecting period, sustainable collecting, mode of preparation and consumption. Analysis of results showed that most of the edible plants are used mainly by the tribals both during normal and difficult times. Safe conservation and sustainable use and in depth study of wild edible plants diversity is essential to the continued use of this diversity for meeting the present and future food needs of the tribals and rural people living nearby during the periods of food scarcity. The documentation of these wild edible plants is significant in the current era to update the indigenous know-how. Detailed investigation on the chemical composition and nutritive qualities/toxic effects of all the wild edible plant species becomes the need of the hour.

Summary

A total of 242 species of edible plants belonging to 180 genera and 82 families used by the tribal communities of East Godavari district were identified. Most of the edible parts are fruits followed by leaves, seeds, tubers, flowers, tender stem, branches and root. Most of the edible plants are used mainly by the tribals both during normal and difficult times. Safe conservation and sustainable use and in depth study of wild edible plants diversity is essential to the continued use of this diversity for meeting the present and future food needs of the tribals during the periods of food scarcity.

Acknowledgements

The authors are very much grateful to the tribal people of the East Godavari district for sharing their valuable information on herbal medicine.

References

Arinathan, V., Mohan, V. R. and Maruthupandian, A. 2009. Little known wild edible seeds of Western Ghats, Tamil Nadu. J. Non-Timber Forest Products **16:** 119-124.

Arinathan, V., Mohan, V. R. and Maruthupandian, A. 2011. Wild edible unripe fruits used by the Palliyars of Western Ghats, Tamil Nadu. J. Non-Timber Forest Products **18:** 149-152.

Binu Thomas, Aravindhan, V., Maharajan, M. and Rajendran, A. 2010. Wild edible roots and tubers and their contribution to the food security of Shoal Naikas tribe of Kerala, India. J. Non-Timber Forest Products **17:** 449-451.

Jain, S. K. 1963. Studies in Indian Ethnobotany-Less known uses of fifty common plants from the tribal areas of Madhya Pradesh. Bull. Bot. Surv. India. **5:** 223-226.

Jain, S. K. 1964. Wild plant foods of the tribals of Bastar. Khadi Gramodyog. **10:** 557-561.

Krishna Prasad, V., Rajagopal, T., Yogesh Kumar and Badarinath, K.V.S. 1999. Food plants of Konda Reddis of Rampa Agency, East Godavari district, Andhra Pradesh-A case study. Ethnobotany **11:** 92-96.

Kumar, S. and Kumari, B. 2007. Some less-known "Sags" (leafy vegetables) utilized by the tribals and others of Panch Pargana area in Jharkhand. Ethnobotany **19:** 62-66.

Maliya, S. D. 2007. Traditional fruit and leaf therapy among Tharus and indigenous people of district Bahraich, India. Ethnobotany **19:** 131-133.

Pratap Aditya, Somya Yadav, Anil Kumar and Yadav, D. K. 2009. Indigenous knowledge of wild plant resources used as food in Gaya district, Bihar. J. Non-Timber Forest Products. **16:** 85-88.

Radhakrishna, K., Pandurangan, A. G., Pushpangadan, P. and Sasidharan, A. 1996. Less known ethnomedicinal plants of Kerala State and their conservation. Ethnobotany. **8:** 82-84.

Reddy, K.N., Chiranjibi Pattanaik, Reddy, C.S. and Raju, V. S. 2007. Traditional knowledge on wild food plants in Andhra Pradesh. Indian J. Trad. Knowl. **6:** 223-229.

Singh, R. S. and Singh, N. I. 1985. A preliminary ethnobotanical studies on wild edible plants in the markets of Manipur-1. J. Econ. Taxon. Bot. **6:** 699-703.

Singh, U. and Singh, B. 1992. Tropical grain legume as important human foods. J. Econ. Taxon. Bot. **46:** 310-361.

Singh, B., Rethy, P., Hazarika, H., Gajurel, P. R. and Kagyung, R. 2009. Wild edible plants and its commercial potential: A case study from Itanagar area of Arunachal Pradesh. J. Non-Timber Forest Products **16:** 139-144.

Subramanian, A., Mohan, V.R., and Maruthupandian, A. 2011. Little known wild edible plants used by Valains from Madurai district, Tamil Nadu. J. Non-Timber Forest Products **18:** 333-336.

Sudhakar, A and Vedavathy, S. 1999. Wild edible plants used by the tribals of Chittoor district (Andhra Pradesh), India. J. Econ. Taxon. Bot. **23:** 321-329.

Young, V. R. and Pellett, P. L. 1994. Plants proteins in relation to human protein and aminoacid nutrition. Amer. J. Clin. Nutr. **59:** 1203-1212.

Index to Scientific Names

(Italicized names indicate synonymy)

www.ingramcontent.com/pod-product-compliance
Ingram Content Group UK Ltd.
Pitfield, Milton Keynes, MK11 3LW, UK
UKHW022214051026
15624UKWH00001B/153

9 788170 359722